Clustering Methodology for Symbolic Data

## Wiley Series in Computational Statistics

Consulting Editors:

**Paolo Giudici**
University of Pavia, Italy

**Geof H. Givens**
Colorado State University, USA

Wiley Series in Computational Statistics is comprised of practical guides and cutting edge research books on new developments in computational statistics. It features quality authors with a strong applications focus. The texts in the series provide detailed coverage of statistical concepts, methods and case studies in areas at the interface of statistics, computing, and numerics. With sound motivation and a wealth of practical examples, the books show in concrete terms how to select and to use appropriate ranges of statistical computing techniques in particular fields of study. Readers are assumed to have a basic understanding of introductory terminology. The series concentrates on applications of computational methods in statistics to fields of bioinformatics, genomics, epidemiology, business, engineering, finance and applied statistics.

# Clustering Methodology for Symbolic Data

*Lynne Billard*
University of Georgia, USA

*Edwin Diday*
CEREMADE, Université Paris-Dauphine, Université PSL, Paris, France

This edition first published 2020
© 2020 John Wiley & Sons Ltd

The right of Lynne Billard and Edwin Diday to be identified as the authors of this work has been asserted in accordance with law.

*Registered Offices*
John Wiley & Sons, Inc., 111 River Street, Hoboken, NJ 07030, USA
John Wiley & Sons Ltd, The Atrium, Southern Gate, Chichester, West Sussex, PO19 8SQ, UK

*Editorial Office*
9600 Garsington Road, Oxford, OX4 2DQ, UK

For details of our global editorial offices, customer services, and more information about Wiley products visit us at www.wiley.com.

Wiley also publishes its books in a variety of electronic formats and by print-on-demand. Some content that appears in standard print versions of this book may not be available in other formats.

*Library of Congress Cataloging-in-Publication Data*
Names: Billard, L. (Lynne), 1943- author. | Diday, E., author.
Title: Clustering methodology for symbolic data / Lynne Billard (University
    of Georgia), Edwin Diday (CEREMADE, Université Paris-Dauphine, Université PSL, Paris,
France).
Description: Hoboken, NJ : Wiley, 2020. | Includes bibliographical references
    and index. |
Identifiers: LCCN 2019011642 (print) | LCCN 2019018340 (ebook) | ISBN
    9781119010388 (Adobe PDF) | ISBN 9781119010395 (ePub) | ISBN 9780470713938
    (hardcover)
Subjects: LCSH: Cluster analysis. | Multivariate analysis.
Classification: LCC QA278.55 (ebook) | LCC QA278.55 .B55 2019 (print) | DDC
    519.5/3–dc23
LC record available at https://lccn.loc.gov/2019011642

Cover Design: Wiley
Cover Image: © Lynne Billard
Background: © Iuliia_Syrotina_28/Getty Images

Set in 10/12pt WarnockPro by SPi Global, Chennai, India
Printed and bound in Singapore by Markono Print Media Pte Ltd

10  9  8  7  6  5  4  3  2  1

# Contents

# 1

# Introduction

The theme of this volume centers on clustering methodologies for data which allow observations to be described by lists, intervals, histograms, and the like (referred to as "symbolic" data), instead of single point values (traditional "classical" data). Clustering techniques are frequent participants in exploratory data analyses when the goal is to elicit identifying classes in a data set. Often these classes are in and of themselves the goal of an analysis, but they can also become the starting point(s) of subsequent analyses. There are many texts available which focus on clustering for classically valued observations. This volume aims to provide one such outlet for symbolic data.

With the capabilities of the modern computer, large and extremely large data sets are becoming more routine. What is less routine is how to analyze these data. Data sets are becoming so large that even with the increased computational power of today, direct analyses through the myriad of classical procedures developed over the past century alone are not possible; for example, from Stirling's formula, the number of partitions of a data set of only 50 units is approximately $1.85 \times 10^{47}$. As a consequence, subsets of aggregated data are determined for subsequent analyses. Criteria for how and the directions taken in these aggregations would typically be driven by the underlying scientific questions pertaining to the nature and formation of the data sets at hand. Examples abound. Data streams may be aggregated into blocks of data or communications networks may have different patterns in phone usage across age groups and/or regions, studies of network traffic across different networks will inevitably involve symbolic data, satellite observations are aggregated into (smaller) sub-regional measurements, and so on. The list is endless. There are many different approaches and motivations behind the aggregations. The aggregated observations are perforce lists, intervals, histograms, etc., and as such are examples of symbolic data. Indeed, Schweizer (1984) anticipated this progress with his claim that "distributions are the numbers of the future".

In its purest, simplest form, symbolic data can be defined as taking values as hypercubes or as Cartesian products of distributions in $p$-dimensional space

*Clustering Methodology for Symbolic Data*, First Edition. Lynne Billard and Edwin Diday.
© 2020 John Wiley & Sons Ltd. Published 2020 by John Wiley & Sons Ltd.

$\mathbb{R}^p$, in contrast to classical observations whose values are points in $\mathbb{R}^p$. Classical data are well known, being the currency of statistical analyses since the subject began. Symbolic data and their analyses are, however, relatively new and owe their origin to the seminal work of Diday (1987).

More specifically, observations may be multi-valued or lists (of categorical values). To illustrate, consider a text-mining document. The original database may consist of thousands or millions of text files characterized by a number (e.g., 6000) of key words. These words or sets of words can be aggregated into categories of words such as "themes" (e.g., telephone enquiries may be aggregated under categories of accounts, new accounts, discontinued service, broken lines, and so forth, with each of these consisting of its own sub-categories). Thus, a particular text message may contain the specific key words $Y =$ {home-phone, monthly contract, ...} from the list of possible key words $\mathcal{Y} =$ {two -party line, billing, local service, international calls, connections, home, monthly contract, ...}. Or, the color of the bird species rainbow lorikeet is $Y =$ {green, yellow, red, blue} with $Y$ taking values from the list of colors $\mathcal{Y} =$ {black, blue, brown, green, white, red, yellow, ..., (possible colors), ... }. An aggregation of drivers by city census tract may produce a list of automobile ownership for one particular residential tract as $Y =$ {Ford, Renaullt, Volvo, Jeep} from $\mathcal{Y} =$ {..., (possible car models), ... }. As written, these are examples of non-modal observations. If the end user also wants to know proportional car ownership, say, then aggregation of the census tract classical observations might produce the modal list-valued observation $Y =$ {Holden, .2; Falcon, .25; Renault, .5; Volvo, .05} indicating 20% of the drivers own a Holden car, 50% own a Renault, and so forth.

Interval-valued observations, as the name suggests, are characterized as taking values across an interval $Y = [a, b]$ from $\mathcal{Y} \equiv \mathbb{R}$. There are endless examples. Stock prices have daily low and high values; temperatures have daily (or monthly, or yearly, ...) minimum and maximum values. Observations within (or even between adjacent) pixels in a functional magnetic resonance imaging (fMRI) data set (from measurements of $p$ different stimuli, say) are aggregated to produce a range of values across the separate pixels. In their study of face recognition features, Leroy et al. (1990) aggregated pixel values to obtain interval measurements. At the current time, more methodology is available for interval-valued data sets than for other types of symbolic observations, so special attention will be paid to these data.

Another frequently occurring type of symbolic data is the histogram-valued observation. These observations correspond to the traditional histogram that pertains when classical observations are summarized into a histogram format. For example, consider the height ($Y$) of high-school students. Rather than retain the values for each individual, a histogram is calculated to make an analysis of height characteristics of school students across the 1000 schools in the state. Thus, at a particular school, it may be that the heights, in inches, are $Y = \{[50, 60), 0.12; [60, 65), 0.33; [65, 72), 0.45; [72, 80], 0.1\}$, where the relative frequency of students being 60–65 inches tall is 0.33 or 33%. More generally, rather than the sub-interval having a relative frequency, as in this example, other weights may pertain.

These lists, intervals, and histograms are just some of the many possible formats for symbolic data. Chapter 2 provides an introduction to symbolic data. A key question relates to how these data arrive in practice. Clearly, many symbolic data sets arise naturally, especially species data sets, such as the bird colors illustrated herein. However, most symbolic data sets will emerge from the aggregation of the massively large data sets generated by the modern computer. Accordingly, Chapter 2 looks briefly at this generation process. This chapter also considers the calculations of basic descriptive statistics, such as sample means, variances, covariances, and histograms, for symbolic data. It is noted that classical observations are special cases. However, it is also noted that symbolic data have internal variations, unlike classical data (for which this internal variation is zero). Bock and Diday (2000a), Billard and Diday (2003, 2006a), Diday and Noirhomme-Fraiture (2008), the reviews of Noirhomme-Fraiture and Brito (2011) and Diday (2016), and the non-technical introduction in Billard (2011) provide a wide coverage of symbolic data and some of the current methodologies.

As for classical statistics since the subject began, observations are realizations of some underlying random variable. Symbolic observations are also realizations of those same (standard, so to speak) random variables, the difference being that realizations are symbolic-valued instead of numerical or categorical point-valued. Thus, for example, the parameters of a distribution of the random variable, such as $Y \sim N(\mu, \Sigma)$, are still points, e.g., $\mu = (0, \ldots, 0)$ and $\Sigma = I$. This feature is especially evident when calculating descriptive statistics, e.g., the sample mean of interval observations (see section 2.4). That is, the output sample mean of intervals is a point, and is not an interval such as might be the case when interval arithmetic is employed. Indeed, as for classical statistics, standard classical arithmetic is in force (i.e., we do not use intervals or histograms or related arithmetics). In that same vein, aggregated observations are still distributed according to that underlying distribution (e.g., normally distributed); however, it is assumed that those normally distributed observations are uniformly spread across the interval, or sub-intervals for histogram valued data. Indeed, this is akin to the "group" data histogram problems of elementary applied statistics courses. While this uniform spread assumption exists in almost all symbolic data analytic procedures, relaxation to some other form of spread could be possible.

The starting premise of the clustering methodologies presupposes the data are already in a symbolic format, therefore the philosophical concepts involved behind the formation of symbolic data are by and large not included in this volume. The reader should be aware, however, that there are many issues that might be considered between the initial data values (classical or otherwise) and the symbolic values of the data set to be analysed. This is particularly applicable when symbolic data emerge as a result of aggregating larger data sets. These principles are covered extensively in Bock and Diday (2000a), Billard and Diday (2006a), and Diday (2016).

Most clustering methodologies depend in some manner on dissimilarity and/or distance measures. The basic concepts underlying dissimilarity and/or distance measures are described in Chapter 3 along with some of their properties. Chapter 3 also presents dissimilarity/distance measures for non-modal symbolic data, i.e., for non-modal list multi-valued data and for interval-valued data. Chapter 4 considers such measures for modal observations, i.e., for modal list multi-valued data and for modal interval-valued (better known as histogram-valued) data. In most of the relevant literature, it is assumed that all variables are of the same type, e.g., all interval-valued. However, that is not always a necessary restriction. Therefore, the case of mixed type valued variables is illustrated on occasions, mainly in Chapters 6–8.

Chapter 5 reviews clustering procedures in general, with the primary focus on classical approaches. Clustering procedures are heavily computational and so started to emerge for classical data sets in the 1950s with the appearance of computers. Contemporary computers ensure these methods are even more accessible and even more in demand.

Broadly, clustering procedures can be categorized as organizing the entire data set $\Omega$ into non-overlapping but exhaustive partitions or into building hierarchical trees. The most frequent class of partitioning algorithm is the $k$-means algorithm or its variants usually based on cluster means or centroid values, including versions of the $k$-medoids algorithm which is typically based on dissimilarity or distance measures, and the more general dynamical partitioning method. Mixture distributions are also part of the partitioning paradigm.

There are two types of hierarchical tree constructions. The first approach is when the hierarchical tree is constructed from the top down divisively whereby the first cluster contains the entire data set $\Omega$. At each step, a cluster is divided into two sub-clusters, with the division being dictated by some criteria, such as producing new clusters which attain a reduced sum of squares of the observations within clusters and/or between the clusters according to some collective measure of the cluster diagnostics. Alternatively, hierarchical trees can be built from the bottom up when the starting point consists of clusters of one only observation which are successively merged until reaching the top of the tree, with this tree-top cluster containing all observations in $\Omega$. In this case, several criteria exist for the selection of which clusters are to be merged at each stage, e.g., nearest neighbor, farthest neighbor, Ward's minimum variance, among other criteria. An extension of the standard non-overlapping clusters of hierarchies is the agglomerative pyramidal methodology, which allows observations to belong to at most two distinct clusters.

These methods are extended to symbolic data in Chapter 6 for partitioning methods. Chapter 7 considers divisive hierarchies using either a monothetic algorithm or a polythetic algorithm. Because of the unique structure of symbolic data (e.g., they are not points), it becomes necessary to introduce new concepts (e.g., association measures) in order to develop algorithms for symbolic data. In Chapter 8, agglomerative methods are described for both hierarchies and pyramids. In each chapter, these constructions are illustrated for the different types of symbolic data: modal and non-modal list multi-valued data, interval-valued data, histogram-valued data and sometimes for mixed-valued data. As for classical methods, what becomes evident very quickly is that there is a plethora of available algorithms. These algorithms are in turn based on an extensive array of underlying criteria such as distance or dissimilarity matrices, with many different ways of calculating said matrices, and a further array of possible reallocation and starting and stopping rules.

At the end of a clustering process – be this a partitioning, a divisive hierarchy, an agglomerative hierarchy, or even a principal component analysis or aggregation of observations in some form – it is often the case that calculation of the cluster profile into a summarizing form is desired. Indeed, the field of symbolic data owes its origins to output data sets of clustering classical data when Diday (1987) recognised that summarizing obtained clusters by single point values involved a loss of critical details, especially a loss of information relating to the variation of the observations within a given cluster. This loss is especially significant if the clustering procedure is designed to produce outputs for future analyses. For example, suppose a process produces a cluster with two interval observations for a given variable of $Y = [1, 5]$ and $Y = [2, 8]$. One summary of these observations might be the interval $[1, 8]$ or the interval $[1.5, 6.5]$, among other possibilities. Chapters 5–8 contain numerous examples of output clusters. Rather than calculate a cluster representation value for each cluster for all these examples, the principle of this representation calculation is illustrated for some output clusters obtained in Chapter 6 (see section 6.7).

Likewise, by the same token, for all the hierarchy trees – those built by divisive algorithms or those built by agglomerative methods be these pure hierarchies or pyramidal hierarchies – tree heights (as measured along the $y$-axis) can be calculated. How this is done is illustrated for some trees in Chapter 7 (see section 7.4). An additional feature of Chapter 8 is the consideration of logical rules applied to a data set. This is particularly relevant when the data set being analysed was a result of aggregation of large classical data sets, though this can also be a factor in the aggregation of symbolic data sets. Thus, for example, apparent interval-valued observations may in fact be histogram-valued after appropriate logical rules are invoked (as in the data of Example 8.12).

All of these aspects can be applied to all examples in Chapters 5–8. We leave as exercises for the reader to establish output representations, tree heights, and rules where applicable for the data sets used and clusters obtained throughout.

Finally, all data sets used herein are available at <http://www.stat.uga.edu/faculty/LYNNE/Lynne.html>. The source reference will identify the table number where the data were first used. Sometimes the entire data set is used in the examples, sometimes only a portion is used. It is left as exercises for the reader to re-do those examples with the complete data sets and/or with different portions of them. Some algorithms used in symbolic analyses are contained in the SODAS (Symbolic Official Data Analysis System) package and can be downloaded from <www.ceremade.dauphine.fr/%7Etouati/sodas-pagegarde.htm>; an expanded package, SODAS2, can be downloaded from <http://www.assoproject.be>. Details of the use of these SODAS packages can be found in Bock and Diday (2000a) and Diday and Noirhomme-Fraiture (2008), respectively.

Many researchers in the field have indirectly contributed to this book through their published work. We hope we have done justice to those contributions. Of course, no book, most especially including this one, can provide an extensive detailed coverage of all applicable material; space limitations alone dictate that selections have of necessity had to be made.

# 2

# Symbolic Data: Basics

In this chapter, we describe what symbolic data are, how they may arise, and their different formulations. Some data are naturally symbolic in format, while others arise as a result of aggregating much larger data sets according to some scientific question(s) that generated the data sets in the first place. Thus, section 2.2.1 describes non-modal multi-valued or lists of categorical data, with modal multi-valued data in section 2.2.2; lists or multi-valued data can also be called simply categorical data. Section 2.2.3 considers interval-valued data, with modal interval data more commonly known as histogram-valued data in section 2.2.4. We begin, in section 2.1, by considering the distinctions and similarities between individuals, classes, and observations. How the data arise, such as by aggregation, is discussed in section 2.3. Basic descriptive statistics are presented in section 2.4. Except when necessary for clarification purposes, we will write "interval-valued data" as "interval data" for simplicity; likewise, for the other types of symbolic data.

It is important to remember that symbolic data, like classical data, are just different manifestations of sub-spaces of the $p$-dimensional space $\mathbb{R}^p$ always dealing with the same random variables. A classical datum is a point in $\mathbb{R}^p$, whereas a symbolic value is a hypercube or a Cartesian product of distributions in $\mathbb{R}^p$. Thus, for example, the $p = 2$-dimensional random variable $(Y_1, Y_2)$ measuring height and weight (say) can take a classical value $Y_1 = 68$ inches and $Y_2 = 70$ kg, or it may take a symbolic value with $Y_1 = [65, 70]$ and $Y_2 = [70, 74]$ interval values which form a rectangle or a hypercube in the plane. That is, the random variable is itself unchanged, but the realizations of that random variable differ depending on the format. However, it is also important to recognize that since classical values are special cases of symbolic values, then regardless of analytical technique, classical analyses and symbolic analyses should produce the same results when applied to those classical values.

*Clustering Methodology for Symbolic Data*, First Edition. Lynne Billard and Edwin Diday.
© 2020 John Wiley & Sons Ltd. Published 2020 by John Wiley & Sons Ltd.

## 2.1 Individuals, Classes, Observations, and Descriptions

In classical statistics, we talk about having a random sample of $n$ observations $Y_1, \ldots, Y_n$ as outcomes for a random variable $Y$. More precisely, we say $Y_i$ is the observed value for individual $i$, $i = 1, \ldots, n$. A particular observed value may be $Y = 3$, say. We could equivalently say the description of the $i$th individual is $Y_i = 3$. Usually, we think of an individual as just that, a single individual. For example, our data set of $n$ individuals may record the height $Y$ of individuals, Bryson, Grayson, Ethan, Coco, Winston, Daisy, and so on. The "individual" could also be an inanimate object such as a particular car model with $Y$ describing its capacity, or some other measure relating to cars. On the other hand, the "individual" may represent a class of individuals. For example, the data set consisting of $n$ individuals may be $n$ classes of car models, Ford, Renault, Honda, Volkswagen, Nova, Volvo, … , with $Y$ recording the car's speed over a prescribed course, etc. However individuals may be defined, the realization of $Y$ for that individual is a single point value from its domain $\mathcal{Y}$.

If the random variable $Y$ takes quantitative values, then the domain (also called the range or observation space) is $\mathcal{Y}$ taking values on the real line $\mathbb{R}$, or a subset of $\mathbb{R}$ such as $\mathbb{R}^+$ if $Y$ can only take non-negative or zero values. When $Y$ takes qualitative values, then a classically valued observation takes one of two possible values such as {Yes, No} or coded to $\mathcal{Y} = \{0, 1\}$, for example. Typically, if there are several categories of possible values, e.g., bird colors with domain $\mathcal{Y} = \{$red, blue, green, white,…$\}$, a classical analysis will include a different random variable for each category and then record the presence (Yes) or absence (No) of each category. When there are $p$ random variables, then the domain of $\mathbf{Y} = (Y_1, \ldots, Y_p)$ is $\mathcal{Y} = \mathcal{Y}_1 \times \cdots \times \mathcal{Y}_p$.

In contrast, when the data are symbolic-valued, the observations $Y_1, \ldots, Y_m$ are typically realizations that emerge after aggregating observed values for the random variable $Y$ across some specified class or category of interest (see section 2.3). Thus, for example, observations may refer now to $m$ classes, or categories, of age×income, or to $m$ species of dogs, and so on. Thus, the class Boston (say) has a June temperature range of $[58°\text{F}, 75°\text{F}]$. In the language of symbolic analysis, the individuals are ground-level or order-one individuals and the aggregations – classes – are order-two individuals or "objects" (see, e.g., Diday (1987, 2016), Bock and Diday (2000a,b), Billard and Diday (2006a), or Diday and Noirhomme-Fraiture (2008)).

On the other hand, suppose Gracie's pulse rate $Y$ is the interval $Y = [62, 66]$. Gracie is a single individual and a classical value for her pulse rate might be $Y = 63$. However, this interval of values would result from the collection, or aggregation, of Gracie's classical pulse rate values over some specified time period. In the language of symbolic data, this interval represents the pulse rate of the class "Gracie". However, this interval may be the result of aggregating the classical point values of all individuals named "Gracie" in some larger data base. That is, some symbolic realizations may relate to one single individual, e.g., Gracie, whose pulse rate may be measured as [62, 66] over time, or to a set of all those Gracies of interest. The context should make it clear which situation prevails.

In this book, symbolic realizations for the observation $u$ can refer interchangeably to the description $Y_u$ of classes or categories or "individuals" $u$, $u = 1, \ldots, m$, that is, simply, $u$ will be the unit (which is itself a class, category, individual, or observation) that is described by $Y_u$. Furthermore, in the language of symbolic data, the realization of $Y_u$ is referred to as the "description" $d$ of $Y_u$, $d(Y(u))$. For simplicity, we write simply $Y_u$, $u = 1, \ldots, m$.

## 2.2 Types of Symbolic Data

### 2.2.1 Multi-valued or Lists of Categorical Data

We have a random variable $Y$ whose realization is the set of values $\{Y_1, \ldots, Y_{s'}\}$ from the set of possible values or categories $\mathcal{Y} = \{Y_1, \ldots, Y_s\}$, where $s$ and $s'$ with $s' \leq s$ are finite. Typically, for a symbolic realization, $s' > 1$, whereas for a classical realization, $s' = 1$. This realization is called a list (of $s'$ categories from $\mathcal{Y}$) or a multi-valued realization, or even a multi-categorical realization, of $Y$. Formally, we have the following definition.

**Definition 2.1** Let the $p$-dimensional random variable $\mathbf{Y} = (Y_1, \ldots, Y_p)$ take possible values from the list of possible values in its domain $\mathcal{Y} = \mathcal{Y}_1 \times \cdots \times \mathcal{Y}_p$ with $\mathcal{Y}_j = \{Y_{j1}, \ldots, Y_{js_j}\}$, $j = 1, \ldots, p$. In a random sample of size $m$, the realization $\mathbf{Y}_u$ is a **list** or **multi-valued** observation whenever

$$\mathbf{Y}_u = (\{Y_{ujk_j}; \ k_j = 1, \ldots, s_{uj}\}, \ j = 1, \ldots, p), \quad u = 1, \ldots, m. \quad (2.2.1)$$

□

Notice that, in general, the number of categories $s_{uj}$ in the actual realization differs across realizations (i.e., $s_{uj} \neq s_j$), $u = 1, \ldots, m$, and across variables $Y_j$ (i.e., $s_{uj} \neq s_u$), $j = 1, \ldots, p$.

**Example 2.1** Table 2.1 (second column) shows the list of major utilities used in a set of seven regions. Here, the domain for the random variable $Y =$ utility is $\mathcal{Y} = \{\text{coal, oil, wood, electricity, gas}, \ldots, \text{(possible utilities)}, \ldots \}$. For example, for the fifth region ($u = 5$), $Y_5 = \{\text{gas, oil, other}\}$. The third region, $u = 3$, has a single utility (coal), i.e., the utility usage for this region is a classical realization. Thus, we write $Y_3 = \{\text{coal}\}$. If a region were uniquely identified by its utility usage, and we were to try to identify a region $u = 7$ (say) by a single usage, such as electricity, then it could be mistaken for region six, which is quite a

**Table 2.1** List or multi-valued data: regional utilities (Example 2.1)

| Region $u$ | Major utility | Cost |
|---|---|---|
| 1 | {electricity, coal, wood, gas} | [190, 230] |
| 2 | {electricity, oil, coal} | [21.5, 25.5] |
| 3 | {coal} | [40, 53] |
| 4 | {other} | 15.5 |
| 5 | {gas, oil, other} | [25, 30] |
| 6 | {electricity} | 46.0 |
| 7 | {electricity, coal} | [37, 43] |

different region and is clearly not the same as region seven. That is, a region cannot in general be described by a single classical value, but only by the list of types of utilities used in that region, i.e., as a symbolic list or multi-valued value. □

While list data mostly take qualitative values that are verbal descriptions of an outcome, such as the types of utility usages in Example 2.1, quantitative values such as coded values {1, 2, … } may be the recorded value. These are not necessarily the same as ordered categorical values such as $\mathcal{Y}$ = {small, medium, large}. Indeed, a feature of categorical values is that there is no prescribed ordering of the listed realizations. For example, for the seventh region ($u = 7$) in Table 2.1, the description {electricity, coal} is exactly the same description as {coal, electricity}, i.e., the same region. This feature does not carry over to quantitative values such as histograms (see section 2.2.4).

### 2.2.2 Modal Multi-valued Data

Modal lists or modal multi-valued data (sometimes called modal categorical data) are just list or multi-valued data but with each realized category occurring with some specified weight such as an associated probability. Examples of non-probabilistic weights include the concepts of capacities, possibilities, and necessities (see Billard and Diday, 2006a, Chapter 2; see also Definitions 2.6–2.9 in section 2.2.5). In this section and throughout most of this book, it is assumed that the weights are probabilities; suitable adjustment for other weights is left to the reader.

**Definition 2.2**   Let the $p$-dimensional random variable $\mathbf{Y} = (Y_1, \ldots, Y_p)$ take possible values from the list of possible values in its domain $\mathcal{Y} = \mathcal{Y}_1 \times \cdots \times \mathcal{Y}_p$ with $\mathcal{Y}_j = \{Y_{j1}, \ldots, Y_{js_j}\}$, $j = 1, \ldots, p$. In a random sample of size $m$, the realization $\mathbf{Y}_u$ is a **modal list**, or **modal multi-valued** observation whenever

$$\mathbf{Y}_u = (\{Y_{ujk_j}, p_{ujk_j}; \ k_j = 1, \dots, s_{uj}\}, \ j = 1, \dots, p), \ \sum_{k_j=1}^{s_{uj}} p_{ujk_j} = 1, \ u = 1, \dots, m.$$

(2.2.2)

□

Without loss of generality, we can write the number of categories from $\mathcal{Y}_j$ for the random variable $Y_j$, as $s_{uj} = s_j$, for all $u = 1, \dots, m$, by simply giving unrealized categories ($Y_{jk'}$, say) the probability $p_{ujk'} = 0$. Furthermore, the non-modal multi-valued realization of Eq. (2.2.1) can be written as a modal multi-valued observation of Eq. (2.2.2) by assuming actual realized categories from $\mathcal{Y}_j$ occur with equal probability, i.e., $p_{ujk_j} = 1/s_{uj}$ for $k_j = 1, \dots, s_{uj}$, and unrealized categories occur with probability zero, for each $j = 1, \dots, p$.

**Example 2.2** A study of ascertaining deaths attributable to smoking, for $m = 8$ regions, produced the data of Table 2.2. Here, $Y =$ cause of death from smoking, with domain $\mathcal{Y} = \{$death from smoking, death from lung cancer, death from respiratory diseases induced by smoking$\}$ or simply $\mathcal{Y} = \{$smoking, lung cancer, respiratory$\}$. Thus, for region $u = 1$, smoking caused lung cancer deaths in 18.4% of the smoking population, 18.8% of all smoking deaths were from respiratory diseases, and 62.8% of smoking deaths were from other smoking-related causes, i.e., $(p_{11}, p_{12}, p_{13}) = (0.628, 0.184, 0.188)$, where for simplicity we have dropped the $j = 1 = p$ subscript. Data for region $u = 7$ are limited to $p_{71} = 0.648$. However, we know that $p_{72} + p_{73} = 0.352$. Hence, we can assume that $p_{72} = p_{73} = 0.176$. On the other hand, for region $u = 8$, the categories $\{$lung cancer, respiratory$\}$ did not occur, thus the associated probabilities are $p_{82} = p_{83} = 0.0$. □

**Table 2.2** Modal multi-valued data: smoking deaths ($p_{uk}$) (Example 2.2)

| Region | Proportion $p_{uk}$ of smoking deaths attributable to: | | |
|---|---|---|---|
| $u$ | Smoking | Lung cancer | Respiratory |
| 1 | 0.628 | 0.184 | 0.188 |
| 2 | 0.623 | 0.202 | 0.175 |
| 3 | 0.650 | 0.197 | 0.153 |
| 4 | 0.626 | 0.209 | 0.165 |
| 5 | 0.690 | 0.160 | 0.150 |
| 6 | 0.631 | 0.204 | 0.165 |
| 7 | 0.648 | | |
| 8 | 1.000 | | |

### 2.2.3  Interval Data

A classical realization for quantitative data takes a point value on the real line $\mathbb{R}$. An interval-valued realization takes values from a subset of $\mathbb{R}$. This is formally defined as follows.

**Definition 2.3**  Let the $p$-dimensional random variable $\mathbf{Y} = (Y_1, \ldots, Y_p)$ take quantitative values from the space $\mathbb{R}^p$. A random sample of $m$ realizations takes **interval** values when

$$\mathbf{Y}_u = ([a_{u1}, b_{u1}], \ldots, [a_{up}, b_{up}]), \quad u = 1, \ldots, m, \qquad (2.2.3)$$

where $a_{uj} \leq b_{uj}, j = 1, \ldots, p, u = 1, \ldots, m$, and where the intervals may be open or closed at either end (i.e., $[a, b), (a, b], [a, b]$, or $(a, b)$).  □

**Example 2.3**  Table 2.1 (right-hand column) gives the cost (in \$) of the regional utility usage of Example 2.1. Thus, for example, the cost in the first region, $u = 1$, ranges from 190 to 230. Clearly, a particular household in this region has its own cost, 199, say. However, not all households in this region have the same costs, as illustrated by these values. On the other hand, the recorded costs for households in region $u = 6$ is the classical value 46.0 (or the interval [46, 46]).  □

There are numerous examples of naturally occurring symbolic data sets. One such scenario exists in the next example.

**Example 2.4**  The data of Table 2.3 show the minimum and maximum temperatures recorded at six weather stations in China for the months of January ($Y_1$) and July ($Y_2$) in 1988. Also shown is the elevation ($Y_3$), which is a classical value. The complete data set is extracted from <http://dss.ucar.edu/datasets/ds578.5> and is a multivariate time series with temperatures (in °C) for many

**Table 2.3**  Interval data: weather stations (Example 2.4)

| Station | $Y_1$ = January | $Y_2$ = July | $Y_3$ = Elevation |
|---|---|---|---|
| $u$ | $[a_{u1}, b_{u1}]$ | $[a_{u2}, b_{u2}]$ | $a_{u3}$ |
| 1 | $[-18.4, -7.5]$ | $[17.0, 26.5]$ | 4.82 |
| 2 | $[-23.4, -15.5]$ | $[12.9, 23.0]$ | 14.78 |
| 3 | $[-8.4, 9.0]$ | $[10.8, 23.2]$ | 73.16 |
| 4 | $[10.0, 17.7]$ | $[24.2, 33.8]$ | 2.38 |
| 5 | $[11.5, 17.7]$ | $[25.8, 33.5]$ | 1.44 |
| 6 | $[11.8, 19.2]$ | $[25.6, 32.6]$ | 0.02 |

stations for all months over the years 1974–1988 (see also Billard, 2014). Thus, we see that, in July 1988, station $u = 3$ enjoyed temperatures from a low of 10.8°C to a high of 23.2°C, i.e., $Y_{32} = [a_{32}, b_{32}] = [10.8, 23.2]$. □

### 2.2.4 Histogram Data

Histogram data usually result from the aggregation of several values of quantitative random variables into a number of sub-intervals. More formally, we have the following definition.

**Definition 2.4**  Let the $p$-dimensional random variable $\mathbf{Y} = (Y_1, \ldots, Y_p)$ take quantitative values from the space $\mathbb{R}^p$. A random sample of $m$ realizations takes **histogram** values when, for $u = 1, \ldots, m$,

$$\mathbf{Y}_u = (\{[a_{ujk_j}, b_{ujk_j}), p_{ujk_j}; \ k_j = 1, \ldots, s_{uj}\}, \quad j = 1, \ldots, p), \ \sum_{k_j=1}^{s_{uj}} p_{ujk} = 1,$$

$$(2.2.4)$$

where $a_{ujk_j} \leq b_{ujk_j}, j = 1, \ldots, p, u = 1, \ldots, m.$  □

Usually, histogram sub-intervals are closed at the left end and open at the right end except for the last sub-interval, which is closed at both ends. Furthermore, note that the number of histogram sub-intervals $s_{uj}$ differs across $u$ and across $j$. For the special case that $s_{uj} = 1$, and hence $p_{uj1} = 1$ for all $u = 1, \ldots, m$, the histogram is an interval.

**Example 2.5**  Table 2.4 shows a histogram-valued data set of $m = 10$ observations. Here, the random variable is $Y$ = flight time for airlines to fly from several departure cities into one particular hub city airport. There were approximately 50000 flights recorded. Rather than a single flight, interest was on performance for specific carriers. Accordingly, the aggregated values by airline carrier were obtained and the histograms of those values were calculated in the usual way. Notice that the number of histogram sub-intervals $s_u$ varies across $u = 1, \ldots, m$; also, the sub-intervals $[a_{uk}, b_{uk})$ can differ for $u = 1, \ldots, m$, reflecting, in this case, different flight distances depending on flight routes and the like. Thus, for example, $Y_7 = \{[10, 50), 0.117; [50, 90), 0.476; [90,130), 0.236; [130,170], 0.171\}$ (data extracted from Falduti and Taibaly, 2004). □

In the context of symbolic data methodology, the starting data are already in a histogram format. All data, including histogram data, can themselves be aggregated to form histograms (see section 2.4.4).

**Table 2.4** Histogram data: flight times (Example 2.5)

| Airline | $Y$ = Flight time |
|---|---|
| $u$ | $\{[a_{uk}, b_{uk}), p_{uk}; \; k = 1 \dots, s_u\}$ |
| 1 | $\{[40, 100), 0.082; [100, 180), 0.530; [180, 240), 0.172; [240, 320), 0.118; [320, 380], 0.098\}$ |
| 2 | $\{[40, 90), 0.171; [90, 140), 0.285; [140, 190), 0.351; [190, 240), 0.022; [240, 290], 0.171\}$ |
| 3 | $\{[35, 70), 0.128; [70, 135), 0.114; [135, 195), 0.424; [195, 255], 0.334\}$ |
| 4 | $\{[20, 40), 0.060; [40, 60), 0.458; [60, 80), 0.259; [80, 100), 0.117; [100, 120], 0.106\}$ |
| 5 | $\{[200, 250), 0.164; [250, 300), 0.395; [300, 350), 0.340; [350, 400], 0.101\}$ |
| 6 | $\{[25, 50), 0.280; [50, 75), 0.301; [75, 100), 0.250; [100, 125], 0.169\}$ |
| 7 | $\{[10, 50), 0.117; [50, 90), 0.476; [90, 130), 0.236; [130, 170], 0.171\}$ |
| 8 | $\{[10, 50), 0.069; [50, 90), 0.368; [90, 130), 0.514; [130, 170], 0.049\}$ |
| 9 | $\{[20, 35), 0.066; [35, 50), 0.337; [50, 65), 0.281; [65, 80), 0.208; [80, 95], 0.108\}$ |
| 10 | $\{[20, 40), 0.198; [40, 60), 0.474; [60, 80), 0.144; [80, 100), 0.131; [100, 120], 0.053\}$ |

### 2.2.5 Other Types of Symbolic Data

A so-called mixed data set is one in which not all of the $p$ variables take the same format. Instead, some may be interval data, some histograms, some lists, etc.

**Example 2.6** To illustrate a mixed-valued data set, consider the data of Table 2.5 for a random sample of joggers from each group of $m = 10$ body types. Joggers were timed to run a specific distance. The pulse rates ($Y_1$) of joggers at the end of their run were measured and are shown as interval values

**Table 2.5** Mixed-valued data: joggers (Example 2.6)

| Group $u$ | $Y_1$ = Pulse rate | $Y_2$ = Running time |
|---|---|---|
| 1 | [73, 114] | $\{[5.3, 6.2), 0.3; [6.2, 7.1), 0.5; [7.1, 8.3], 0.2\}$ |
| 2 | [70, 100] | $\{[5.5, 6.9), 0.4; [6.7, 8.0), 0.4; [8.0, 9.0], 0.2\}$ |
| 3 | [69, 91] | $\{[5.1, 6.6), 0.4; [6.6, 7.4), 0.4; [7.4, 7.8], 0.2\}$ |
| 4 | [59, 89] | $\{[3.7, 5.8), 0.6; [5.8, 6.3], 0.4\}$ |
| 5 | [61, 87] | $\{[4.5, 5.9), 0.4; [5.9, 6.2], 0.6\}$ |
| 6 | [69, 95] | $\{[4.1, 6.1), 0.5; [6.1, 6.9], 0.5\}$ |
| 7 | [65, 78] | $\{[2.4, 4.8), 0.3; [4.8, 5.7), 0.5; [5.7, 6.2], 0.2\}$ |
| 8 | [58, 83] | $\{[2.1, 5.4), 0.2; [5.4, 6.0), 0.5; [6.0, 6.9], 0.3\}$ |
| 9 | [79, 103] | $\{[4.8, 6.5), 0.3; [6.5, 7.4); 0.5; [7.4, 8.2], 0.2\}$ |
| 10 | [40, 60] | $\{[3.2, 4.1), 0.6; [4.1, 6.7], 0.4\}$ |

for each group. For the first group, the pulse rates fell across the interval $Y_{11} = [73, 114]$. These intervals are themselves simple histograms with $s_{uj} = 1$ for all $u = 1, \ldots, 10$.

In addition, the histogram of running times $(Y_2)$ for each group was calculated, as shown. Thus, for the first group $(u = 1)$, 30% of the joggers took 5.3 to 6.2 time units to run the course, 50% took 6.2 to 7.1, and 20% took 7.1 to 8.3 units of time to complete the run, i.e., we have $Y_{12} = \{[5.3, 6.2), 0.3; [6.2, 7.1), 0.5; [7.1, 8.3], 0.2\}$. On the other hand, half of those in group $u = 6$ ran the distance in under 6.1 time units and half took more than 6.1 units of time.                                                                         □

Other types of symbolic data include probability density functions or cumulative distributions, as in the observations in Table 2.6(a), or models such as the time series models for the observations in Table 2.6(b).

The modal multi-valued data of section 2.2.2 and the histogram data of section 2.2.4 use probabilities as the weights of the categories and the histogram sub-intervals; see Eqs. (2.2.2) and (2.2.4), respectively. While these weights are the most common seen by statistical analysts, there are other possible weights. First, let us define a more general weighted modal type of observation. We take the number of variables to be $p = 1$; generalization to $p \geq 1$ follows readily.

**Definition 2.5**   Let the random sample of size $m$ be realizations of the random variable $Y$ taking values from its domain $\mathcal{Y} = \{\eta_1, \ldots, \eta_S\}$. Then, $Y_u$ are **modal-valued** observations when they can be written in the form

$$Y_u = \{\eta_{uk}, \pi_{uk}; \ k = 1, \ldots, s_u\}, \qquad u = 1, \ldots, m, \tag{2.2.5}$$

where $\pi_{uk}$ is the weight associated with the category $\eta_{uk}$.                                   □

**Table 2.6**  Some other types of symbolic data

|     | $u$ | Description of $u$ |
|-----|-----|--------------------|
| (a) | 1   | Distributed as a normal $N_p(\boldsymbol{\mu}, \boldsymbol{\Sigma})$ |
|     | 2   | Distributed as a normal $N(0, \sigma^2)$ |
|     | 3   | Distributed as exponential $(\beta)$ |
|     | .   | ... |
| (b) | 5   | Follows an AR(1) time-series model |
|     | 6   | Follows a MA($q$) time-series model |
|     | 7   | Is a first-order Markov chain |
|     | .   | ... |

Thus, for a modal list or multi-valued observation of Definition 2.2, the category $Y_{uk} \equiv \eta_{uk}$ and the probability $p_{uk} \equiv \pi_{uk}$; $k = 1, \ldots, s_u$. Likewise, for a histogram observation of Definition 2.4, the sub-interval $[a_{uk}, b_{uk}) \equiv \eta_{uk}$ occurs with relative frequency $p_{uk}$, which corresponds to the weight $\pi_{uk}$, $k = 1, \ldots, s_u$. Note, however, that in Definition 2.5 the condition $\sum_{k=1}^{s_u} \pi_{uk} = 1$ does not necessarily hold, unlike pure modal multi-valued and histogram observations (see Eqs. (2.2.2) and (2.2.4), respectively). Thus, in these two cases, the weights $\pi_k$ are probabilities or relative frequencies. The following definitions relate to situations when the weights do not necessarily sum to one. As before, $s_u$ can differ from observation to observation.

**Definition 2.6** Let the random variable $Y$ take values in its domain $\mathcal{Y} = \{\eta_1, \ldots, \eta_S\}$. The **capacity** of the category $\eta_k$ is the probability that at least one observation from the set of observations $\Omega = (Y_1, \ldots, Y_m)$ includes the category $\eta_k$, $k = 1, \ldots, S$. □

**Definition 2.7** Let the random variable $Y$ take values in its domain $\mathcal{Y} = \{\eta_1, \ldots, \eta_S\}$. The **credibility** of the category $\eta_k$ is the probability that all observations in the set of observations $\Omega = (Y_1, \ldots, Y_m)$ include the category $\eta_k$, $k = 1, \ldots, S$. □

**Definition 2.8** Let the random variable $Y$ take values in its domain $\mathcal{Y} = \{\eta_1, \ldots, \eta_S\}$. Let $C_1$ and $C_2$ be two subsets of $\Omega = (Y_1, \ldots, Y_m)$. A **possibility** measure is the mapping $\pi$ from $\Omega$ to $[0, 1]$ with $\pi(\Omega) = 1$ and $\pi(\phi) = 0$ where $\phi$ is the empty set, such that for all subsets $C_1$ and $C_2$, $\pi(C_1 \cup C_2) = \max\{\pi(C_1), \pi(C_2)\}$. □

**Definition 2.9** Let the random variable $Y$ take values in its domain $\mathcal{Y} = \{\eta_1, \ldots, \eta_S\}$. Let $C$ be a subset of the set of observations $\Omega = (Y_1, \ldots, Y_m)$. A **necessity** measure of $C$, $N(C)$, satisfies $N(C) = 1 - \pi(C^c)$, where $\pi$ is the possibility measure of Definition 2.8 and $C^c$ is the complement of the subset $C$. □

**Example 2.7** Consider the random variable $Y =$ utility of Example 2.1 with realizations shown in Table 2.1. Then, the capacity that at least one region uses the utility $\eta =$ coal is 4/7, while the credibility that every region uses both coal and electricity is 3/7. □

**Example 2.8** Suppose a random variable $Y =$ number of bedrooms in a house takes values $y = \{2, 3, 4\}$ with possibilities $\pi(y) = 0.3, 0.4, 0.5$, respectively. Let $C_1$ and $C_2$ be the subsets that there are two and three bedrooms, respectively. Then, the possibility $\pi(C_1 \cup C_2) = \max\{\pi(C_1), \pi(C_2)\} = \max\{0.3, 0.4\} = 0.4$.

Now suppose $C$ is the set of three bedrooms, i.e., $C = \{3\}$. Then the necessity of $C$ is $N(C) = 1 - \pi(C^c) = 1 - \max\{\pi(2), \pi(4)\} = 1 - \max\{.3, .5\} = 0.5$. □

More examples for these cases can be found in Diday (1995) and Billard and Diday (2006a). This book will restrict attention to modal list or multi-valued data and histogram data cases. However, many of the methodologies in the remainder of the book apply equally to any weights $\pi_{uk}, k = 1, \ldots, s_u,\ u = 1, \ldots, m$, including those for capacities, credibilities, possibilities, and necessities.

More theoretical aspects of symbolic data and concepts along with some philosophical aspects can be found in Billard and Diday (2006a, Chapter 2).

## 2.3   How do Symbolic Data Arise?

Symbolic data arise in a myriad of ways. One frequent source results when aggregating larger data sets according to some criteria, with the criteria usually driven by specific operational or scientific questions of interest.

For example, a medical data set may consist of millions of observations recording a slew of medical information for each individual for every visit to a healthcare facility since the year 1990 (say). There would be records of demographic variables (such as age, gender, weight, height, ...), geographical information (such as street, city, county, state, country of residence, etc.), basic medical tests results (such as pulse rate, blood pressure, cholesterol level, glucose, hemoglobin, hematocrit, ...), specific aliments (such as whether or not the patient has diabetes, a heart condition and if so what, i.e. mitral value syndrome, congestive heart failure, arrhythmia, diverticulitis, myelitis, etc.). There would be information as to whether the patient had a heart attack (and the prognosis) or cancer symptoms (such as lung cancer, lymphoma, brain tumor, etc.). For given aliments, data would be recorded indicating when and what levels of treatments were applied and how often, and so on. The list of possible symptoms is endless. The pieces of information would in analytic terms be the variables (for which the number $p$ is also large), while the information for each individual for each visit to the healthcare facility would be an observation (where the number of observations $n$ in the data set can be extremely large). Trying to analyze this data set by traditional classical methods is likely to be too difficult to manage.

It is unlikely that the user of this data set, whether s/he be a medical insurer or researcher or maybe even the patient him/herself, is particularly interested in the data for a particular visit to the care provider on some specific date. Rather, interest would more likely center on a particular disease (angina, say), or respiratory diseases in a particular location (Lagos, say), and so on. Or, the focus may be on age × gender classes of patients, such as 26-year-old men or 35-year-old women, or maybe children (aged 17 years and under) with leukemia, again the list is endless. In other words, the interest is on characteristics between different groups of individuals (also called classes or categories, but these categories should not be confused with the categories that make up the lists or multi-valued types of data of sections 2.2.1 and 2.2.2).

However, when the researcher looks at the accumulated data for a specific group, 50-year-old men with angina living in the New England district (say), it is unlikely all such individuals weigh the same (or have the same pulse rate, or the same blood pressure measurement, etc.). Rather, thyroid measurements may take values along the lines of, e.g., 2.44, 2.17, 1.79, 3.23, 3.59, 1.67, .... These values could be aggregated into an interval to give [1.67, 3.59] or they could be aggregated as a histogram realization (especially if there are many values being aggregated). In general, aggregating all the observations which satisfy a given group/class/category will perforce give realizations that are symbolic valued. In other words, these aggregations produce the so-called second-level observations of Diday (1987). As we shall see in section 2.4, taking the average of these values for use in a (necessarily) classical methodology will give an answer certainly, but also most likely that answer will not be correct.

Instead of a medical insurer's database, an automobile insurer would aggregate various entities (such as pay-outs) depending on specific classes, e.g., age $\times$ gender of drivers or type of car (Volvo, Renault, Chevrolet, …), including car type by age and gender, or maybe categories of drivers (such as drivers of red convertibles). Statistical agencies publish their census results according to groups or categories of households. For example, salary data are published as ranges such as \$40,000–50,000, i.e., the interval [40, 50] in 1000s of \$.

Let us illustrate this approach more concretely through the following example.

**Example 2.9** Suppose a demographer had before her a massively large data set of hundreds of thousands of observations along the lines of Table 2.7. The data set contains, for each household, the county in which the household is located (coded to $c = 1, 2, …$), along with the recorded variables: $Y_1$ = weekly income (in \$) with domain $\mathcal{Y}_1 = \mathbb{R}^+ = Y_1 \geq 0$, $Y_2$ = age of the head of household (in years) with domain being positive integers, $Y_3$ = children under the age of 18 who live at home with domain $\mathcal{Y}_3 = \{$yes, no$\}$, $Y_4$ = house tenure with domain $\mathcal{Y}_4 = \{$owner occupied, renter occupied$\}$, $Y_5$ = type of energy used in the home with domain $\mathcal{Y}_5 = \{$gas, electric, wood, oil, other$\}$, and $Y_6$ = driving distance to work with $\mathcal{Y}_6 = \mathbb{R}^+ = Y_6 \geq 0$. Data for the first 51 households are shown.

Suppose interest is in the energy usage $Y_5$ within each county. Aggregating these household data for energy across the entire county produces the histograms displayed in Table 2.8. Thus, for example, in the first county 45.83% ($p_{151} = 0.4583$) of the households use gas, while 37.5% ($p_{152} = 0.375$) use electric energy, and so on. This realization could also be written as

**Table 2.7** Household information (Example 2.9)

| County u | Income $Y_1$ | Age $Y_2$ | Child $Y_3$ | Tenure $Y_4$ | Energy $Y_5$ | Distance $Y_6$ |
|---|---|---|---|---|---|---|
| 1 | 801 | 44 | Yes | Owner | Gas | 10 |
| 2 | 752 | 20 | No | Owner | Gas | 6 |
| 1 | 901 | 31 | No | Owner | Electric | 9 |
| 1 | 802 | 47 | Yes | Owner | Gas | 15 |
| 1 | 901 | 42 | No | Renter | Electric | 12 |
| 2 | 750 | 43 | No | Owner | Electric | 4 |
| 1 | 798 | 42 | Yes | Owner | Gas | 8 |
| 1 | 901 | 43 | Yes | Renter | Electric | 9 |
| 2 | 748 | 30 | No | Owner | Gas | 4 |
| 2 | 747 | 32 | No | Owner | Electric | 6 |
| 1 | 901 | 77 | No | Renter | Other | 10 |
| 2 | 751 | 66 | No | Renter | Oil | 6 |
| 1 | 899 | 45 | No | Renter | Wood | 8 |
| 1 | 897 | 48 | No | Renter | Wood | 9 |
| 1 | 804 | 41 | Yes | Owner | Gas | 11 |
| 1 | 797 | 43 | Yes | Owner | Gas | 1 |
| 2 | 750 | 52 | No | Owner | Electric | 4 |
| 2 | 748 | 32 | No | Owner | Electric | 7 |
| 1 | 907 | 42 | No | Renter | Electric | 12 |
| 2 | 748 | 64 | No | Renter | Wood | 6 |
| 1 | 799 | 22 | No | Owner | Gas | 8 |
| 2 | 754 | 51 | No | Owner | Electric | 3 |
| 1 | 799 | 24 | No | Owner | Gas | 11 |
| 1 | 897 | 35 | No | Renter | Electric | 10 |
| 2 | 749 | 38 | No | Owner | Electric | 5 |
| 1 | 802 | 39 | Yes | Owner | Gas | 8 |
| 2 | 751 | 37 | No | Owner | Electric | 5 |
| 2 | 747 | 29 | No | Owner | Electric | 6 |

(Continued)

Table 2.7 (Continued)

| County $u$ | Income $Y_1$ | Age $Y_2$ | Child $Y_3$ | Tenure $Y_4$ | Energy $Y_5$ | Distance $Y_6$ |
|---|---|---|---|---|---|---|
| 1 | 800 | 39 | Yes | Owner | Gas | 8 |
| 1 | 899 | 32 | No | Owner | Electric | 12 |
| 2 | 703 | 46 | Yes | Owner | Gas | 4 |
| 2 | 753 | 48 | No | Owner | Electric | 5 |
| 1 | 898 | 32 | Yes | Renter | Electric | 10 |
| 2 | 748 | 43 | No | Owner | Electric | 5 |
| 1 | 896 | 29 | No | Owner | Electric | 11 |
| 1 | 802 | 39 | Yes | Owner | Gas | 11 |
| 2 | 746 | 37 | No | Owner | Electric | 4 |
| 2 | 752 | 54 | No | Owner | Electric | 6 |
| 1 | 901 | 59 | No | Renter | Oil | 11 |
| 3 | 674 | 41 | Yes | Renter | Gas | 10 |
| 2 | 749 | 27 | No | Owner | Gas | 6 |
| 2 | 749 | 60 | No | Renter | Wood | 4 |
| 2 | 750 | 33 | No | Renter | Gas | 4 |
| 1 | 800 | 21 | No | Owner | Electric | 6 |
| 2 | 747 | 36 | No | Owner | Electric | 4 |
| 1 | 802 | 40 | Yes | Owner | Gas | 10 |
| 2 | 699 | 43 | Yes | Owner | Gas | 5 |
| 2 | 696 | 39 | Yes | Owner | Gas | 5 |
| 2 | 747 | 24 | No | Renter | Gas | 6 |
| 2 | 752 | 67 | No | Owner | Wood | 5 |
| 2 | 749 | 50 | No | Owner | Electric | 4 |
| ... | ... | ... | ... | ... | ... | ... |

**Table 2.8** Aggregated households (Example 2.9)

| Class | $Y_3$ = Children | $Y_5$ = Energy |
|---|---|---|
| County 1 | {no, 0.5417; yes, 0.4583} | {gas, 0.4583; electric, 0.375; wood, 0.0833; oil, 0.0417; other, 0.0417} |
| Owner | {no, 0.4000; yes, 0.6000} | {gas, 0.7333; electric, 0.2667} |
| Renter | {no, 0.7778; yes, 0.2222} | {electric, 0.5556; wood, 0.2222; oil, 0.111; other, 0.1111} |
| County 2 | {no, 0.8846; yes, 0.1154} | {gas, 0.3077; electric, 0.5385; wood, 0.1154; oil, 0.0385} |
| Owner | {no, 0.8571; yes, 0.1429} | {gas, 0.2857; electric, 0.6667; wood, 0.0476} |
| Renter | {no, 1.000} | {gas, 0.4000; wood, 0.4000; oil, 0.2000} |

$Y_{15}$ = {gas, 0.4583; electric, 0.3750; wood, 0.0833; oil, 0.0417; other, 0.0417}. Of those who are home owners, 68.75% use gas and 56.25% use electricity, with no households using any other category of energy. These values are obtained in this case by aggregating across the class of county × tenure. The aggregated energy usage values for both counties as well as those for all county × tenure classes are shown in Table 2.8.

This table also shows the aggregated values for $Y_3$ which indicate whether or not there are children under the age of 18 years living at home. Aggregation by county shows that, for counties $u = 1$ and $u = 2$, respectively, $Y_3$ takes values $Y_{13}$ = {no, 0.5417; yes, 0.4583} and $Y_{23}$ = {no, 0.8846; yes, 0.1154}, respectively. We can also show that $Y_{14}$ = {owner, 0.625; renter, 0.375} and $Y_{24}$ = {owner, 0.8077; renter, 0.1923}, for home tenure $Y_4$.

Both $Y_3$ and $Y_5$ are modal multi-valued observations. Had the aggregated household values been simply identified by categories only, then we would have non-modal multi-valued data, e.g., energy $Y_5$ for owner occupied households in the first county may have been recorded simply as {gas, electric}. In this case, any subsequent analysis would assume that gas and electric occurred with equal probability, to give the realization {gas, 0.5; electric, 0.5; wood, 0; oil, 0; other, 0}.

Let us now consider the quantitative variable $Y_6$ = driving distance to work. The histograms obtained by aggregating across all households for each county are shown in Table 2.9. Notice in particular that the numbers of histogram sub-intervals $s_{u6}$ differ for each county $u$: here, $s_{16} = 3$, and $s_{26} = 2$. Notice also that within each histogram, the sub-intervals are not necessarily of equal length: here, e.g., for county $u = 2$, $[a_{261}, b_{261}) = [3, 5)$, whereas $[a_{262}, b_{262}) = [6, 7]$. Across counties, histograms do not necessarily have the same sub-intervals: here, e.g., $[a_{161}, b_{161}) = [1, 5)$, whereas $[a_{261}, b_{261}) = [3, 5)$.

**Table 2.9** Aggregated households (Example 2.9)

| Class | $Y_6$ = Distance to work |
|---|---|
| County 1 | $\{[1,5), 0.042; [6,10), 0.583; [11,15], 0.375\}$ |
| Owner | $\{[1,5), 0.0625; [6,8), 0.3125; [9,12), 0.5625, [13,15], 0.0625\}$ |
| Renter | $\{[8,9), 0.375; [10,12], 0.625\}$ |
| County 2 | $\{[3,5), 0.654; [6,7], 0.346\}$ |
| Owner | $\{[3,5), 0.714; [6,7], 0.286\}$ |
| Renter | $[4,6]$ |

The corresponding histograms for county × tenure classes are also shown in Table 2.9. We see that for renters in the second county, this distance is aggregated to give the interval $Y_6 = [4, 6]$, a special case of a histogram. □

Most symbolic data sets will arise from these types of aggregations usually of large data sets but it can be aggregation of smaller data sets. A different situation can arise from some particular scientific question, regardless of the size of the data set. We illustrate this via a question regarding hospitalizations of cardiac patients, described more fully in Quantin *et al.* (2011).

**Example 2.10** Cardiologists had long suspected that the survival rate of patients who presented with acute myocardial infarction (AMI) depended on whether or not the patients first went to a cardiology unit and the types of hospital units to which patients were subsequently moved. However, analyses of the raw classical data failed to show this as an important factor to survival. In the Quantin *et al.* (2011) study, patient pathways were established covering a variety of possible pathways. For example, one pathway consisted of admission to one unit (such as intensive care, or cardiology, etc.) before being sent home, while another pathway consisted of admission to an intensive care unit at one hospital, then being moved to a cardiology unit at the same or a different hospital, and then sent home. Each patient could be identified as having followed a specific pathway over the course of treatment, thus the class/group/category was the "pathway." The recorded observed values for a vast array of medical variables were aggregated across those patients within each pathway.

As a simple case, let the data of Table 2.10 be the observed values for $Y_1$ = age and $Y_2$ = smoker for eight patients admitted to three different hospitals. The domain of $Y_1$ is $\mathbb{R}^+$. The smoking multi-valued variable records if the patient does not smoke, is a light smoker, or is a heavy smoker. Suppose the domain

**Table 2.10** Cardiac patients (Example 2.8)

| Patient | Hospital | Age | Smoker |
|---------|----------|-----|--------|
| Patient 1 | Hospital 1 | 74 | Heavy |
| Patient 2 | Hospital 1 | 78 | Light |
| Patient 3 | Hospital 2 | 69 | No |
| Patient 4 | Hospital 2 | 73 | Heavy |
| Patient 5 | Hospital 2 | 80 | Light |
| Patient 6 | Hospital 1 | 70 | Heavy |
| Patient 7 | Hospital 1 | 82 | Heavy |
| Patient 8 | Hospital 3 | 76 | No |
| ... | ... | ... | ... |

**Table 2.11** Hospital pathways (Example 2.8)

| Pathway | Age | Smoker |
|---------|-----|--------|
| Hospital 1 | [70, 82] | {light,1/4; heavy, 3/4} |
| Hospital 2 | [69, 80] | {no, light, heavy} |
| Hospital 3 | [76, 76] | {no} |
| ... | ... | ... |

for $Y_2$ is written as $\mathcal{Y}_2 = \{$no, light, heavy$\}$. Let a pathway be described as a one-step pathway corresponding to a particular hospital, as shown. Thus, for example, four patients collectively constitute the pathway corresponding to the class "Hospital 1"; likewise for the pathways "Hospital 2" and "Hospital 3". Then, observations by pathways are the symbolic data obtained by aggregating classical values for patients who make up a pathway. The age values were aggregated into intervals and the smoking values were aggregated into list values, as shown in Table 2.11. The aggregation of the single patient in the "Hospital 3" pathway ($u = 3$) is the classically valued observation $Y_3 = (Y_{31}, Y_{32}) = ([76, 76], \{no\})$.

The analysis of the Quantin *et al.* (2011) study, based on the pathways symbolic data, showed that pathways were not only important but were in fact the most important factor affecting survival rates, thus corroborating what the cardiologists felt all along. □

There are numerous other situations which perforce are described by symbolic data. Species data are examples of naturally occurring symbolic data. Data with minimum and maximum values, such as the temperature data of

Table 2.4, also occur as a somewhat natural way to record measurements of interest. Many stockmarket values are reported as high and low values daily (or weekly, monthly, annually). Pulse rates may more accurately be recorded as $64 \pm 2$, i.e., $[62, 66]$ rather than the midpoint value of 64; blood pressure values are notorious for "bouncing around", so that a given value of say 73 for diastolic blood pressure may more accurately be $[70, 80]$. Sensitive census data, such as age, may be given as $[30, 40]$, and so on. There are countless examples.

A question that can arise after aggregation has occurred deals with the handling of outlier values. For example, suppose data aggregated into intervals produced an interval with specific values $\{9, 25, 26, 26.4, 27, 28.1, 29, 30\}$. Or, better yet, suppose there were many many observations between 25 and 30 along with the single value 9. In mathematical terms, our interval, after aggregation, can be formally written as $[a,b]$, where

$$a = \min_{i \in \mathcal{X}}\{x_i\}, \quad b = \max_{i \in \mathcal{X}}\{x_i\}, \tag{2.3.1}$$

where $\mathcal{X}$ is the set of all $x_i$ values aggregated into the interval $[a, b]$. In this case, we obtain the interval $[9, 30]$. However, intuitively, we conclude that the value 9 is an outlier and really does not belong to the aggregations in the interval $[25, 30]$. Suppose instead of the value 9, we had a value 21, which, from Eq. (2.3.1), gives the interval $[21, 30]$. Now, it may not be at all clear if the value 21 is an outlier or if it truly belongs to the interval of aggregated values. Since most analyses involving interval data assume that observations within an interval are uniformly spread across that interval, the question becomes one of testing for uniformity across those intervals. Stéphan (1998), Stéphan et al. (2000), and Cariou and Billard (2015) have developed tests of uniformity, gap tests and distance tests, to help address this issue. They also give some reduction algorithms to achieve the deletion of genuine outliers.

## 2.4 Descriptive Statistics

In this section, basic descriptive statistics, such as sample means, sample variances and covariances, and histograms, for the differing types of symbolic data are briefly described. For quantitative data, these definitions implicitly assume that within each interval, or sub-interval for histogram observations, observations are uniformly spread across that interval. Expressions for the sample mean and sample variance for interval data were first derived by Bertrand and Goupil (2000). Adjustments for non-uniformity can be made. For list multi-valued data, the sample mean and sample variance given herein are simply the respective classical values for the probabilities associated with each of the corresponding categories in the variable domain.

## 2.4.1 Sample Means

**Definition 2.10** Let $Y_u$, $u = 1, \ldots, m$, be a random sample of size $m$, with $Y_u$ taking modal list multi-valued values $Y_u = \{Y_{uk}, p_{uk}; k = 1, \ldots, s\}$ from the domain $\mathcal{Y} = \{Y_1, \ldots, Y_s\}$ (as defined in Definition 2.2). Then, the **sample mean** for **list, multi-valued** observations is given by

$$\bar{Y} = \{Y_k, \bar{p}_k; k = 1, \ldots, s\}, \quad \bar{p}_k = \frac{1}{m} \sum_{u=1}^{m} p_{uk}, \tag{2.4.1}$$

where, without loss of generality, it is assumed that the number of categories from $\mathcal{Y}$ contained in $Y_u$ is $s_u = s$ for all observations by suitably setting $p_{uk} = 0$ where appropriate, and where, for non-modal list observations, it is assumed that each category that occurs has the probability $p_{uk} = 1/m$ and those that do not occur have probability $p_{uk} = 0$ (see section 2.2.2). □

**Example 2.11** Consider the deaths attributable to smoking data of Table 2.2. It is easy to show, by applying Eq. (2.4.1), that

$$\bar{Y} = \{\text{smoking}, 0.687; \text{lung cancer}, 0.167; \text{respiratory}, 0.146\},$$

where we have assumed in observation $u = 7$ that the latter two categories have occurred with equal probability, i.e., $p_{72} = p_{72} = 0.176$ (see Example 2.2). □

**Definition 2.11** Let $Y_u$, $u = 1, \ldots, m$, be a random sample of size $m$, with $Y_u = [a_u, b_u]$ taking interval values (as defined in Definition 2.3). Then, the **interval sample mean** $\bar{Y}$ is given by

$$\bar{Y} = \frac{1}{2m} \sum_{u=1}^{m} (a_u + b_u). \tag{2.4.2}$$

□

**Definition 2.12** Let $Y_u$, $u = 1, \ldots, m$, be a random sample of size $m$, with $Y_u$ taking histogram values (as defined in Definition 2.4), $Y_u = \{[a_{uk}, b_{uk}), p_{uk}; k = 1, \ldots, s_u\}$, $u = 1, \ldots, m$. Then, the **histogram sample mean** $\bar{Y}$ is

$$\bar{Y} = \frac{1}{2m} \sum_{u=1}^{m} \sum_{k=1}^{s_u} (a_{uk} + b_{uk}) p_{uk}. \tag{2.4.3}$$

□

**Example 2.12** Take the joggers data of Table 2.5 and Example 2.6. Consider pulse rate $Y_1$. Applying Eq. (2.4.2) gives

$$\bar{Y}_1 = [(73 + 114) + \cdots + (40 + 60)]/(2 \times 10) = 77.150.$$

Likewise, for the histogram values for running time $Y_2$, from Eq. (2.4.3), we have

$$\bar{Y}_2 = [\{(5.3 + 6.2) \times 0.3 + \cdots + (7.1 + 8.3) \times 0.2\} + \ldots$$
$$+ \{3.2 + 4.1) \times 0.6 + (4.1 + 6.7) \times 0.4\}]/(2 \times 10) = 5.866.$$

□

### 2.4.2 Sample Variances

**Definition 2.13** Let $Y_u$, $u = 1, \ldots, m$, be a random sample of size $m$, with $Y_u$ taking modal list or multi-valued values $Y_u = \{Y_{uk}, p_{uk}, k = 1, \ldots, s\}$ from the domain $\mathcal{Y} = \{Y_1, \ldots, Y_s\}$. Then, the **sample variance** for **list, multi-valued** data is given by

$$S^2 = \{Y_k, S_k^2; k = 1, \ldots, s\}, \quad S_k^2 = \frac{1}{m-1} \sum_{u=1}^{m} (p_{uk} - \bar{p}_k)^2, \tag{2.4.4}$$

where $\bar{p}_k$ is given in Eq. (2.4.1) and where, as in Definition 2.10, without loss of generality, we assume all possible categories from $\mathcal{Y}$ occur with some probabilities being zero as appropriate. □

**Example 2.13** For the smoking deaths data of Table 2.2, by applying Eq. (2.4.4) and using the sample mean $\bar{p} = (0.687, 0.167, 0.146)$ from Example 2.11, we can show that the sample variance $S^2$ and standard deviation $S$ are, respectively, calculated as

$$S^2 = \{\text{smoking}, 0.0165; \text{lung cancer}, 0.0048; \text{respiratory}, 0.0037\},$$
$$S = \{\text{smoking}, 0.128; \text{lung cancer}, 0.069; \text{respiratory}, 0.061\}.$$

□

**Definition 2.14** Let $Y_u$, $u = 1, \ldots, m$, be a random sample of size $m$, with $Y_u = [a_u, b_u]$ taking interval values with sample mean $\bar{Y}$ as defined in Eq. (2.4.2). Then, the **interval sample variance** $S^2$ is given by

$$S^2 = \frac{1}{3m} \sum_{u=1}^{m} [(a_u - \bar{Y})^2 + (a_u - \bar{Y})(b_u - \bar{Y}) + (b_u - \bar{Y})^2]. \tag{2.4.5}$$

□

Let us consider Eq. (2.4.5) more carefully. For these observations, it can be shown that the total sum of squares (SS), Total SS, i.e., $mS^2$, can be written as

$$mS^2 = \sum_{u=1}^{m} [(a_u + b_u)/2 - \bar{Y}]^2 + \sum_{u=1}^{m} [(a_u - \bar{Y}_u)^2 + (a_u - \bar{Y}_u)(b_u - \bar{Y}_u)$$
$$+ (b_u - \bar{Y}_u)^2]/3, \tag{2.4.6}$$

where $\bar{Y}$ is the overall mean of Eq. (2.4.2), and where the sample mean of the observation $Y_u$ is

$$\bar{Y}_u = (a_u + b_u)/2, \ u = 1, \ldots, m. \tag{2.4.7}$$

The term inside the second summation in Eq. (2.4.6) equals $S^2$ given in Eq. (2.4.5) when $m = 1$. That is, this is a measure of the internal variation, the internal variance, of the single observation $Y_u$. When summed over all such observations, $u = 1, \ldots, m$, we obtain the internal variation of all $m$ observations; we call this the Within SS. To illustrate, suppose we have a single observation $Y = [7, 13]$. Then, substituting into Eq. (2.4.5), we obtain the sample variance as $S^2 = 3 \neq 0$, i.e., interval observations each contain internal variation. The first term in Eq. (2.4.6) is the variation of the interval midpoints across all observations, i.e., the Between SS.

Hence, we can write

$$\text{Total SS} = \text{Between SS} + \text{Within SS} \tag{2.4.8}$$

where

$$\text{Between SS} = \sum_{u=1}^{m} [(a_u + b_u)/2 - \bar{Y}]^2, \tag{2.4.9}$$

$$\text{Within SS} = \sum_{u=1}^{m} [(a_u - \bar{Y}_u)^2 + (a_u - \bar{Y}_u)(b_u - \bar{Y}_u) + (b_u - \bar{Y}_u)^2]/3. \tag{2.4.10}$$

By assuming that values across an interval are uniformly spread across the interval, we see that the Within SS can also be obtained from

$$\text{Within SS} = \sum_{u=1}^{m} (b_u - a_u)^2/12.$$

Therefore, researchers, who upon aggregation of sets of classical data restrict their analyses to the average of the symbolic observation (such as interval means) are discarding important information; they are ignoring the internal variations (i.e., the Within SS) inherent to their data.

When the data are classically valued, with $Y_u = a_u \equiv [a_u, a_u]$, then $\bar{Y}_u = a_u$ and hence the Within SS of Eq. (2.4.10) is zero and the Between SS of Eq. (2.4.9) is the same as the Total SS for classical data. Hence, the sample variance of Eq. (2.4.5) for interval data reduces to its classical counterpart for classical point data, as it should.

**Definition 2.15** Let $Y_u$, $u = 1, \ldots, m$, be a random sample of size $m$, with $Y_u$ taking histogram values, $Y_u = \{[a_{uk}, b_{uk}), p_{uk}; k = 1, \ldots, s_u\}$, $u = 1, \ldots, m$, and let the sample mean $\bar{Y}$ be as defined in Eq. (2.4.3). Then, the **histogram sample variance** $S^2$ is

$$S^2 = \frac{1}{3m} \sum_{u=1}^{m} \sum_{k=1}^{s_u} \{[(a_{uk} - \bar{Y})^2 + (a_{uk} - \bar{Y})(b_{uk} - \bar{Y}) + (b_{uk} - \bar{Y})^2]p_{uk}\}.$$

(2.4.11)

□

As for intervals, we can show that the total variation for histogram data consists of two parts, as in Eq. (2.4.8), where now its components are given, respectively, by

$$\text{Between SS} = \sum_{u=1}^{m} \sum_{k=1}^{s_u} \{[(a_{uk} + b_{uk})/2 - \bar{Y}]^2 p_{uk}\},$$

(2.4.12)

$$\text{Within SS} = \sum_{u=1}^{m} \sum_{k=1}^{s_u} \{[(a_{uk} - \bar{Y}_u)^2 + (a_{uk} - \bar{Y}_u)(b_{uk} - \bar{Y}_u) + (b_{uk} - \bar{Y}_u)^2]p_{uk}\}/3$$

(2.4.13)

with

$$\bar{Y}_u = \sum_{k=1}^{s_u} p_{uk}(a_{uk} + b_{uk})/2, \ u = 1, \dots, m.$$

(2.4.14)

It is readily seen that for the special case of interval data, where now $s_u = 1$ and hence $p_{u1} = 1$ for all $u = 1, \dots, m$, the histogram sample variance of Eq. (2.4.11) reduces to the interval sample variance of Eq. (2.4.5).

**Example 2.14**  Consider the joggers data of Table 2.5. From Example 2.12, we know that the sample means are, respectively, $\bar{Y}_1 = 77.150$ for pulse rate and $\bar{Y}_2 = 5.866$ for running time. Then applying Eqs. (2.4.5) and (2.4.11), respectively, to the interval data for pulse rates and the histogram data for running times, we can show that the sample variances and hence the sample standard deviations are, respectively,

$$S_1^2 = 197.611; \quad S_1 = 14.057; \quad S_2^2 = 1.458; \quad S_2 = 1.207.$$

□

### 2.4.3 Sample Covariance and Correlation

When the number of variables $p \geq 2$, it is of interest to obtain measures of how these variables depend on each other. One such measure is the covariance. We note that for modal data it is necessary to know the corresponding probabilities for the pairs of each cross-sub-intervals in order to calculate the covariances. This is not an issue for interval data since there is only one possible cross-interval/rectangle for each observation.

**Definition 2.16** Let $\mathbf{Y}_u = (Y_1, Y_2)$, $u = 1, \ldots, m$, be a random sample of interval data with $Y_{uj} = [a_{uj}, b_{uj}]$, $j = 1, 2$. Then, the **sample covariance** between $Y_1$ and $Y_2$, $S_{12}$, is given by

$$S_{12} = \frac{1}{6m} \sum_{u=1}^{m} [2(a_{u1} - \bar{Y}_1)(a_{u2} - \bar{Y}_2) + (a_{u1} - \bar{Y}_1)(b_{u2} - \bar{Y}_2)$$
$$+ (b_{u1} - \bar{Y}_1)(a_{u2} - \bar{Y}_2) + 2(b_{u1} - \bar{Y}_1)(b_{u2} - \bar{Y}_2)]. \tag{2.4.15}$$

□

As for the variance, we can show that the sum of products (SP) satisfies

$$mS_{12} = \text{Total SP} = \text{Between SP} + \text{Within SP} \tag{2.4.16}$$

where

$$\text{Between SP} = \sum_{u=1}^{m} [(a_{u1} + b_{u1})/2 - \bar{Y}_1][(a_{u2} + b_{u2})/2 - \bar{Y}_2], \tag{2.4.17}$$

$$\text{Within SP} = \frac{1}{6} \sum_{u=1}^{m} [2(a_{u1} - \bar{Y}_{u1})(a_{u2} - \bar{Y}_{u2}) + (a_{u1} - \bar{Y}_{u1})(b_{u2} - \bar{Y}_{u2})$$
$$+ (b_{u1} - \bar{Y}_{u1})(a_{u2} - \bar{Y}_{u2}) + 2(b_{u1} - \bar{Y}_{u1})(b_{u2} - \bar{Y}_{u2})] \tag{2.4.18}$$

$$= \sum_{u=1}^{m} (b_{u1} - a_{u1})(b_{u2} - a_{u2})/12,$$

$$\bar{Y}_{uj} = \frac{1}{2}(a_{uj} + b_{uj}), \ j = 1, 2, \tag{2.4.19}$$

with $\bar{Y}_j$, $j = 1, 2$, obtained from Eq. (2.4.2).

**Example 2.15** Consider the $m = 6$ minimum and maximum temperature observations for the variables $Y_1 = $ January and $Y_2 = $ July of Table 2.3 (and Example 2.4). From Eq. (2.4.2), we calculate the sample means $\bar{Y}_1 = -0.40$ and $\bar{Y}_2 = 23.09$. Then, from Eq. (2.4.15), we have

$$S_{12} = \frac{1}{6 \times 6} [2(-18.4 - (-0.4))(17.0 - 23.09) + (-18.4 - (-0.4))(26.5 - 23.09)$$
$$+ (-7.5 - (-0.04))(17.0 - 23.09) + 2((-7.5 - (-0.04))(26.5 - 23.09)]$$
$$+ \ldots$$
$$+ [2(11.8 - (-0.4))(25.6 - 23.09) + (11.8 - (-0.4))(32.6 - 23.09)$$
$$+ (19.2 - (-0.04))(25.6 - 23.09) + 2((19.2 - (-0.04))(32.6 - 23.09)]$$
$$= 69.197.$$

We can also calculate the respective standard deviations, $S_1 = 14.469$ and $S_2 = 6.038$, from Eq. (2.4.5). Hence, the correlation coefficient (see Definition 2.18 and Eq. (2.4.24)) is

$$Corr(Y_1, Y_2) = \frac{S_{12}}{S_1 S_2} = \frac{69.197}{14.469 \times 6.038} = 0.792.$$

□

**Definition 2.17** Let $Y_{uj}$, $u = 1, \ldots, m$, $j = 1, 2$, be a random sample of size $m$, with $Y_u$ taking joint histogram values, $(Y_{u1}, Y_{u2}) = \{[a_{u1k_1}, b_{u1k_1}), [a_{u2k_2}, b_{u2k_2}), p_{uk_1 k_2}; k_j = 1, \ldots, s_{uj}, j = 1, 2\}$, $u = 1, \ldots, m$, where $p_{uk_1 k_2}$ is the relative frequency associated with the rectangle $[a_{u1k_1}, b_{u1k_1}) \times [a_{u2k_2}, b_{u2k_2})$, and let the sample means $\bar{Y}_j$ be as defined in Eq. (2.4.3) for $j = 1, 2$. Then, the **histogram sample covariance**, $Cov(Y_1, Y_2) = S_{12}$, is

$$S_{12} = \frac{1}{6m} \sum_{u=1}^{m} \sum_{k_1=1}^{s_{u1}} \sum_{k_2=1}^{s_{u2}} [2(a_{u1k_1} - \bar{Y}_1)(a_{u2k_2} - \bar{Y}_2) + (a_{u1k_1} - \bar{Y}_1)(b_{u2k_2} - \bar{Y}_2)$$

$$+ (b_{u1k_1} - \bar{Y}_1)(a_{u2k_2} - \bar{Y}_2) + 2(b_{u1k_1} - \bar{Y}_1)(b_{u2k_2} - \bar{Y}_2)]p_{uk_1 k_2}.$$

(2.4.20)

□

As for the variance, we can show that Eq. (2.4.16) holds where now

$$\text{Between SP} = \sum_{u=1}^{m} \sum_{k_1=1}^{s_{u1}} \sum_{k_2=1}^{s_{u2}} [(a_{u1k_1} + b_{u1k_1})/2 - \bar{Y}_1]$$

$$\times [(a_{u2k_2} + b_{u2k_2})/2 - \bar{Y}_2]p_{uk_1 k_2}, \qquad (2.4.21)$$

$$\text{Within SP} = \frac{1}{6} \sum_{u=1}^{m} \sum_{k_1=1}^{s_{u1}} \sum_{k_2=1}^{s_{u2}} [2(a_{u1k_1} - \bar{Y}_{u1})(a_{u2k_2} - \bar{Y}_{u2})$$

$$+ (a_{u1k_1} - \bar{Y}_{u1})(b_{u2k_2} - \bar{Y}_{u2}) + (b_{u1k_1} - \bar{Y}_{u1})(a_{u2k_2} - \bar{Y}_{u2})$$

$$+ 2(b_{u1k_1} - \bar{Y}_{u1})(b_{u2k_2} - \bar{Y}_{u2})]p_{uk_1 k_2} \qquad (2.4.22)$$

$$= \sum_{u=1}^{m} \sum_{k_1=1}^{s_{u1}} \sum_{k_2=1}^{s_{u2}} (b_{u1k_1} - a_{u1k_1})(b_{u2k_2} - a_{u2k_2})p_{uk_1 k_2}/12,$$

$$\bar{Y}_{uj} = \frac{1}{2} \sum_{k_1=1}^{s_{u1}} \sum_{k_2=1}^{s_{u2}} (a_{ujk_j} + b_{ujk_j})p_{uk_1 k_2}, \; j = 1, 2, \qquad (2.4.23)$$

with $\bar{Y}_j$, $j = 1, 2$, obtained from Eq. (2.4.3).

**Definition 2.18** The Pearson (1895) product-moment **correlation coefficient** between two variables $Y_1$ and $Y_2$, $r_{sym}(Y_1, Y_2)$, for symbolic-valued observations is given by

$$r_{sym}(Y_1, Y_2) = \frac{Cov(Y_1, Y_2)}{S_{Y_1} S_{Y_2}}. \qquad (2.4.24)$$

□

**Example 2.16** Table 2.12 gives the joint histogram observations for the random variables $Y_1$ = flight time (AirTime) and $Y_2$ = arrival delay time (ArrDelay) in minutes for airlines traveling into a major airport hub. The original values were aggregated across airline carriers into the histograms shown in these tables. (The data of Table 2.4 and Example 2.5 deal only with the single variable $Y_1$ = flight time. Here, we need the joint probabilities $p_{uk_1 k_2}$.)

Let us take the first $m = 5$ airlines only. Then, from Eq. (2.4.3), we have the sample means as $\bar{Y}_1 = 36.448$ and $\bar{Y}_2 = 3.384$. From Eq. (2.4.20), the covariance function, $Cov(Y_1, Y_2)$, is calculated as

$$
\begin{aligned}
S_{12} = \frac{1}{6 \times 5} \{ & [2(25 - 36.448)(-40 - 3.384) + (25 - 36.448)(-20 - 3.384) \\
& + (50 - 36.448)(-40 - 3.384) + 2(50 - 36.448)(-20 - 3.384)]0.0246 \\
& + \dots \\
& + [2(100 - 36.448)(35 - 3.384) + (100 - 36.448)(60 - 3.384) \\
& + (120 - 36.448)(35 - 3.384) + 2(120 - 36.448)(60 - 3.384)]0.0056 \} \\
= & \ 119.524.
\end{aligned}
$$

Likewise, from Eq. (2.4.11), the sample variances for $Y_1$ and $Y_2$ are, respectively, $S_1^2 = 1166.4738$ and $S_s^2 = 280.9856$; hence, the standard deviations are, respectively, $S_1 = 34.154$ and $S_2 = 16.763$. Therefore, the sample correlation function, $Corr(Y_1, Y_2)$, is

$$Corr(Y_1, Y_2) = \frac{S_{12}}{S_1 S_2} = \frac{119.524}{34.154 \times 16.763} = 0.209.$$

Similarly, covariances and hence correlation functions for the variable pairs $(Y_1, Y_3)$ and $(Y_2, Y_3)$, where $Y_3$ = departure delay time (DepDelay) in minutes, can be obtained from Table 2.13 and Table 2.14, respectively, and are left to the reader. □

### 2.4.4 Histograms

Brief descriptions of the construction of a histogram based on interval data and on histogram data, respectively, are presented here. More complete details and examples can be found in Billard and Diday (2006a).

Table 2.12 Airlines joint histogram $(Y_1, Y_2)$ (Example 2.16). $Y_1$ = flight time in minutes, $Y_2$ = arrival delay time in minutes

| | u = 1 | | | u = 2 | | | u = 3 | | | u = 4 | | | u = 5 | | |
|---|---|---|---|---|---|---|---|---|---|---|---|---|---|---|---|
| $[a_{u1k}, b_{u1k}]$ | $[a_{u1k}, b_{u1k}]$ | $[a_{u2k}, b_{u2k}]$ | $p_{uk_1k_2}$ | $[a_{u1k}, b_{u1k}]$ | $[a_{u2k}, b_{u2k}]$ | $p_{uk_1k_2}$ | $[a_{u1k}, b_{u1k}]$ | $[a_{u2k}, b_{u2k}]$ | $p_{uk_1k_2}$ | $[a_{u1k}, b_{u1k}]$ | $[a_{u2k}, b_{u2k}]$ | $p_{uk_1k_2}$ | $[a_{u1k}, b_{u1k}]$ | $[a_{u2k}, b_{u2k}]$ | $p_{uk_1k_2}$ |
| [25, 50] | [10, 50) | [−40, −20) | 0.0246 | [10, 50) | [−30, −10) | 0.0113 | [10, 50) | [−50, −20) | 0.0143 | [20, 35) | [−35, −15) | 0.0062 | [20, 40) | [−30, −15) | 0.0808 |
| | | [−20, 0) | 0.1068 | | [−10, 10) | 0.0676 | | [−20, 0) | 0.0297 | | [−15, 10) | 0.0412 | | [−15, 5) | 0.0874 |
| | | [0, 25) | 0.0867 | | [10, 30) | 0.0218 | | [0, 30) | 0.0132 | | [10, 35) | 0.0075 | | [5, 35) | 0.0220 |
| | | [25, 50) | 0.0293 | | [30, 50] | 0.0166 | | [30, 80] | 0.0116 | | [35, 60] | 0.0106 | | [35, 60] | 0.0080 |
| | | [50, 75] | 0.0328 | | | | | | | | | | | | |
| [50, 75] | [50, 90) | [−40, −20) | 0.0215 | [50, 90) | [−30, −10) | 0.0689 | [50, 90) | [−50, −20) | 0.0388 | [35, 50) | [−35, −15) | 0.0301 | [40, 60) | [−30, −15) | 0.0714 |
| | | [−20, 0) | 0.1013 | | [−10, 10) | 0.2293 | | [−20, 0) | 0.1725 | | [−15, 10) | 0.1950 | | [−15, 5) | 0.2836 |
| | | [0, 25) | 0.0921 | | [10, 30) | 0.0976 | | [0, 30) | 0.1047 | | [10, 35) | 0.0674 | | [5, 35) | 0.0933 |
| | | [25, 50) | 0.0398 | | [30, 50] | 0.0802 | | [30, 80] | 0.0521 | | [35, 60] | 0.0443 | | [35, 60] | 0.0255 |
| | | [50, 75] | 0.0463 | | | | | | | | | | | | |
| [75, 100] | [90, 130) | [−40, −20) | 0.0070 | [90, 130) | [−30, −10) | 0.0336 | [90, 130) | [−50, −20) | 0.0535 | [50, 65) | [−35, −15) | 0.0182 | [60, 80) | [−30, −15) | 0.0172 |
| | | [−20, 0) | 0.0677 | | [−10, 10) | 0.1011 | | [−20, 0) | 0.1943 | | [−15, 10) | 0.1503 | | [−15, 5) | 0.0747 |
| | | [0, 25) | 0.0925 | | [10, 30) | 0.0562 | | [0, 30) | 0.1726 | | [10, 35) | 0.0700 | | [5, 35) | 0.0390 |
| | | [25, 50) | 0.0377 | | [30, 50] | 0.0449 | | [30, 80] | 0.0941 | | [35, 60] | 0.0430 | | [35, 60] | 0.0130 |
| | | [50, 75] | 0.0449 | | | | | | | | | | | | |
| [100, 125] | [130, 170] | [−40, −20) | 0.0123 | [130, 170] | [−30, −10) | 0.0344 | [130, 170] | [−50, −20) | 0.0023 | [65, 80) | [−35, −15) | 0.0306 | [80, 100) | [−30, −15) | 0.0288 |
| | | [−20, 0) | 0.0420 | | [−10, 10) | 0.0711 | | [−20, 0) | 0.0097 | | [−15, 10) | 0.1126 | | [−15, 5) | 0.0626 |
| | | [0, 25) | 0.0558 | | [10, 30) | 0.0418 | | [0, 30) | 0.0186 | | [10, 35) | 0.0381 | | [5, 35) | 0.0314 |
| | | [25, 50) | 0.0258 | | [30, 50] | 0.0235 | | [30, 80] | 0.0181 | | [35, 60] | 0.0270 | | [35, 60] | 0.0083 |
| | | [50, 75] | 0.0330 | | | | | | | [80, 95] | [−35, −15) | 0.0027 | [100, 120] | [−30, −15) | 0.0045 |
| | | | | | | | | | | | [−15, 10) | 0.0585 | | [−15, 5) | 0.0210 |
| | | | | | | | | | | | [10, 35) | 0.0301 | | [5, 35) | 0.0217 |
| | | | | | | | | | | | [35, 60] | 0.0164 | | [35, 60] | 0.0057 |

(Continued)

**Table 2.12** (Continued)

| $[a_{u1k}, b_{u1k}]$ | $u = 6$ | | | $u = 7$ | | | $u = 8$ | | | $u = 9$ | | | $u = 10$ | | |
|---|---|---|---|---|---|---|---|---|---|---|---|---|---|---|---|
| | $[a_{u2k}, b_{u2k}]$ | $[a_{u1k}, b_{u1k}]$ | $p_{uk_1k_2}$ | $[a_{u2k}, b_{u2k}]$ | $[a_{u1k}, b_{u1k}]$ | $p_{uk_1k_2}$ | $[a_{u2k}, b_{u2k}]$ | $[a_{u1k}, b_{u1k}]$ | $p_{uk_1k_2}$ | $[a_{u1k}, b_{u1k}]$ | $[a_{u2k}, b_{u2k}]$ | $p_{uk_1k_2}$ | $[a_{u1k}, b_{u1k}]$ | $[a_{u2k}, b_{u2k}]$ | $p_{uk_1k_2}$ |
| [40,100) | [-50,-20) | [40,90) | 0.0079 | [-40,-20) | [35,70) | 0.0011 | [-35,-15) | [20,40) | 0.0152 | [200,250) | [-45,-20) | 0.0113 | [200,250) | [-50,-30) | 0.0120 |
| | [-20,10) | | 0.0421 | [-20,0) | | 0.0675 | [-15,5) | | 0.0675 | | [-20,0) | 0.0249 | | [-30,0) | 0.0737 |
| | [10,40) | | 0.0147 | [0,20) | | 0.0660 | [5,25) | | 0.0262 | | [0,30) | 0.0117 | | [0,30) | 0.0556 |
| | [40,80) | | 0.0095 | [20,40] | | 0.0363 | [25,60] | | 0.0193 | | [30,60] | 0.0120 | | [30,60] | 0.0226 |
| | [80,130] | [90,140) | 0.0076 | [-40,-20) | [70,135) | 0.0119 | [-35,-15) | [40,60) | 0.0179 | [250,300) | [-45,-20) | 0.0652 | [250,300) | [-50,-30) | 0.0226 |
| [100,180) | [-50,-20) | | 0.0293 | [-20,0) | | 0.1259 | [-15,5) | | 0.0537 | | [-20,0) | 0.2258 | | [-30,0) | 0.1368 |
| | [-20,10) | | 0.2793 | [0,20) | | 0.1008 | [5,25) | | 0.0207 | | [0,30) | 0.0964 | | [0,30) | 0.1729 |
| | [10,40) | | 0.1216 | [20,40] | | 0.0463 | [25,60] | | 0.0220 | | [30,60] | 0.0709 | | [30,60] | 0.0632 |
| | [40,80) | [140,190) | 0.0568 | [-40,-20) | [135,195) | 0.0197 | [-35,-15) | [60,80) | 0.0758 | [300,350) | [-45,-20) | 0.0180 | [300,350) | [-50,-30) | 0.0135 |
| | [80,130] | | 0.0435 | [-20,0) | | 0.1173 | [-15,5) | | 0.1860 | | [-20,0) | 0.1096 | | [-30,0) | 0.1398 |
| [180,240) | [-50,-20) | | 0.0095 | [0,20) | | 0.1251 | [5,25) | | 0.0978 | | [0,30) | 0.0791 | | [0,30) | 0.1474 |
| | [-20,10) | | 0.0798 | [20,40] | | 0.0886 | [25,60] | | 0.0647 | | [30,60] | 0.0520 | | [30,60] | 0.0391 |
| | [10,40) | [190,240) | 0.0473 | [-40,-20) | [195,255] | 0.0019 | [-35,-15) | [80,100) | 0.0468 | [350,400] | [-45,-20) | 0.0050 | [350,400] | [-50,-30) | 0.0195 |
| | [40,80) | | 0.0189 | [-20,0) | | 0.0038 | [-15,5) | | 0.1708 | | [-20,0) | 0.0239 | | [-30,0) | 0.0481 |
| | [80,130] | | 0.0161 | [0,20) | | 0.0048 | [5,25) | | 0.0826 | | [0,30) | 0.0472 | | [0,30) | 0.0331 |
| [240,320) | [-50,-20) | | 0.0202 | [20,40] | | 0.0119 | [25,60] | | 0.0331 | | [30,60] | 0.0413 | | [30,60] | |
| | [-20,10) | [240,290) | 0.0656 | [-40,-20) | | 0.0314 | | [100,120] | | | [-45,-20) | 0.0050 | | | |
| | [10,40) | | 0.0211 | [-20,0) | | 0.0868 | | | | | [-20,0) | 0.0195 | | | |
| | [40,80) | | 0.0065 | [0,20) | | 0.0400 | | | | | [0,30) | 0.0378 | | | |
| | [80,130] | | 0.0046 | [20,40] | | 0.0130 | | | | | [30,60] | 0.0435 | | | |
| [320,380) | [-50,-20) | | 0.0082 | | | | | | | | | | | | |
| | [-20,10) | | 0.0446 | | | | | | | | | | | | |
| | [10,40) | | 0.0314 | | | | | | | | | | | | |
| | [40,80) | | 0.0077 | | | | | | | | | | | | |
| | [80,130] | | 0.0063 | | | | | | | | | | | | |

**Table 2.13** Airlines joint histogram $(Y_1, Y_3)$ (Example 2.16). $Y_1$ = flight time in minutes, $Y_3$ = departure delay time in minutes

| | u = 1 | | | u = 2 | | | u = 3 | | | u = 4 | | | u = 5 | | |
|---|---|---|---|---|---|---|---|---|---|---|---|---|---|---|---|
| $[a_{u1k}, b_{u1k}]$ | $[a_{u3k}, b_{u3k}]$ | $p_{uk_1k_3}$ | $[a_{u1k}, b_{u1k}]$ | $[a_{u3k}, b_{u3k}]$ | $p_{uk_1k_3}$ | $[a_{u1k}, b_{u1k}]$ | $[a_{u3k}, b_{u3k}]$ | $p_{uk_1k_3}$ | $[a_{u1k}, b_{u1k}]$ | $[a_{u3k}, b_{u3k}]$ | $p_{uk_1k_3}$ | $[a_{u1k}, b_{u1k}]$ | $[a_{u3k}, b_{u3k}]$ | $p_{uk_1k_3}$ | |
| [25, 50] | [−20, −5] | 0.0835 | [10, 50] | [−20, −5] | 0.0227 | [10, 50] | [−30, −5] | 0.0165 | [20, 35] | [−18, −5] | 0.0106 | [20, 40] | [−20, −5] | 0.0636 |
| | [−5, 0) | 0.0767 | | [−5, 0) | 0.0532 | | [−5, 0) | 0.0114 | | [−5, 0) | 0.0319 | | [−5, 0) | 0.0924 |
| | [0, 10) | 0.0381 | | [0, 20) | 0.0292 | | [0, 15) | 0.0242 | | [0, 20) | 0.0111 | | [0, 60] | 0.0423 |
| | [10, 50) | 0.0547 | | [20, 50] | 0.0122 | | [15, 50) | 0.0089 | | [20, 120] | 0.0120 | [40, 60] | [−20, −5] | 0.0978 |
| | [50, 150] | 0.0273 | [50, 90] | [−20, −5] | 0.1238 | | [50, 150] | 0.0079 | [35, 50] | [−18, −5] | 0.0550 | | [−5, 0) | 0.2361 |
| [50, 75] | [−20, −5] | 0.0703 | | [−5, 0) | 0.1844 | [50, 90] | [−30, −5] | 0.0410 | | [−5, 0) | 0.1543 | | [0, 60] | 0.1399 |
| | [−5, 0) | 0.0789 | | [0, 20) | 0.1024 | | [−5, 0) | 0.0896 | | [0, 20) | 0.0718 | [60, 80] | [−20, −5] | 0.0312 |
| | [0, 10) | 0.0459 | | [20, 50] | 0.0654 | | [0, 15) | 0.1572 | | [20, 120] | 0.0559 | | [−5, 0) | 0.0527 |
| | [10, 50) | 0.0681 | [90, 130] | [−20, −5] | 0.0436 | | [15, 50) | 0.0454 | [50, 65] | [−18, −5] | 0.0417 | | [0, 60] | 0.0600 |
| | [50, 150] | 0.0377 | | [−5, 0) | 0.0889 | | [50, 150] | 0.0347 | | [−5, 0) | 0.1246 | [80, 100] | [−20, −5] | 0.0255 |
| [75, 100] | [−20, −5] | 0.0525 | | [0, 20) | 0.0628 | [90, 130] | [−30, −5] | 0.0583 | | [0, 20) | 0.0705 | | [−5, 0) | 0.0546 |
| | [−5, 0) | 0.0550 | | [20, 50] | 0.0405 | | [−5, 0) | 0.0896 | | [20, 120] | 0.0448 | | [0, 60] | 0.0510 |
| | [0, 10) | 0.0472 | [130, 170] | [−20, −5] | 0.0471 | | [0, 15) | 0.2377 | [65, 80] | [−18, −5] | 0.0470 | [100, 120] | [−20, −5] | 0.0116 |
| | [10, 50) | 0.0595 | | [−5, 0) | 0.0615 | | [15, 50) | 0.0753 | | [−5, 0) | 0.0971 | | [−5, 0) | 0.0201 |
| | [50, 150] | 0.0357 | | [0, 20) | 0.0436 | | [50, 150] | 0.0535 | | [0, 20) | 0.0368 | | [0, 60] | 0.0213 |
| [100, 125] | [−20, −5] | 0.0451 | | [20, 50] | 0.0187 | [130, 170] | [−30, −5] | 0.0066 | | [20, 120] | 0.0275 | | | |
| | [−5, 0) | 0.0377 | | | | | [−5, 0) | 0.0068 | [80, 95] | [−18, −5] | 0.0230 | | | |
| | [0, 10) | 0.0297 | | | | | [0, 15) | 0.0181 | | [−5, 0) | 0.0505 | | | |
| | [10, 50) | 0.0316 | | | | | [15, 50) | 0.0088 | | [0, 20) | 0.0230 | | | |
| | [50, 150] | 0.0248 | | | | | [50, 150] | 0.0084 | | [20, 120] | 0.0111 | | | |

*(Continued)*

Table 2.13 (Continued)

| | $u=6$ | | | $u=7$ | | | $u=8$ | | | $u=9$ | | | $u=10$ | | |
|---|---|---|---|---|---|---|---|---|---|---|---|---|---|---|---|
| $[a_{u1k}, b_{u1k}]$ | $[a_{u1k}, b_{u1k}]$ | $[a_{u3k}, b_{u3k}]$ | $p_{uk_1k_3}$ | $[a_{u1k}, b_{u1k}]$ | $[a_{u3k}, b_{u3k}]$ | $p_{uk_1k_3}$ | $[a_{u1k}, b_{u1k}]$ | $[a_{u3k}, b_{u3k}]$ | $p_{uk_1k_3}$ | $[a_{u1k}, b_{u1k}]$ | $[a_{u3k}, b_{u3k}]$ | $p_{uk_1k_3}$ | $[a_{u1k}, b_{u1k}]$ | $[a_{u3k}, b_{u3k}]$ | $p_{uk_1k_3}$ |
| [40,100) | [40,90) | [−20,−5) | 0.0115 | [35,70) | [−20,−5) | 0.0214 | [20,40) | [−15,−5) | 0.0496 | [20,40) | [−20,−5) | 0.0072 | [200,250) | [−16,−5) | 0.0135 |
| | | [−5,0) | 0.0227 | | [−5,0) | 0.0394 | | [−5,0) | 0.0551 | | [−5,0) | 0.0110 | | [−5,0) | 0.0526 |
| | | [0,10) | 0.0151 | | [0,10) | 0.0499 | | [0,40] | 0.0234 | | [0,10) | 0.0154 | | [0,10) | 0.0511 |
| | | [10,50) | 0.0188 | | [10,50) | 0.0534 | [40,60) | [−15,−5) | 0.0413 | | [10,50) | 0.0186 | | [10,30] | 0.0466 |
| | | [50,150] | 0.0136 | | [50,100] | 0.0069 | | [−5,0) | 0.0427 | | [50,100] | 0.0076 | [250,300) | [−16,−5) | 0.0662 |
| [100,180) | [90,140) | [−20,−5) | 0.0714 | [70,135) | [−20,−5) | 0.0252 | | [0,40] | 0.0303 | [40,60) | [−20,−5) | 0.0976 | | [−5,0) | 0.1368 |
| | | [−5,0) | 0.1640 | | [−5,0) | 0.0648 | [60,80) | [−15,−5) | 0.0661 | | [−5,0) | 0.1411 | | [0,10) | 0.1143 |
| | | [0,10) | 0.1195 | | [0,10) | 0.0884 | | [−5,0) | 0.1887 | | [0,10) | 0.0951 | | [10,30] | 0.0782 |
| | | [10,50) | 0.1121 | | [10,50) | 0.0905 | | [0,40] | 0.1694 | | [10,50) | 0.0784 | [300,350) | [−16,−5) | 0.0541 |
| | | [50,150] | 0.0634 | | [50,100] | 0.0159 | [80,100) | [−15,−5) | 0.1019 | | [50,100] | 0.0460 | | [−5,0) | 0.0887 |
| [180,240) | [140,190) | [−20,−5) | 0.0194 | [135,195) | [−20,−5) | 0.0358 | | [−5,0) | 0.1639 | [60,80) | [−20,−5) | 0.0611 | | [0,10) | 0.1158 |
| | | [−5,0) | 0.0435 | | [−5,0) | 0.0897 | | [0,40] | 0.0675 | | [−5,0) | 0.0721 | | [10,30] | 0.0812 |
| | | [0,10) | 0.0468 | | [0,10) | 0.1167 | | | | | [0,10) | 0.0444 | [350,400) | [−16,−5) | 0.0256 |
| | | [10,50) | 0.0377 | | [10,50) | 0.0947 | | | | | [10,50) | 0.0450 | | [−5,0) | 0.0361 |
| | | [50,150] | 0.0241 | | [50,100] | 0.0138 | | | | | [50,100] | 0.0359 | | [0,10) | 0.0271 |
| [240,320) | [190,240] | [−20,−5) | 0.0216 | [195,255] | [−20,−5) | 0.0019 | | | | [80,100) | [−20,−5) | 0.0148 | | [10,30] | 0.0120 |
| | | [−5,0) | 0.0432 | | [−5,0) | 0.0044 | | | | | [−5,0) | 0.0202 | | | |
| | | [0,10) | 0.0278 | | [0,10) | 0.0090 | | | | | [0,10) | 0.0302 | | | |
| | | [10,50) | 0.0170 | | [10,50) | 0.0063 | | | | | [10,50) | 0.0265 | | | |
| | | [50,150] | 0.0084 | | [50,100] | 0.0008 | | | | | [50,100] | 0.0258 | | | |
| [320,380] | [240,290] | [−20,−5) | 0.0104 | [240,290] | [−20,−5) | 0.0289 | | | | [100,120) | [−20,−5) | 0.0192 | | | |
| | | [−5,0) | 0.0292 | | [−5,0) | 0.0633 | | | | | [−5,0) | 0.0183 | | | |
| | | [0,10) | 0.0295 | | [0,10) | 0.0413 | | | | | [0,10) | 0.0195 | | | |
| | | [10,50) | 0.0210 | | [10,50) | 0.0314 | | | | | [10,50) | 0.0214 | | | |
| | | [50,150] | 0.0082 | | [50,100] | 0.0063 | | | | | [50,100] | 0.0274 | | | |

**Table 2.14** Airlines joint histogram $(Y_2, Y_3)$ (Example 2.16). $Y_2$ = arrival delay time in minutes, $Y_3$ = departure delay time in minutes

| | u = 1 | | | u = 2 | | | u = 3 | | | u = 4 | | | u = 5 | | |
|---|---|---|---|---|---|---|---|---|---|---|---|---|---|---|---|
| | $[a_{u2k}, b_{u2k}]$ | $[a_{u3k}, b_{u3k}]$ | $p_{uk_2k_3}$ | $[a_{u2k}, b_{u2k}]$ | $[a_{u3k}, b_{u3k}]$ | $p_{uk_2k_3}$ | $[a_{u2k}, b_{u2k}]$ | $[a_{u3k}, b_{u3k}]$ | $p_{uk_2k_3}$ | $[a_{u2k}, b_{u2k}]$ | $[a_{u3k}, b_{u3k}]$ | $p_{uk_2k_3}$ | $[a_{u2k}, b_{u2k}]$ | $[a_{u3k}, b_{u3k}]$ | $p_{uk_2k_3}$ |
| | [-40, -20) | [-20, -5) | 0.0484 | [-50, -20) | [-20, -5) | 0.0667 | [-35, -15) | [-30, -5) | 0.0469 | [-18, -5) | [-18, -5) | 0.0434 | [-30, -15) | [-20, -5) | 0.1011 |
| | | [-5, 0) | 0.0139 | | [-5, 0) | 0.0693 | | [-5, 0) | 0.0343 | | [-5, 0) | 0.0443 | | [-5, 0) | 0.0912 |
| | | [0, 10] | 0.0031 | | [0, 20] | 0.0122 | | [0, 15) | 0.0275 | [-5, 0) | [-18, -5) | 0.1157 | | [0, 60] | 0.0104 |
| | [-20, 0) | [-20, -5) | 0.1349 | [-20, 0) | [-20, -5) | 0.1369 | | [15, 50] | 0.0002 | | [-5, 0) | 0.3409 | [-15, 5) | [-20, -5) | 0.1101 |
| | | [-5, 0) | 0.1277 | | [-5, 0) | 0.2345 | [-15, 10) | [-30, -5) | 0.0606 | | [0, 20) | 0.1006 | | [-5, 0) | 0.2935 |
| | | [0, 10) | 0.0476 | | [0, 20) | 0.0968 | | [-5, 0) | 0.1156 | | [20, 120] | 0.0004 | | [0, 60] | 0.1257 |
| | | [10, 50] | 0.0076 | | [20, 50] | 0.0009 | | [0, 15) | 0.2191 | [0, 20) | [-18, -5) | 0.0155 | [5, 35) | [-20, -5) | 0.0163 |
| | [0, 25) | [-20, -5) | 0.0601 | [0, 30) | [-20, -5) | 0.0283 | | [15, 50) | 0.0107 | | [-5, 0) | 0.0665 | | [-5, 0) | 0.0640 |
| | | [-5, 0) | 0.0898 | | [-5, 0) | 0.0702 | | [50, 150] | 0.0129 | | [0, 20) | 0.0962 | | [0, 60] | 0.1271 |
| | | [0, 10) | 0.0865 | | [0, 20) | 0.0920 | [10, 35) | [-30, -5) | 0.0447 | | [20, 120] | 0.0350 | [35, 60] | [-20, -5) | 0.0021 |
| | | [10, 50) | 0.0904 | | [20, 50] | 0.0270 | | [-5, 0) | 0.1714 | [20, 120] | [-18, -5) | 0.0027 | | [-5, 0) | 0.0071 |
| | | [50, 150] | 0.0002 | [30, 50] | [-20, -5) | 0.0052 | | [0, 15) | 0.0787 | | [-5, 0) | 0.0066 | | [0, 60] | 0.0513 |
| | [25, 50) | [-20, -5) | 0.0064 | | [-5, 0) | 0.0139 | | [15, 50) | 0.0014 | | [0, 20) | 0.0164 | | | |
| | | [-5, 0) | 0.0129 | | [0, 20) | 0.0371 | | [50, 150] | 0.0020 | | [20, 120] | 0.1157 | | | |
| | | [0, 10) | 0.0197 | | [20, 50] | 0.1090 | [35, 60] | [-30, -5) | 0.0029 | | | | | | |
| | | [10, 50) | 0.0824 | | | | | [-5, 0) | 0.0191 | | | | | | |
| | | [50, 150] | 0.0113 | | | | | [0, 15) | 0.0488 | | | | | | |
| | [50, 75) | [-20, -5) | 0.0016 | | | | | [15, 50] | 0.1030 | | | | | | |
| | | [-5, 0) | 0.0039 | | | | | | | | | | | | |
| | | [0, 10) | 0.0041 | | | | | | | | | | | | |
| | | [10, 50) | 0.0334 | | | | | | | | | | | | |
| | | [50, 150] | 0.1140 | | | | | | | | | | | | |

(Continued)

**Table 2.14** (Continued)

| | u = 6 | | | u = 7 | | | u = 8 | | | u = 9 | | | u = 10 | |
|---|---|---|---|---|---|---|---|---|---|---|---|---|---|---|
| $[a_{u2k}, b_{u2k}]$ | $[a_{u3k}, b_{u3k}]$ | $p_{uk_2k_3}$ | $[a_{u2k}, b_{u2k}]$ | $[a_{u3k}, b_{u3k}]$ | $p_{uk_2k_3}$ | $[a_{u2k}, b_{u2k}]$ | $[a_{u3k}, b_{u3k}]$ | $p_{uk_2k_3}$ | $[a_{u2k}, b_{u2k}]$ | $[a_{u3k}, b_{u3k}]$ | $p_{uk_2k_3}$ | $[a_{u2k}, b_{u2k}]$ | $[a_{u3k}, b_{u3k}]$ | $p_{uk_2k_3}$ |
| [-50, -20] | [-20, -5] | 0.0267 | [-40, -20] | [-20, -5] | 0.0220 | [-35, -15] | [-15, -5] | 0.0689 | [-45, -20] | [-20, -5] | 0.0652 | [-50, -30] | [-16, -5] | 0.0135 |
| | [-5, 0) | 0.0331 | | [-5, 0) | 0.0314 | | [-5, 0) | 0.0813 | | [-5, 0) | 0.0318 | | [-5, 0) | 0.0165 |
| | [0, 10) | 0.0145 | | [0, 10) | 0.0107 | | [0, 40] | 0.0055 | | [0, 10) | 0.0076 | | [0, 10) | 0.0165 |
| | [10, 50] | 0.0008 | | [10, 50] | 0.0019 | [-15, 5) | [-15, -5) | 0.1377 | [-20, 0) | [-20, -5) | 0.1090 | | [10, 30] | 0.0015 |
| [-20, 10) | [-20, -5) | 0.0904 | [-20, 0) | [-20, -5) | 0.0691 | | [-5, 0) | 0.2562 | | [-5, 0) | 0.1729 | [-30, 0) | [-16, -5) | 0.0917 |
| | [-5, 0) | 0.2129 | | [-5, 0) | 0.1601 | | [0, 40] | 0.0840 | | [0, 10) | 0.1043 | | [-5, 0) | 0.1398 |
| | [0, 10) | 0.1545 | | [0, 10) | 0.1446 | [5, 25) | [-15, -5) | 0.0510 | | [10, 50] | 0.0176 | | [0, 10) | 0.1023 |
| | [10, 50] | 0.0533 | | [10, 50] | 0.0275 | | [-5, 0) | 0.0978 | [0, 30) | [-20, -5) | 0.0230 | | [10, 30] | 0.0361 |
| | [50, 150] | 0.0003 | | [50, 150] | 0.0199 | | [0, 40] | 0.0785 | | [-5, 0) | 0.0523 | [0, 30) | [-16, -5) | 0.0496 |
| [10, 40) | [-20, -5) | 0.0147 | [0, 20) | [-20, -5) | 0.0627 | [25, 60] | [-15, -5) | 0.0014 | | [0, 10) | 0.0828 | | [-5, 0) | 0.1353 |
| | [-5, 0) | 0.0486 | | [-5, 0) | 0.1224 | | [-5, 0) | 0.0152 | | [10, 50] | 0.1128 | | [0, 10) | 0.1564 |
| | [0, 10) | 0.0604 | | [0, 10) | 0.1310 | | [0, 40] | 0.1226 | | [50, 100] | 0.0013 | | [10, 30] | 0.0827 |
| | [10, 50] | 0.1069 | | [10, 50] | 0.0006 | | | | [30, 60] | [-20, -5) | 0.0028 | [30, 60] | [-16, -5) | 0.0045 |
| | [50, 150] | 0.0055 | | [50, 150] | 0.0023 | | | | | [-5, 0) | 0.0057 | | [-5, 0) | 0.0226 |
| [40, 80) | [-20, -5) | 0.0025 | [20, 40] | [-20, -5) | 0.0075 | | | | | [0, 10) | 0.0101 | | [0, 10) | 0.0331 |
| | [-5, 0) | 0.0068 | | [-5, 0) | 0.0275 | | | | | [10, 50] | 0.0595 | | [10, 30] | 0.0977 |
| | [0, 10) | 0.0074 | | [0, 10) | 0.1157 | | | | | [50, 100] | 0.1414 | | | |
| | [10, 50] | 0.0378 | | [10, 50] | 0.0430 | | | | | | | | | |
| | [50, 150] | 0.0448 | | | | | | | | | | | | |
| [80, 130] | [-20, -5) | 0.0002 | | | | | | | | | | | | |
| | [-5, 0) | 0.0013 | | | | | | | | | | | | |
| | [0, 10) | 0.0019 | | | | | | | | | | | | |
| | [10, 50] | 0.0077 | | | | | | | | | | | | |
| | [50, 150] | 0.0670 | | | | | | | | | | | | |

**Definition 2.19** Let $Y_u$, $u = 1, \ldots, m$, be a random sample of interval observations with $Y_u = [a_u, b_u] \equiv Z(u)$. Then the **histogram** of these data is the set $\{(p_g, I_g), g = 1, \ldots, r\}$, where $I_g = [a_{hg}, b_{hg})$, $g = 1, \ldots, r-1$, and $[a_{hr}, b_{hr}]$ are the histogram sub-intervals, and $p_g$ is the relative frequency for the sub-interval $I_g$, with

$$p_g = f_g/m, \quad f_g = \sum_{u=1}^{m} \frac{||Z(u) \cap I_g||}{||Z(u)||}, \tag{2.4.25}$$

where $||A||$ is the length of $A$. □

**Example 2.17** Consider the $Y_1 =$ pulse rate interval data of Table 2.5 considered in Example 2.6. Then, a histogram of these data is, from Eq. (2.4.25),

$$\{[40, 55), 0.075; [55, 70), 0.191; [70, 85), 0.451; [85, 100), 0.236; [100, 115], 0.047\}.$$

□

**Definition 2.20** Let $Y_u$, $u = 1, \ldots, m$, be a random sample of histogram observations with $Y_u = \{[a_{uk}, b_{uk}), p_{uk}; k = 1, \ldots, s_u\}$. Then the **histogram** of these data is the set $\{(p_g, I_g), g = 1, \ldots, r\}$, where $I_g = [a_{hg}, b_{hg}), g = 1, \ldots, r-1$, and $[a_{hr}, b_{hr}]$ are the histogram sub-intervals, and $p_g$ is the relative frequency for the sub-interval $I_g$, with

$$p_g = f_g/m, \quad f_g = \sum_{u=1}^{m} \sum_{k=1}^{s_u} \frac{||Z(k; u) \cap I_g||}{||Z(k; u)||} p_{uk}, \tag{2.4.26}$$

where $Z(k; u) = [a_{uk}, b_{uk})$ is the $k$th, $k = 1, \ldots, s_u$, sub-interval of the observation $Y_u$. □

**Example 2.18** Consider the $Y_2 =$ running time histogram data of Table 2.5 considered in Example 2.6. We can show that, from Eq. (2.4.26), a histogram of these histogram observations is

$$\{[1.5, 3.5), 0.042; [3.5, 5.5), 0.281; [5.5, 7.5), 0.596; [7.5, 9.5], 0.081\}.$$

□

## 2.5 Other Issues

There are very few theoretical results underpinning the methodologies pertaining to symbolic data. Some theory justifying the weights associated with modal valued observations, such as capacities, credibilities, necessities, possibilities, and probabilities (briefly described in Definitions 2.6–2.9), can be found in Diday (1995) and Diday and Emilion (2003). These concepts include the union and interception probabilities of Chapter 4, and are the only choice which gives Galois field sets. Their results embody *Choquet* (1954) capacities. Other Galois field theory supporting classification and clustering ideas can be found in Brito and Polaillon (2005).

The descriptive statistics described in section 2.4 are empirically based and are usually moment estimators for the underlying means, variances, and covariances. Le-Rademacher and Billard (2011) have shown that these estimators for the mean and variance of interval data in Eqs. (2.4.2) and (2.4.5), respectively, are the maximum likelihood estimators under reasonable distributional assumptions; likewise, Xu (2010) has shown the moment estimator for the covariance in Eq. (2.4.15) is also the maximum likelihood estimator. These derivations involve separating out the overall distribution from the internal distribution within the intervals, and then invoking conditional moment theory in conjunction with standard maximum likelihood theory. The current work assumed the overall distribution to be a normal distribution with the internal variations following appropriately defined conjugate distributions. Implicit in the formulation of these estimators for the mean, variance, and covariance is the assumption that the points inside a given interval are uniformly spread across the intervals. Clearly, this uniformity assumption can be changed. Le-Rademacher and Billard (2011), Billard (2008), Xu (2010), and Billard et al. (2016) discuss how these changes can be effected, illustrating with an internal triangular distribution. There is a lot of foundational work that still needs to be done here.

By and large, however, methodologies and statistics seem to be intuitively correct when they correspond to their classical counterparts. However, to date, they are not generally rigorously justified theoretically. One governing validity criterion is that, crucially and most importantly, methods developed for symbolic data must produce the corresponding classical results when applied to the special case of classical data.

## Exercises

**2.1**  Show that the histogram data of Table 2.4 for $Y =$ flight time have the sample statistics $\bar{Y} = 122.216$, $S_Y^2 = 7309.588$ and $S_Y = 85.486$.

**2.2**  Refer to Example 2.16 and use the data of Tables 2.12–2.14 for all $m = 10$ airlines for $Y_1 =$ AirTime, $Y_2 =$ ArrDelay, and $Y_3 =$ DepDelay. Show that the sample statistics are $\bar{Y}_1 = 122.216$, $\bar{Y}_2 = 6.960$, $\bar{Y}_3 = 9.733$, $S_{Y_1} = 85.496$, $S_{Y_2} = 25.367$, $S_{Y_3} = 26.712$, $\mathrm{Cov}(Y_1, Y_2) = 110.250$, $\mathrm{Cov}(Y_1, Y_3) = -61.523$, $\mathrm{Cov}(Y_2, Y_3) = 466.482$, $\mathrm{Corr}(Y_1, Y_2) = 0.051$, $\mathrm{Corr}(Y_1, Y_3) = -0.027$, $\mathrm{Corr}(Y_2, Y_3) = 0.688$.

**2.3** Consider the airline data with joint distributions in Tables 2.15–2.17.

(a) Using these tables, calculate the sample statistics $\bar{Y}_j$, $S_j$, $\text{Cov}(Y_{j_1}, Y_{j_2})$ and $\text{Corr}(Y_{j_1}, Y_{j_2})$, $j, j_1, j_2 = 1, 2, 3$.

(b) How do the statistics of (a) differ from those of Exercise 2.2, if at all? If there are differences, what do they tell us about different aggregations?

# Appendix

**Table 2.15** Airlines joint histogram $(Y_1, Y_2)$ (Exercise 2.5.3). $Y_1$ = flight time in minutes, $Y_2$ = arrival delay time in minutes

**$u = 1$**

| $[a_{u1k}, b_{u1k}]$ | $[a_{u2k}, b_{u2k}]$ | $p_{uk_1 k_2}$ |
|---|---|---|
| [10, 50] | [-40, -20] | 0.0246 |
|  | [-20, 0] | 0.1068 |
|  | [0, 25] | 0.0867 |
|  | [25, 50] | 0.0293 |
|  | [50, 75] | 0.0328 |
| [50, 90] | [-40, -20] | 0.0215 |
|  | [-20, 0] | 0.1013 |
|  | [0, 25] | 0.0921 |
|  | [25, 50] | 0.0398 |
|  | [50, 75] | 0.0463 |
| [90, 130] | [-40, -20] | 0.0070 |
|  | [-20, 0] | 0.0677 |
|  | [0, 25] | 0.0925 |
|  | [25, 50] | 0.0377 |
|  | [50, 75] | 0.0449 |
| [130, 170] | [-40, -20] | 0.0123 |
|  | [-20, 0] | 0.0420 |
|  | [0, 25] | 0.0558 |
|  | [25, 50] | 0.0258 |
|  | [50, 75] | 0.0330 |

**$u = 2$**

| $[a_{u1k}, b_{u1k}]$ | $[a_{u2k}, b_{u2k}]$ | $p_{uk_1 k_2}$ |
|---|---|---|
| [10, 50] | [-30, -10] | 0.0113 |
|  | [-10, 10] | 0.0676 |
|  | [10, 30] | 0.0218 |
|  | [30, 50] | 0.0166 |
| [50, 90] | [-30, -10] | 0.0689 |
|  | [-10, 10] | 0.2293 |
|  | [10, 30] | 0.0976 |
|  | [30, 50] | 0.0802 |
| [90, 130] | [-30, -10] | 0.0336 |
|  | [-10, 10] | 0.1011 |
|  | [10, 30] | 0.0562 |
|  | [30, 50] | 0.0449 |
| [130, 170] | [-30, -10] | 0.0344 |
|  | [-10, 10] | 0.0711 |
|  | [10, 30] | 0.0418 |
|  | [30, 50] | 0.0235 |

**$u = 3$**

| $[a_{u1k}, b_{u1k}]$ | $[a_{u2k}, b_{u2k}]$ | $p_{uk_1 k_2}$ |
|---|---|---|
| [10, 50] | [-50, -30] | 0.0032 |
|  | [-30, 0] | 0.0408 |
|  | [0, 30] | 0.0132 |
|  | [30, 60] | 0.0116 |
| [50, 90] | [-50, -30] | 0.0088 |
|  | [-30, 0] | 0.2025 |
|  | [0, 30] | 0.1047 |
|  | [30, 60] | 0.0521 |
| [90, 130] | [-50, -30] | 0.0157 |
|  | [-30, 0] | 0.2320 |
|  | [0, 30] | 0.1726 |
|  | [30, 60] | 0.0941 |
| [130, 170] | [-50, -30] | 0.0005 |
|  | [-30, 0] | 0.0114 |
|  | [0, 30] | 0.0186 |
|  | [30, 60] | 0.0181 |

**$u = 4$**

| $[a_{u1k}, b_{u1k}]$ | $[a_{u2k}, b_{u2k}]$ | $p_{uk_1 k_2}$ |
|---|---|---|
| [20, 35] | [-35, -15] | 0.0062 |
|  | [-15, 10] | 0.0412 |
|  | [10, 35] | 0.0075 |
|  | [35, 60] | 0.0106 |
| [35, 50] | [-35, -15] | 0.0301 |
|  | [-15, 10] | 0.1950 |
|  | [10, 35] | 0.0674 |
|  | [35, 60] | 0.0443 |
| [50, 65] | [-35, -15] | 0.0182 |
|  | [-15, 10] | 0.1503 |
|  | [10, 35] | 0.0700 |
|  | [35, 60] | 0.0430 |
| [65, 80] | [-35, -15] | 0.0306 |
|  | [-15, 10] | 0.1126 |
|  | [10, 35] | 0.0381 |
|  | [35, 60] | 0.0270 |
| [80, 95] | [-35, -15] | 0.0027 |
|  | [-15, 10] | 0.0585 |
|  | [10, 35] | 0.0301 |
|  | [35, 60] | 0.0164 |

**$u = 5$**

| $[a_{u1k}, b_{u1k}]$ | $[a_{u2k}, b_{u2k}]$ | $p_{uk_1 k_2}$ |
|---|---|---|
| [20, 40] | [-30, -15] | 0.0808 |
|  | [-15, 10] | 0.0940 |
|  | [10, 35] | 0.0154 |
|  | [35, 60] | 0.0080 |
| [40, 60] | [-30, -15] | 0.0714 |
|  | [-15, 10] | 0.3155 |
|  | [10, 35] | 0.0614 |
|  | [35, 60] | 0.0255 |
| [60, 80] | [-30, -15] | 0.0172 |
|  | [-15, 10] | 0.0846 |
|  | [10, 35] | 0.0291 |
|  | [35, 60] | 0.0130 |
| [80, 100] | [-30, -15] | 0.0288 |
|  | [-15, 10] | 0.0709 |
|  | [10, 35] | 0.0232 |
|  | [35, 60] | 0.0083 |
| [100, 120] | [-30, -15] | 0.0045 |
|  | [-15, 10] | 0.0279 |
|  | [10, 35] | 0.0149 |
|  | [35, 60] | 0.0057 |

*(Continued)*

**Table 2.15** (Continued)

| u = 6 | | | u = 7 | | | u = 8 | | | u = 9 | | | u = 10 | | |
|---|---|---|---|---|---|---|---|---|---|---|---|---|---|---|
| $[a_{u1k}, b_{u1k}]$ | $[a_{u2k}, b_{u2k}]$ | $p_{uk_1k_2}$ | $[a_{u1k}, b_{u1k}]$ | $[a_{u2k}, b_{u2k}]$ | $p_{uk_1k_2}$ | $[a_{u1k}, b_{u1k}]$ | $[a_{u2k}, b_{u2k}]$ | $p_{uk_1k_2}$ | $[a_{u1k}, b_{u1k}]$ | $[a_{u2k}, b_{u2k}]$ | $p_{uk_1k_2}$ | $[a_{u1k}, b_{u1k}]$ | $[a_{u2k}, b_{u2k}]$ | $p_{uk_1k_2}$ |
| [40,100] | [−50,−20] | 0.0079 | [40,90] | [−40,−20] | 0.0011 | [35,70] | [−35,−15] | 0.0193 | [20,40] | [−50,−30] | 0.0022 | [200,250] | [−50,−30] | 0.0120 |
|  | [−20,10] | 0.0421 |  | [−20,0] | 0.0675 |  | [−15,5] | 0.0634 |  | [−30,0] | 0.0340 |  | [−30,0] | 0.0737 |
|  | [10,40] | 0.0147 |  | [0,20] | 0.0660 |  | [5,25] | 0.0262 |  | [0,30] | 0.0117 |  | [0,30] | 0.0556 |
|  | [40,70] | 0.0074 |  | [20,40] | 0.0363 |  | [25,45] | 0.0096 |  | [30,60] | 0.0120 |  | [30,60] | 0.0226 |
|  | [70,100] | 0.0046 | [90,140] | [−40,−20] | 0.0119 |  | [45,65] | 0.0179 | [40,60] | [−50,−30] | 0.0110 | [250,300] | [−50,−30] | 0.0226 |
|  | [100,130] | 0.0050 |  | [−20,0] | 0.1259 | [70,135] | [−35,−15] | 0.0537 |  | [−30,0] | 0.2800 |  | [−30,0] | 0.1368 |
| [100,180] | [−50,−20] | 0.0293 |  | [0,20] | 0.1008 |  | [−15,5] | 0.0207 |  | [0,30] | 0.0964 |  | [0,30] | 0.1729 |
|  | [−20,10] | 0.2793 |  | [20,40] | 0.0463 |  | [5,25] | 0.0138 |  | [30,60] | 0.0709 |  | [30,60] | 0.0632 |
|  | [10,40] | 0.1216 | [140,190] | [−40,−20] | 0.0197 |  | [25,45] | 0.0083 | [60,80] | [−50,−30] | 0.0013 | [300,350] | [−50,−30] | 0.0135 |
|  | [40,70] | 0.0462 |  | [−20,0] | 0.1173 |  | [45,65] | 0.0813 |  | [−30,0] | 0.1263 |  | [−30,0] | 0.1398 |
|  | [70,100] | 0.0238 |  | [0,20] | 0.1251 | [135,195] | [−35,−15] | 0.1804 |  | [0,30] | 0.0791 |  | [0,30] | 0.1474 |
|  | [100,130] | 0.0303 |  | [20,40] | 0.0886 |  | [−15,5] | 0.0978 |  | [30,60] | 0.0520 |  | [30,60] | 0.0391 |
| [180,240] | [−50,−20] | 0.0095 | [190,240] | [−40,−20] | 0.0019 |  | [5,25] | 0.0234 | [80,100] | [−50,−30] | 0.0003 | [350,400] | [−50,−30] | 0.0000 |
|  | [−20,10] | 0.0798 |  | [−20,0] | 0.0038 |  | [25,45] | 0.0413 |  | [−30,0] | 0.0287 |  | [−30,0] | 0.0195 |
|  | [10,40] | 0.0473 |  | [0,20] | 0.0048 |  | [45,65] | 0.0565 |  | [0,30] | 0.0472 |  | [0,30] | 0.0481 |
|  | [40,70] | 0.0156 |  | [20,40] | 0.0119 | [195,255] | [−35,−15] | 0.1612 |  | [30,60] | 0.0413 |  | [30,60] | 0.0331 |
|  | [70,100] | 0.0079 | [240,290] | [−40,−20] | 0.0314 |  | [−15,5] | 0.0826 | [100,120] | [−50,−30] | 0.0013 |  |  |  |
|  | [100,130] | 0.0115 |  | [−20,0] | 0.0868 |  | [5,25] | 0.0207 |  | [−30,0] | 0.0233 |  |  |  |
| [240,320] | [−50,−20] | 0.0202 |  | [0,20] | 0.0400 |  | [25,45] | 0.0124 |  | [0,30] | 0.0378 |  |  |  |
|  | [−20,10] | 0.0656 |  | [20,40] | 0.0130 |  |  |  |  | [30,60] | 0.0435 |  |  |  |
|  | [10,40] | 0.0211 |  |  |  |  |  |  |  |  |  |  |  |  |
|  | [40,70] | 0.0062 |  |  |  |  |  |  |  |  |  |  |  |  |
|  | [70,100] | 0.0013 |  |  |  |  |  |  |  |  |  |  |  |  |
|  | [100,130] | 0.0036 |  |  |  |  |  |  |  |  |  |  |  |  |
| [320,380] | [−50,−20] | 0.0082 |  |  |  |  |  |  |  |  |  |  |  |  |
|  | [−20,10] | 0.0446 |  |  |  |  |  |  |  |  |  |  |  |  |
|  | [10,40] | 0.0314 |  |  |  |  |  |  |  |  |  |  |  |  |
|  | [40,70] | 0.0065 |  |  |  |  |  |  |  |  |  |  |  |  |
|  | [70,100] | 0.0030 |  |  |  |  |  |  |  |  |  |  |  |  |
|  | [100,130] |  |  |  |  |  |  |  |  |  |  |  |  |  |

**Table 2.16** Airlines joint histogram $(Y_1, Y_3)$ (Exercise 2.5.3). $Y_1$ = flight time in minutes, $Y_3$ = departure delay time in minutes

| | u = 1 | | | u = 2 | | | u = 3 | | | u = 4 | | | u = 5 | | |
|---|---|---|---|---|---|---|---|---|---|---|---|---|---|---|---|
| $[a_{u1k}, b_{u1k})$ | $[a_{u3k}, b_{u3k})$ | $p_{uk_1k_3}$ | $[a_{u1k}, b_{u1k})$ | $[a_{u3k}, b_{u3k})$ | $p_{uk_1k_3}$ | $[a_{u1k}, b_{u1k})$ | $[a_{u3k}, b_{u3k})$ | $p_{uk_1k_3}$ | $[a_{u1k}, b_{u1k})$ | $[a_{u3k}, b_{u3k})$ | $p_{uk_1k_3}$ | $[a_{u1k}, b_{u1k})$ | $[a_{u3k}, b_{u3k})$ | $p_{uk_1k_3}$ |
| [25, 50) | [−20, −5) | 0.0835 | [10, 50) | [−20, −5) | 0.0227 | [35, 50) | [−30, −5) | 0.0165 | [20, 35) | [−18, −5) | 0.0106 | [20, 40) | [−20, −5) | 0.0636 |
| | [−5, 0) | 0.0767 | | [−5, 0) | 0.0532 | | [−5, 0) | 0.0114 | | [−5, 0) | 0.0319 | | [−5, 0) | 0.0924 |
| | [0, 10) | 0.0381 | | [0, 20) | 0.0292 | | [0, 10) | 0.0224 | | [0, 20) | 0.0111 | | [0, 60] | 0.0423 |
| | [10, 50) | 0.0547 | | [20, 50) | 0.0122 | | [10, 50) | 0.0107 | | [20, 120] | 0.0120 | [40, 60) | [−20, −5) | 0.0978 |
| | [50, 150] | 0.0273 | [50, 90) | [−20, −5) | 0.1238 | | [50, 150] | 0.0079 | [50, 65) | [−18, −5) | 0.0550 | | [−5, 0) | 0.2361 |
| [50, 75) | [−20, −5) | 0.0703 | | [−5, 0) | 0.1844 | [50, 90) | [−30, −5) | 0.0410 | | [−5, 0) | 0.0718 | | [0, 60] | 0.1399 |
| | [−5, 0) | 0.0789 | | [0, 20) | 0.1024 | | [−5, 0) | 0.0896 | | [0, 20) | 0.0559 | [60, 80) | [−20, −5) | 0.0312 |
| | [0, 10) | 0.0459 | | [20, 50) | 0.0654 | | [0, 10) | 0.1447 | | [20, 120] | 0.0417 | | [−5, 0) | 0.0527 |
| | [10, 50) | 0.0681 | [90, 130) | [−20, −5) | 0.0436 | | [10, 50) | 0.0580 | [65, 80) | [−18, −5) | 0.1246 | | [0, 60] | 0.0600 |
| | [50, 150] | 0.0377 | | [−5, 0) | 0.0889 | | [50, 150] | 0.0347 | | [−5, 0) | 0.0705 | [80, 100) | [−20, −5) | 0.0255 |
| [75, 100) | [−20, −5) | 0.0525 | | [0, 20) | 0.0628 | [90, 130) | [−30, −5) | 0.0583 | | [0, 20) | 0.0448 | | [−5, 0) | 0.0546 |
| | [−5, 0) | 0.0550 | | [20, 50) | 0.0405 | | [−5, 0) | 0.0896 | | [20, 120] | 0.0470 | | [0, 60] | 0.0510 |
| | [0, 10) | 0.0472 | [130, 170) | [−20, −5) | 0.0471 | | [0, 10) | 0.2165 | [80, 95] | [−18, −5) | 0.0971 | [100, 120) | [−20, −5) | 0.0116 |
| | [10, 50) | 0.0595 | | [−5, 0) | 0.0615 | | [10, 50) | 0.0966 | | [−5, 0) | 0.0368 | | [−5, 0) | 0.0201 |
| | [50, 150] | 0.0357 | | [0, 20) | 0.0436 | | [50, 150] | 0.0535 | | [0, 20) | 0.0275 | | [0, 60] | 0.0213 |
| [100, 125] | [−20, −5) | 0.0451 | | [20, 50) | 0.0187 | [130, 170) | [−30, −5) | 0.0066 | | [20, 120] | 0.0230 | | | |
| | [−5, 0) | 0.0377 | | | | | [−5, 0) | 0.0068 | | [−5, 0) | 0.0505 | | | |
| | [0, 10) | 0.0297 | | | | | [0, 10) | 0.0161 | | [0, 20) | 0.0230 | | | |
| | [10, 50) | 0.0316 | | | | | [10, 50) | 0.0107 | | [20, 120] | 0.0111 | | | |
| | [50, 150] | 0.0248 | | | | | [50, 150] | 0.0084 | | | | | | |

(Continued)

**Table 2.16** (Continued)

| u = 6 | | | u = 7 | | | u = 8 | | | u = 9 | | | u = 10 | | |
|---|---|---|---|---|---|---|---|---|---|---|---|---|---|---|
| $[a_{u1k}, b_{u1k}]$ | $[a_{u3k}, b_{u3k}]$ | $p_{uk_1k_3}$ | $[a_{u3k}, b_{u3k}]$ | $[a_{u1k}, b_{u1k}]$ | $p_{uk_1k_3}$ | $[a_{u1k}, b_{u1k}]$ | $[a_{u3k}, b_{u3k}]$ | $p_{uk_1k_3}$ | $[a_{u3k}, b_{u3k}]$ | $[a_{u1k}, b_{u1k}]$ | $p_{uk_1k_3}$ | $[a_{u1k}, b_{u1k}]$ | $[a_{u3k}, b_{u3k}]$ | $p_{uk_1k_3}$ |
| [40,100) | [-18,-5) | 0.0115 | [-20,-5) | [40,90) | 0.0214 | [35,70) | [-15,-5) | 0.0496 | [-20,-5) | [20,40) | 0.0072 | [200,250) | [-16,-5) | 0.0135 |
| | [-5,0) | 0.0227 | [-5,0) | | 0.0394 | | [-5,0) | 0.0551 | [-5,0) | | 0.0110 | | [-5,0) | 0.0526 |
| | [0,10) | 0.0151 | [0,10) | | 0.0499 | | [0,40] | 0.0234 | [0,10) | | 0.0154 | | [0,10) | 0.0511 |
| | [10,50) | 0.0188 | [10,50) | | 0.0534 | [70,135) | [-15,-5) | 0.0413 | [10,50) | | 0.0186 | | [10,30] | 0.0466 |
| | [50,150] | 0.0136 | [50,100] | | 0.0069 | | [-5,0) | 0.0427 | [50,100] | | 0.0076 | [250,300) | [-16,-5) | 0.0662 |
| [100,180) | [-18,-5) | 0.0714 | [-20,-5) | [90,140) | 0.0252 | | [0,40] | 0.0303 | [-20,-5) | [40,60) | 0.0976 | | [-5,0) | 0.1368 |
| | [-5,0) | 0.1640 | [-5,0) | | 0.0648 | [135,195) | [-15,-5) | 0.0661 | [-5,0) | | 0.1411 | | [0,10) | 0.1143 |
| | [0,10) | 0.1195 | [0,10) | | 0.0884 | | [-5,0) | 0.1887 | [0,10) | | 0.0951 | | [10,30] | 0.0782 |
| | [10,50) | 0.1121 | [10,50) | | 0.0905 | | [0,40] | 0.1694 | [10,50) | | 0.0784 | [300,350) | [-16,-5) | 0.0541 |
| | [50,150] | 0.0634 | [50,100] | | 0.0159 | [195,255) | [-15,-5) | 0.1019 | [50,100] | | 0.0460 | | [-5,0) | 0.0887 |
| [180,240) | [-18,-5) | 0.0194 | [-20,-5) | [140,190) | 0.0358 | | [-5,0) | 0.1639 | [-20,-5) | [60,80) | 0.0611 | | [0,10) | 0.1158 |
| | [-5,0) | 0.0435 | [-5,0) | | 0.0897 | | [0,40] | 0.0675 | [-5,0) | | 0.0721 | | [10,30] | 0.0812 |
| | [0,10) | 0.0468 | [0,10) | | 0.1167 | | | | [0,10) | | 0.0444 | [350,400) | [-16,-5) | 0.0256 |
| | [10,50) | 0.0377 | [10,50) | | 0.0947 | | | | [10,50) | | 0.0450 | | [-5,0) | 0.0361 |
| | [50,150] | 0.0241 | [50,100] | | 0.0138 | | | | [50,100] | | 0.0359 | | [0,10) | 0.0271 |
| [240,320) | [-18,-5) | 0.0216 | [-20,-5) | [190,240) | 0.0019 | | | | [-20,-5) | [80,100) | 0.0148 | | [10,30] | 0.0120 |
| | [-5,0) | 0.0432 | [-5,0) | | 0.0044 | | | | [-5,0) | | 0.0202 | | | |
| | [0,10) | 0.0278 | [0,10) | | 0.0090 | | | | [0,10) | | 0.0302 | | | |
| | [10,50) | 0.0170 | [10,50) | | 0.0063 | | | | [10,50) | | 0.0265 | | | |
| | [50,150] | 0.0084 | [50,100] | | 0.0008 | | | | [50,100] | | 0.0258 | | | |
| [320,380) | [-18,-5) | 0.0104 | [-20,-5) | [240,290) | 0.0289 | | | | [-20,-5) | [100,120) | 0.0192 | | | |
| | [-5,0) | 0.0292 | [-5,0) | | 0.0633 | | | | [-5,0) | | 0.0183 | | | |
| | [0,10) | 0.0295 | [0,10) | | 0.0413 | | | | [0,10) | | 0.0195 | | | |
| | [10,50) | 0.0210 | [10,50) | | 0.0314 | | | | [10,50) | | 0.0214 | | | |
| | [50,150] | 0.0082 | [50,100] | | 0.0063 | | | | [50,100] | | 0.0274 | | | |

**Table 2.17** Airlines joint histogram $(Y_2, Y_3)$ (Exercise 2.5.3), $Y_2$ = arrival delay time in minutes, $Y_3$ = departure delay time in minutes

| $[a_{u2k}, b_{u2k}]$ | u = 1 | | u = 2 | | | u = 3 | | | u = 4 | | | u = 5 | | |
|---|---|---|---|---|---|---|---|---|---|---|---|---|---|---|
| | $[a_{u3k}, b_{u3k}]$ | $p_{uk_2k_3}$ | $[a_{u2k}, b_{u2k}]$ | $[a_{u3k}, b_{u3k}]$ | $p_{uk_2k_3}$ | $[a_{u2k}, b_{u2k}]$ | $[a_{u3k}, b_{u3k}]$ | $p_{uk_2k_3}$ | $[a_{u2k}, b_{u2k}]$ | $[a_{u3k}, b_{u3k}]$ | $p_{uk_2k_3}$ | $[a_{u2k}, b_{u2k}]$ | $[a_{u3k}, b_{u3k}]$ | $p_{uk_2k_3}$ |
| [-40, -20) | [-20, -8) | 0.0277 | [-30, -5) | [-20, -5) | 0.1116 | [-50, -20) | [-30, -5) | 0.0469 | [-35, -15] | [-18, -5) | 0.0434 | [-30, -15] | [-20, -5) | 0.1011 |
| | [-8, -1) | 0.0328 | | [-5, 0) | 0.1373 | | [-5, 0) | 0.0343 | | [-5, 0] | 0.0443 | | [-5, 0) | 0.0912 |
| | [-1, 15] | 0.0049 | | [0, 20] | 0.0275 | | [0, 15) | 0.0275 | [-15, 10) | [-18, -5] | 0.1157 | | [0, 60] | 0.0104 |
| [-20, 0) | [-20, -8) | 0.0595 | [-5, 10) | [-20, -5) | 0.0920 | | [15, 50] | 0.0002 | | [-5, 0) | 0.3409 | [-15, 5) | [-20, -5) | 0.1101 |
| | [-8, -1) | 0.1837 | | [-5, 0) | 0.1665 | [-20, 0) | [-30, -5) | 0.0606 | | [0, 20] | 0.1006 | | [-5, 0) | 0.2935 |
| | [-1, 15] | 0.0718 | | [0, 20] | 0.0815 | | [-5, 0) | 0.1156 | | [20, 100] | 0.0004 | | [0, 60] | 0.1257 |
| | [15, 60] | 0.0029 | | [20, 50] | 0.0009 | | [0, 15) | 0.2191 | [10, 35) | [-18, -5) | 0.0155 | [5, 35) | [-20, -5) | 0.0163 |
| [0, 20) | [-20, -8) | 0.0273 | [10, 30) | [-20, -5) | 0.0283 | | [15, 50] | 0.0107 | | [-5, 0) | 0.0665 | | [-5, 0) | 0.0640 |
| | [-8, -1) | 0.0960 | | [-5, 0) | 0.0702 | [10, 35) | [-30, -5) | 0.0129 | | [0, 20] | 0.0962 | | [0, 60] | 0.1271 |
| | [-1, 15] | 0.1171 | | [0, 20] | 0.0920 | | [-5, 0) | 0.0447 | | [20, 100] | 0.0350 | [35, 70] | [-20, -5) | 0.0021 |
| | [15, 60] | 0.0457 | | [20, 50] | 0.0270 | | [0, 15) | 0.1714 | [35, 70] | [-18, -5) | 0.0027 | | [-5, 0) | 0.0071 |
| [20, 50) | [-20, -8) | 0.0037 | [30, 50] | [-20, -5) | 0.0052 | | [15, 50] | 0.0787 | | [-5, 0) | 0.0066 | | [0, 60] | 0.0513 |
| | [-8, -1) | 0.0207 | | [-5, 0) | 0.0139 | | [50, 150] | 0.0014 | | [0, 20] | 0.0164 | | | |
| | [-1, 15] | 0.0459 | | [0, 20] | 0.0371 | [35, 85] | [-30, -5) | 0.0020 | | [20, 100] | 0.1157 | | | |
| | [15, 60] | 0.1007 | | [20, 50] | 0.1090 | | [-5, 0) | 0.0025 | | | | | | |
| | [60, 150] | 0.0027 | | | | | [0, 15) | 0.0168 | | | | | | |
| [50, 75] | [-20, -8) | 0.0006 | | | | | [15, 50] | 0.0463 | | | | | | |
| | [-8, -1) | 0.0037 | | | | | [50, 150] | 0.0504 | | | | | | |
| | [-1, 15] | 0.0076 | | | | | [-5, 0) | 0.0004 | | | | | | |
| | [15, 60] | 0.0515 | | | | | [0, 15) | 0.0023 | | | | | | |
| | [60, 150] | 0.0937 | | | | | [15, 50] | 0.0025 | | | | | | |
| | | | | | | | [50, 150] | 0.0526 | | | | | | |

*(Continued)*

**Table 2.17** (Continued)

| u = 6 | | | u = 7 | | | u = 8 | | | u = 9 | | | u = 10 | | |
|---|---|---|---|---|---|---|---|---|---|---|---|---|---|---|
| $[a_{u2k}, b_{u2k}]$ | $[a_{u3k}, b_{u3k}]$ | $p_{uk_2k_3}$ | $[a_{u2k}, b_{u2k}]$ | $[a_{u3k}, b_{u3k}]$ | $p_{uk_2k_3}$ | $[a_{u2k}, b_{u2k}]$ | $[a_{u3k}, b_{u3k}]$ | $p_{uk_2k_3}$ | $[a_{u2k}, b_{u2k}]$ | $[a_{u3k}, b_{u3k}]$ | $p_{uk_2k_3}$ | $[a_{u2k}, b_{u2k}]$ | $[a_{u3k}, b_{u3k}]$ | $p_{uk_2k_3}$ |
| [−50, −20] | [−18, −5] | 0.0267 | [−40, −20] | [−20, −5] | 0.0220 | [−35, −15] | [−15, −5] | 0.0702 | [−45, −20] | [−20, −5] | 0.0652 | [−50, −30] | [−16, −5] | 0.0135 |
| | [−5, 0] | 0.0331 | | [−5, 0] | 0.0314 | | [−5, 0] | 0.0829 | | [−5, 0] | 0.0318 | | [−5, 0] | 0.0165 |
| | [0, 10] | 0.0145 | | [0, 10] | 0.0107 | | [0, 40] | 0.0056 | | [0, 10] | 0.0076 | | [0, 10] | 0.0165 |
| | [10, 50] | 0.0008 | | [10, 50] | 0.0019 | [−15, 5] | [−15, −5] | 0.1320 | [−20, 0] | [−20, −5] | 0.1090 | | [10, 30] | 0.0015 |
| [−20, 10] | [−18, −5] | 0.0904 | [−20, 0] | [−20, −5] | 0.0691 | | [−5, 0] | 0.2528 | | [−5, 0] | 0.1729 | [−30, 0] | [−16, −5] | 0.0917 |
| | [−5, 0] | 0.2129 | | [−5, 0] | 0.1601 | | [0, 40] | 0.0829 | | [0, 10] | 0.1043 | | [−5, 0] | 0.1398 |
| | [0, 10] | 0.1545 | | [0, 10] | 0.1446 | [5, 25] | [−15, −5] | 0.0520 | | [10, 50] | 0.0176 | | [0, 10] | 0.1023 |
| | [10, 50] | 0.0533 | | [10, 50] | 0.0275 | | [−5, 0] | 0.0997 | [0, 30] | [−20, −5] | 0.0230 | | [10, 30] | 0.0361 |
| | [50, 150] | 0.0003 | | [50, 100] | 0.0199 | | [0, 40] | 0.0801 | | [−5, 0] | 0.0523 | [0, 30] | [−16, −5] | 0.0496 |
| [10, 40] | [−18, −5] | 0.0147 | [0, 20] | [−5, 0] | 0.0627 | [25, 60] | [−15, −5] | 0.0014 | | [0, 10] | 0.0828 | | [−5, 0] | 0.1353 |
| | [−5, 0] | 0.0486 | | [0, 10] | 0.1224 | | [−5, 0] | 0.0154 | | [10, 50] | 0.1128 | | [0, 10] | 0.1564 |
| | [0, 10] | 0.0604 | | [10, 50] | 0.1310 | | [0, 40] | 0.1250 | | [50, 100] | 0.0013 | | [10, 30] | 0.0827 |
| | [10, 50] | 0.1069 | | [50, 100] | 0.0006 | | | | [30, 100] | [−20, −5] | 0.0028 | [30, 60] | [−16, −5] | 0.0045 |
| | [50, 150] | 0.0055 | [20, 40] | [−20, −5] | 0.0023 | | | | | [−5, 0] | 0.0057 | | [−5, 0] | 0.0226 |
| [40, 80] | [−18, −5] | 0.0025 | | [−5, 0] | 0.0075 | | | | | [0, 10] | 0.0101 | | [0, 10] | 0.0331 |
| | [−5, 0] | 0.0068 | | [0, 10] | 0.0275 | | | | | [10, 50] | 0.0595 | | [10, 30] | 0.0977 |
| | [0, 10] | 0.0074 | | [10, 50] | 0.1157 | | | | | [50, 100] | 0.1414 | | | |
| | [10, 50] | 0.0378 | | [50, 100] | 0.0430 | | | | | | | | | |
| | [50, 150] | 0.0448 | | | | | | | | | | | | |
| [80, 150] | [−18, −5] | 0.0002 | | | | | | | | | | | | |
| | [−5, 0] | 0.0013 | | | | | | | | | | | | |
| | [0, 10] | 0.0019 | | | | | | | | | | | | |
| | [10, 50] | 0.0077 | | | | | | | | | | | | |
| | [50, 150] | 0.0670 | | | | | | | | | | | | |

# 3

# Dissimilarity, Similarity, and Distance Measures

Many clustering techniques use dissimilarity, similarity, or distance measures as a basis for forming appropriate clusters. Their role in clustering is covered in Chapters 5–8. In this chapter, we first consider some general features of these measures. We then go on to discuss these measures in detail for two types of symbolic data, namely, non-modal multi-valued (i.e., list) data and interval data, and in Chapter 4 we give the details for measures for modal valued observations, both modal multi-valued (or modal list) data and histogram-valued data. In these cases, all variables are assumed to be of the same type (e.g., all interval-valued). However, this is not a necessary restriction, as seen in the mixed data examples of later chapters (e.g., Example 7.14).

## 3.1   Some General Basic Definitions

We start with a set of objects $\Omega = A, B, \ldots$. The aim in clustering is to form clusters such that those objects that are most alike are clustered together and those that are most unalike are in different clusters. Thus, the notion of a dissimilarity measure, $d(A, B)$, is to calculate how much two particular objects, $A$ and $B$, are not alike, or, if we want to calculate how alike they are, then we have a similarity measure, $s(A, B)$. Typically, a dissimilarity measure is inversely proportional to its similarity measure, e.g., $d(A, B) = 1 - s(A, B)$, or $d(A, B) = 1/s(A, B)$. Therefore, we will consider only dissimilarity measures, rather than similarity measures.

**Definition 3.1**   Let $A$ and $B$ be any two objects in the space $\Omega$. Then, a **dissimilarity measure** between $A$ and $B$, $d(A, B)$, is a measure of how dissimilar these objects are and satisfies

(i)  $d(A, B) = d(A, B)$
(ii)  $d(A, B) > 0$ for all $A \neq B$
(iii)  $d(A, A) = 0$ for all $A \in \Omega$. □

*Clustering Methodology for Symbolic Data*, First Edition. Lynne Billard and Edwin Diday.
© 2020 John Wiley & Sons Ltd. Published 2020 by John Wiley & Sons Ltd.

**Definition 3.2** Let $A$ and $B$ be any two objects in the space $\Omega$. Then, a **distance measure**, $d(A, B)$, between $A$ and $B$ is a dissimilarity measure which satisfies properties (i)–(iii) of Definition 3.1 and also satisfies

(iv) $d(A, B) = 0$ implies $A = B$

(v) $d(A, B) \leq d(A, C) + d(B, C)$ for all $A, B, C \in \Omega$.     □

 

Property (i) tells us that a dissimilarity measure is symmetric. Property (v) is known as the triangle property. A distance measure is sometimes called a **metric**.

**Definition 3.3** Let the space $\Omega$ contain the set of objects $\{A_1, \ldots, A_m\}$. The set of dissimilarity measures, or equally the set of distance measures, $d(A_i, A_j)$, $i, j = 1, \ldots, m$, are the elements of the **dissimilarity matrix**, or **distance matrix, D**.     □

There are two special types of dissimilarity measures that are of particular importance: ultrametric dissimilarity when constructing hierarchies and Robinson dissimilarity when constructing pyramids.

**Definition 3.4** Let $A$, $B$, and $C$ be any three objects in the space $\Omega$. Then, an **ultrametric** is a dissimilarity measure satisfying properties (i)–(iii) of Definition 3.1 and (iv)–(v) of Definition 3.2, and also satisfies

(vi) $d(A, B) \leq \max\{d(A, C), d(B, C)\}$ for all $A, B, C \in \Omega$.     □

 

It can be shown that dissimilarity measures $d(\cdot, \cdot)$ for the set $\Omega = \{A, B, C, \ldots, Z\}$ are ultrametric if and only if all the triangles formed by all triplets $(A, B, C) \subset \Omega$ are isosceles triangles with base smaller than the sides. Ultrametric dissimilarities are important when comparing hierarchies as they are in a one-to-one correspondence with hierarchies (Johnson, 1967). Furthermore, an ultrametric matrix allows for the construction of a hierarchy without crossover branches.

**Definition 3.5** Let $A$, $B$, and $C$ be any three objects in the space $\Omega$. A **pyramdial dissimilarity** is such that there exists an order $O$ on the triple $(A, B, C)$ such that

$$\max\{d(A, C), d(B, C)\} \leq d(A, B). \qquad (3.1.1)$$

    □

 

**Definition 3.6** Let **D** be a dissimilarity, or distance, matrix whose elements are monotonically non-decreasing as they move further away by column and by row from the diagonal. Such a matrix is called a **Robinson matrix** (Robinson, 1951). □

Some researchers define a Robinson matrix as having "increasing" rather than "non-decreasing" elements in **D**.

Any ultrametric matrix can be transformed into a Robinson matrix but not all Robinson matrices can be transformed into an ultrametric matrix. Therefore, a set of Robinson matrices contains a set of ultrametric matrices. Robinson matrices are in a one-to-one correspondence with indexed pyramids, and so play an important role in pyramid construction.

**Example 3.1** (i) Suppose we have objects $A$, $B$, and $C$, with dissimilarity (or, equally in this case, distance) measures $d(A, B) = 2$, $d(A, C) = 3$ and $d(B, C) = 3$ as in Figure 3.1(i). Then, it is easy to see that

$$d(A, B) \leq \max\{d(A, C), d(B, C)\},$$

e.g., $d(A, B) = 2 \leq \max\{d(A, C), d(B, C)\} = 3,$

holds for all three comparisons and so are ultrametric dissimilarities.

The dissimilarity matrix, or ultrametric matrix, is

$$\mathbf{D} = \begin{bmatrix} 0 & 2 & 3 \\ 2 & 0 & 3 \\ 3 & 3 & 0 \end{bmatrix}. \tag{3.1.2}$$

From Definition 3.6, we see that the distance matrix **D** of Eq. (3.1.2) is a Robinson matrix, as the elements are non-decreasing in the third column as we move away (upward) from the diagonal although all other movements away from the diagonal do give increasing distance values. On the other hand, if we define such a matrix as having increasing elements as we move away from the diagonal, then this **D** is not a Robinson matrix.

(ii) However, if the dissimilarities measures are $d(A, B) = 2$, $d(A, C) = 1.5$, and $d(B, C) = 1.2$ as in Figure 3.1(ii), we have

$$d(A, B) = 2 > \max\{d(A, C), d(B, C)\} = 1.5;$$

although $\quad d(A, C) = 1.5 < \max\{d(A, B), d(B, C)\} = 2,$

$$d(B, C) = 1.2 < \max\{d(A, B), d(A, C)\} = 2.$$

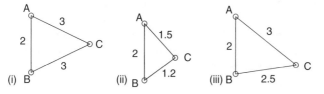

**Figure 3.1** Distances (Example 3.1). (i) Ultrametric (ii) Non-ultrametric (iii) Robinson

Therefore, these are not ultrametric dissimilarities, since condition (vi) does not hold for all three comparisons.

The dissimilarity matrix is not a Robinson matrix since

$$\mathbf{D} = \begin{bmatrix} 0 & 2 & 1.5 \\ 2 & 0 & 1.2 \\ 1.5 & 1.2 & 0 \end{bmatrix}. \tag{3.1.3}$$

(iii) Finally, if the dissimilarity or distance measures are given by $d(A, B) = 2$, $d(A, C) = 3$, and $d(B, C) = 2.5$ as in Figure 3.1(iii), then we see that the distance matrix is a Robinson matrix with

$$\mathbf{D} = \begin{bmatrix} 0 & 2 & 3 \\ 2 & 0 & 2.5 \\ 3 & 2.5 & 0 \end{bmatrix}. \tag{3.1.4}$$

□

It can be the case that a particular ordering of the objects produces a distance matrix that is not a Robinson matrix. However, by re-ordering the objects, a Robinson matrix can be obtained. This order, or transformed order, becomes important when building pyramids.

**Example 3.2**  Suppose there are three objects ordered as $\{A_1, A_2, A_3\}$ with distance measures that produce the distance matrix

$$\mathbf{D} = \begin{bmatrix} 0 & 4 & 9 \\ 4 & 0 & 11 \\ 9 & 11 & 0 \end{bmatrix}. \tag{3.1.5}$$

As structured, it would seem that this distance matrix $\mathbf{D}$ is not a Robinson matrix. However, if we re-ordered the objects according to $\{A_2, A_1, A_3\}$, the distance matrix becomes

$$\mathbf{D} = \begin{bmatrix} 0 & 4 & 11 \\ 4 & 0 & 9 \\ 11 & 9 & 0 \end{bmatrix}. \tag{3.1.6}$$

Now, clearly, the distance matrix in Eq. (3.1.6) is a Robinson matrix.  □

In subsequent sections we consider the calculation of some known dissimilarity or distance measures for a variety of different types of data. However, we first present here some generalized distances which are functions of any dissimilarity or distance measures $d(A, B)$ between any objects $A, B \in \Omega$, be they classical or symbolic valued observations.

**Definition 3.7**    Let $A$ and $B$ be any two objects in the space $\Omega$, and suppose the dissimilarity, or distance, measure between them is $d_j(A,\ B)$ for the variable $Y_j, j = 1, \ldots, p$. The **Minkowski distance of order** $q$, $d^{(q)}(A,\ B)$, is given by

$$d^{(q)}(A,\ B) = \left( \sum_{j=1}^{p} w_j [d_j(A,\ B)]^q \right)^{1/q} \tag{3.1.7}$$

where $w_j \geq 0$ is a weight associated with the variable $Y_j$.    □

Some definitions put the weight $w_j$ inside the squared bracket [·] term in Eq. (3.1.7). The weight $w_j$ in Eq. (3.1.7) can take various forms. A frequently used weight is one designed to standardize the distances by variable to account for differences in scale (see, e.g., the $\mathcal{Y}$-span in the Gowda–Diday dissimilarity of Definition 3.18 below, or the span normalized Euclidean Hausdorff distance of Definition 3.25, or the span normalized Euclidean Ichino–Yaguchi distance of Example 3.13 and Eq. (3.3.26)). Other weights might be used to account for relationships between the variables (e.g., the Mahalanobis distance of Definition 3.11 below takes into account dependencies between variables). There can also be additional "weights" to account for other distortions that might occur; one such case is when the number of variables $p$ is too large (e.g., the $1/p$ weight in the normalized Euclidean distance of Definition 3.10). When there is no weighting incorporated for $Y_j$, then $w_j \equiv 1$ in Eq. (3.1.7).

**Definition 3.8**    Let $A$ and $B$ be any two objects in the space $\Omega$, and suppose the dissimilarity, or distance, measure between them is $d_j(A,\ B)$ for the variable $Y_j, j = 1, \ldots, p$. The **city block distance** is a Minkowski distance of order $q = 1$.    □

This city block distance is also called a **Manhattan distance**. Frequently, the weights take the form $w_j \times c_j, \ j = 1, \ldots, p$, with $\sum_j c_j = 1$, where typically $c_j = 1/p$.

**Definition 3.9**    Let $A$ and $B$ be any two objects in the space $\Omega$, and suppose the dissimilarity, or distance, measure between them is $d_j(A,\ B)$ for the variable $Y_j, j = 1, \ldots, p$. The **Euclidean distance** is a Minkowski distance of order $q = 2$, i.e.,

$$d^{(2)}(A,\ B) = \left( \sum_{j=1}^{p} w_j [d_j(A,\ B)]^2 \right)^{1/2}. \tag{3.1.8}$$

□

While Eq. (3.1.8) is the Euclidean distance for any dissimilarity or distance measures $d_j(A, B)$, $j = 1, \dots, p$, when the data are classical values $\{(x_{ij}, j = 1, \dots, p), \ i = 1, \dots, n\}$, usually the Euclidean distance metric is a (weighted or unweighted) function of the squared distances between the point observations, given by

$$d^{(2)}(A, B) = \left( \sum_{j=1}^{p} w_j (x_{i_1 j} - x_{i_2 j})^2 \right)^{1/2} .$$

**Definition 3.10**  Let $A$ and $B$ be any two objects in the space $\Omega$, and suppose the dissimilarity, or distance, measure between them is $d_j(A, B)$ for the variable $Y_j$, $j = 1, \dots, p$. The **normalized Euclidean distance** is a Minkowski distance of order $q = 2$ given by

$$d^{(2)}(A, B) = \left( \frac{1}{p} \sum_{j=1}^{p} w_j [d_j(A, B)]^2 \right)^{1/2} . \tag{3.1.9}$$

□

**Definition 3.11**  Let $A$ and $B$ be any two objects in the space $\Omega$, and suppose the Euclidean dissimilarity, or distance, measure across all variables $Y_j$, $j = 1, \dots, p$, between them is $d(A, B)$. The **Mahalanobis distance**, Mahalanobis (1936), is given by

$$d^{(Mah)}(A, B) = (\det(\mathbf{\Sigma}))^{1/p} d(A, B) \tag{3.1.10}$$

where $\mathbf{\Sigma}$ is the sample $p \times p$ covariance matrix.  □

**Definition 3.12**  Let $A$ and $B$ be any two $p$-dimensional objects in the space $\Omega = \Omega_1 \times \cdots \times \Omega_p$ taking realizations $\mathbf{Y} = (Y_1, \dots, Y_p)$, and suppose the dissimilarity, or distance, measure between them is $d_j(A, B)$ for the variable $Y_j$, $j = 1, \dots, p$. The **Chebychev distance**, $d^c(A, B)$, is the limit of the unweighted Minkowski distance as the order $q \to \infty$. That is,

$$d^c(A, B) = \lim_{q \to \infty} \left( \sum_{j=1}^{p} [d_j(A, B)]^q \right)^{1/q} = \max_j \{ |d_j(A, B)| \}. \tag{3.1.11}$$

□

**Example 3.3**  Consider the three objects $(A, B, C)$ for which the distances between each pair are as shown in Table 3.1. Also shown are weights by variable, $w_j \equiv$ weight of $Y_j, j = 1, \dots, p = 4$.

Then, the unweighted city block distances are found from Eq. (3.1.7) with $w_j = 1, j = 1, \dots, 4$, and $q = 1$. For the weighted city block distances, let us take

**Table 3.1** Distances (Example 3.3)

| Object pair | Original distances | | | |
|---|---|---|---|---|
| | $d_1(\cdot)$ | $d_2(\cdot)$ | $d_3(\cdot)$ | $d_4(\cdot)$ |
| $(A, B)$ | 2 | 8 | 3 | 2 |
| $(A, C)$ | 3 | 6 | 5 | 3 |
| $(B, C)$ | 3 | 5 | 3 | 3 |
| $w_j$ | 3 | 1 | 2 | 4 |

**Table 3.2** City block distances (Example 3.3)

| Object pair | City block distances | |
|---|---|---|
| | Unweighted | Weighted |
| $(A, B)$ | 15 | 3.75 |
| $(A, C)$ | 17 | 4.25 |
| $(B, C)$ | 14 | 3.50 |

weights $c_j = 1/p = 1/4$. These distances are displayed in Table 3.2. It follows that the weighted city block distance matrix is

$$\mathbf{D}^{(1)} = \begin{bmatrix} 0 & 3.75 & 4.25 \\ 3.75 & 0 & 3.50 \\ 4.25 & 3.50 & 0 \end{bmatrix}. \tag{3.1.12}$$

$\square$

**Example 3.4** For the objects of Example 3.3, let us find some Euclidean distances. The weighted Euclidean distance is found from Eq. (3.1.8), for the object pair $(A, B)$, as

$$d^{(2)}(A, B) = [3 \times 2^2 + 1 \times 8^2 + 2 \times 3^2 + 4 \times 2^2]^{1/2} = 10.488.$$

The unweighted normalized Euclidean distances are found from Eq. (3.1.9) with $w_j = 1$, $j = 1, \ldots, p = 4$. Thus, e.g.,

$$d^{(2)}(B, C) = [(3^2 + 5^2 + 3^2 + 3^2)/4]^{1/2} = 3.606.$$

Suppose now we want to standardize the variable weights to $w_j^* = w_j/\sum_j w_j$, $j = 1, \ldots, 4$, so that $\sum_j w_j^* = 1$. Then, the normalized standardized Euclidean

**Table 3.3** Euclidean distances: $Y_j$-weighted (Example 3.4)

| Object pair | Unweighted | Weighted | Normalized Unweighted | Normalized Weighted | Standardized | Normalized Standardized |
|---|---|---|---|---|---|---|
| $(A, B)$ | 9.000 | 10.488 | 4.500 | 5.244 | 3.317 | 1.658 |
| $(A, C)$ | 8.888 | 12.207 | 4.444 | 6.103 | 3.860 | 1.930 |
| $(B, C)$ | 7.211 | 10.296 | 3.606 | 5.148 | 3.256 | 1.628 |

distances can be found from Eq. (3.1.9). For example, we have the normalized standardized Euclidean distance for the object pair $(A, C)$ as

$$d^{(2)}(A, C) = [(0.3 \times 3^2 + 0.1 \times 6^2 + 0.2 \times 5^2 + 0.4 \times 3^2)/4]^{1/2} = 1.930.$$

The complete set of weighted and unweighted, normalized and standardized Euclidean distances for all three object pairs is shown in Table 3.3. Hence, appropriate distance matrices can be found. For example, the (non-normalized) standardized Euclidean distance matrix is the (Robinson) matrix

$$\mathbf{D}^{(2)} = \begin{bmatrix} 0 & 3.317 & 3.860 \\ 3.317 & 0 & 3.256 \\ 3.860 & 3.256 & 0 \end{bmatrix}. \tag{3.1.13}$$

Finally, the Chebychev distance between $A$ and $B$, for example, is $\max_j d_j(A, B) = 8$. □

**Definition 3.13** Let $\{1, \dots, m\}$ be a set of objects in $\Omega$. Suppose these objects are clustered into the non-overlapping sub-clusters $C_1, \dots, C_K$, with $m_k$ objects in $C_k$, $k = 1, \dots, K$, and $\sum_k m_k = m$. When the distances between objects are Euclidean distances (as in Definition 3.9), **Huygens theorem** gives

$$\sum_{u=1}^{m} d^2(u, g) = \sum_{k=1}^{K} \sum_{u \in C_k} d^2(u, g_k) + \sum_{k=1}^{K} d^2(g_k, g) \tag{3.1.14}$$

where $g$ is the center of gravity of $\Omega$ and $g_k$ is the center of gravity of $C_k$, $k = 1, \dots, K$. The left-hand term is referred to as the total overall variation, and the right-hand terms as the total within-cluster variation and the total between-cluster variation, respectively. □

Typically, these centers of gravity are the means obtained from section 2.4.1 depending on the type of data at hand. A proof of Huygens theorem can be found in Celeux et al. (1989).

## 3.2  Distance Measures: List or Multi-valued Data

Suppose our sample space $\Omega$ consists of objects $A_u$, $u = 1, \ldots, m$, and suppose each object contains lists or multi-valued values for the $p$-dimensional random variable $\mathbf{Y} = (Y_1, \ldots, Y_p)$ in the form

$$\xi(A_u) \equiv \mathbf{Y}_u = (\{Y_{ujk_j}, p_{ujk_j}; \ k_j = 1, \ldots, s_{uj}\}, j = 1, \ldots, p), \quad u = 1, \ldots, m, \tag{3.2.1}$$

where $Y_{ujk}$ is the $k$th $\equiv k_j$th category observed from the set of possible categories $\mathcal{Y}_j = \{Y_{j1}, \ldots, Y_{js_j}\}$ for the variable $Y_j$ for the object $A_u$, and where $p_{ujk_j}$ is the probability or relative frequency associated with $Y_{ujk_j}$, with $\sum_{k_j} p_{ujk_j} = 1$. As written, these are modal-valued realizations. However, we recall from Chapter 2 that when list or multi-valued values are non-modal, it is implied that $p_{ujk_j} \equiv 1/s_{uj}$, for all $(u, j, k_j)$. Also, for the modal case, without loss of generality, we will assume that $s_{uj} = s_j$ for all $u = 1, \ldots, m$, that is, when a given category $Y_{jk_j'}$ in $\mathcal{Y}_j$ does not occur, it has probability $p_{jk_j'} = 0$. In this chapter, we assume our list data are non-modal; modal list data are covered in Chapter 4.

### 3.2.1  Join and Meet Operators for Multi-valued List Data

We will need the concepts of join and meet between two objects, introduced by Ichino (1988).

**Definition 3.14**   Let $A_1$ and $A_2$ be any two objects in $\Omega$ taking multi-valued list values from $\mathcal{Y}$. Then, their **join**, $A_1 \oplus A_2$, is the set of categories from $\mathcal{Y}$ that appears in $A_1$, or $A_2$, or both $A_1$ and $A_2$.                                     □

**Definition 3.15**   Let $A_1$ and $A_2$ be any two objects in $\Omega$ taking multi-valued list values from $\mathcal{Y}$. Then, their **meet**, $A_1 \otimes A_2$, is the set of categories from $\mathcal{Y}$ that appears in both $A_1$ and $A_2$.                                     □

**Definition 3.16**   Let $A_1 = (A_{1j}, \ j = 1, \ldots, p)$ and $A_2 = (A_{2j}, \ j = 1, \ldots, p)$ be any two $p$-dimensional list multi-valued objects in $\Omega = \Omega_1 \times \cdots \times \Omega_p$. Then, the **join** and **meet** of $A_1$ and $A_2$ are the componentwise join and meet, respectively, i.e.,

$$A_1 \oplus A_2 = (A_{11} \oplus A_{12}, \ldots, A_{1p} \oplus A_{1p}), \tag{3.2.2}$$

$$A_1 \otimes A_2 = (A_{11} \otimes A_{12}, \ldots, A_{1p} \otimes A_{1p}). \tag{3.2.3}$$

                                     □

Notice that the join is akin to a union, i.e., $A_1 \oplus A_2 \equiv A_1 \cup A_2$, and the meet compares with the intersection, i.e., $A_1 \otimes A_2 \equiv A_1 \cap A_2$.

**Table 3.4** Multi-valued list data (Example 3.5)

| Household | $Y_1$ Energy type | $Y_2$ Computer type |
| --- | --- | --- |
| 1 | {gas, wood} | {laptop} |
| 2 | {gas, coal, wood} | {laptop, PC} |
| 3 | {gas, coal, wood} | {laptop} |
| 4 | {electricity, wood} | {laptop, PC, other} |
| 5 | {electricity, other} | {PC, other} |

**Example 3.5**  Consider the set of households as given in Table 3.4, with information for the variables $Y_1$ = energy used from $\mathcal{Y}_1$ = {gas, coal, electricity, wood, other} and $Y_2$ = computers used from $\mathcal{Y}_2$ = {laptop, personal computer (PC), other}.

Then, for the objects Household 1 ($A_1$) and Household 2 ($A_2$), we have their join and meet as, respectively,

$$A_1 \oplus A_2 = \{\text{gas, wood, coal}\}, \quad A_1 \otimes A_2 = \{\text{gas, wood}\}, \text{for } Y_1,$$
$$A_1 \oplus A_2 = \{\text{laptop, PC}\}, \qquad A_1 \otimes A_2 = \{\text{laptop}\}, \text{for } Y_2. \qquad \square$$

### 3.2.2 A Simple Multi-valued Distance

A simple distance measure for list or multi-valued observations is adapted from Chavent (2000), where for the non-modal case it is assumed that different categories occur with equal probability and those categories that do not occur have probability zero.

**Definition 3.17**  Let $A_{u_1}$ and $A_{u_2}$ be any two objects in $\Omega = \Omega_1 \times \cdots \times \Omega_p$, $u_1, u_2 = 1, \ldots, m$, taking modal list values of the form in Eq. (3.2.1). Then, a **multi-valued distance measure**, $d(A_{u_1}, A_{u_2})$, between $A_{u_1}$ and $A_{u_2}$ satisfies

$$d^2(A_{u_1}, A_{u_2}) = \sum_{j=1}^{p} \sum_{k_j=1}^{s_j} \left( p \sum_{u=1}^{m} p_{ujk_j} \right)^{-1} (p_{u_1jk_j} - p_{u_2jk_j})^2 \qquad (3.2.4)$$

where for each variable $Y_j$ and each object $A_u$, the probability $p_{ujk_j} = 1/s_{uj}$ if the category $k_j$ occurs and $p_{ujk_j} = 0$ if the category $k_j$ does not occur in object $A_u$, $k_j = 1, \ldots, s_j, j = 1, \ldots, p$, and $u = 1, \ldots, m$. $\qquad \square$

**Example 3.6**  Table 3.5(a) gives a data set of multi-valued observations for $Y_1$ = type of energy used, $Y_2$ = types of computer used, and $Y_3$ = highest educational level attained for households aggregated by region (such as a census tract). The complete sets of possible categories are assumed

**Table 3.5** Multi-valued data (Example 3.6)

| | | (a) Raw data | | |
|---|---|---|---|---|
| Region $u$ | $Y_1$ Energy type | | $Y_2$ Computers | $Y_3$ College |
| 1 | {electricity, gas, wood, other} | | {PC, laptop, none} | {yes, no} |
| 2 | {electricity, wood} | | {PC, laptop, none} | {yes, no} |
| 3 | {electricity, gas} | | {PC, laptop} | {yes} |

| | (b) Modal format | | | | | | | | |
|---|---|---|---|---|---|---|---|---|---|
| | $Y_1$ Energy type | | | | $Y_2$ Computers | | | $Y_3$ College | |
| Region $u$ | Electricity | Gas | Wood | Other | PC | Laptop | None | Yes | No |
| 1 | 0.25 | 0.25 | 0.25 | 0.25 | 0.33 | 0.33 | 0.34 | 0.5 | 0.5 |
| 2 | 0.50 | 0.00 | 0.50 | 0.00 | 0.33 | 0.33 | 0.34 | 0.5 | 0.5 |
| 3 | 0.50 | 0.50 | 0.00 | 0.00 | 0.5 | 0.5 | 0.0 | 1.0 | 0.0 |
| $\sum_u p_{ujk_j}$ | 1.25 | 0.75 | 0.75 | 0.25 | 1.16 | 1.16 | 0.68 | 2.0 | 1.0 |

to be $\mathcal{Y}_1 = \{$electricity, gas, wood, other$\}$, $\mathcal{Y}_2 = \{$PC, laptop, none$\}$, and $\mathcal{Y}_3 = \{$college (yes), college (no)$\}$. Thus, for example, households in Region 1 used all four energy types, all three computer types, and there are occupants with college and some with non-college education. In contrast, households in Region 3 used only gas and electricity energy sources and all household residents had college level education, and so on. Since the aggregation did not retain proportions of the various category types, it is assumed that each type that did occur in the aggregation occurred with equal probability. Therefore, Table 3.5(b) shows the resulting proportion of households for each region across all possible categories in $\mathcal{Y}_j$, $j = 1, 2, 3$, respectively. Thus, for example, in Region 2, 50% of the households used electricity and 50% used wood as their energy source while 50% had a college education and 50% did not. Also shown are the summation terms $\sum_u p_{ujk_j}$ in the denomination of Eq. (3.2.4).

Consider the last two regions $A_2$ and $A_3$. Then, from Eq. (3.2.4) and Table 3.5(b), we have

$$d^2(A_2, A_3) = \frac{1}{3}\left[\frac{(0.5 - 0.5)^2}{1.25} + \frac{(0.0 - 0.5)^2}{0.75} + \frac{(0.5 - 0.0)^2}{0.75} + \frac{(0.0 - 0.0)^2}{0.25}\right.$$

$$+ \frac{(0.33 - 0.5)^2}{1.16} + \frac{(0.33 - 0.5)^2}{1.16} + \frac{(0.34 - 0.0)^2}{0.68}$$

$$\left.+ \frac{(0.5 - 1.0)^2}{2.0} + \frac{(0.5 - 0.0)^2}{1.0}\right] = 0.2776$$

and so $d(A_2, A_3) = 0.527$. Similarly, we can find all the relevant distances. The distance matrix is

$$\mathbf{D} = \begin{bmatrix} 0 & 0.397 & 0.538 \\ 0.397 & 0 & 0.527 \\ 0.538 & 0.527 & 0 \end{bmatrix}. \tag{3.2.5}$$

□

### 3.2.3 Gowda–Diday Dissimilarity

**Definition 3.18** Let $A_{u_1}$ and $A_{u_2}$ be any two objects from the space $\Omega = \Omega_1 \times \cdots \times \Omega_p$ taking non-modal multi-valued list values of the form in Eq. (3.2.1) (i.e., $p_{ujk} = p_{uj} \equiv 1/s_{uj}$). Then, the **Gowda–Diday dissimilarity** (Gowda and Diday, 1991, 1992) between $A_{u_1}$ and $A_{u_2}$ is

$$d(A_{u_1}, A_{u_2}) = \sum_{j=1}^{p} [d_{1j}(A_{u_1}, A_{u_2}) + d_{2j}(A_{u_1}, A_{u_2})] \tag{3.2.6}$$

where

$$d_{1j}(A_{u_1}, A_{u_2}) = (|r_j^1 - r_j^2|)/r_j, \quad j = 1, \dots, p, \tag{3.2.7}$$

$$d_{2j}(A_{u_1}, A_{u_2}) = (r_j^1 + r_j^2 - 2r_j^*)/r_j, \quad j = 1, \dots, p, \tag{3.2.8}$$

with $r_j^k$ being the number of categories from $\mathcal{Y}_j$ actually appearing in $A_u$, $u = u_k, k = 1, 2$ (i.e., $r_j^u \equiv s_{uj}$ of Eq. (3.2.1)), $r_j$ is the number of categories from $\mathcal{Y}_j$ in the join $A_{u_1 j} \oplus A_{u_2 j}$, and $r_j^*$ is the number of categories from $\mathcal{Y}_j$ in the meet $A_{u_1 j} \otimes A_{u_2 j}$ for each variable $Y_j$, $j = 1, \dots, p$. □

The term $d_{1j}(A_{u_1}, A_{u_2})$ gives a measure of the relative sizes of $A_{u_1 j}$ and $A_{u_2 j}$ (also called the **span** component), and the term $d_{2j}(A_{u_1}, A_{u_2})$ gives a measure of the **relative content** of $A_{u_1 j}$ and $A_{u_2 j}$, for each $j = 1, \dots, p$. To normalize these distances to account for differing scales in $\mathcal{Y}_j$, one normalization factor is the term $1/|\mathcal{Y}_j|$ where $|\mathcal{Y}_j|$ is the total number of different categories recorded for $\mathcal{Y}_j$.

**Example 3.7** (i) Consider the data displayed in Table 3.4. Let us find the distances between the first two households $A_1$ and $A_2$. Substituting into Eq. (3.2.7), we obtain for $j = 1$, 2, respectively, the relative sizes or span component as

$$d_{11}(A_1, A_2) = (|2 - 3|)/3 = 0.333,$$

$$d_{12}(A_1, A_2) = (|1 - 2|)/2 = 0.500,$$

and substituting into Eq. (3.2.8), we obtain for $j = 1$, 2, respectively, the relative content component as

$$d_{21}(A_1, A_2) = (2 + 3 - 2 \times 2)/3 = 0.333,$$

$$d_{22}(A_1, A_2) = (1 + 2 - 2 \times 1)/2 = 0.500.$$

**Table 3.6** Gowda–Diday dissimilarities (Example 3.7)

| Object pair | Relative size $Y_1$ | $Y_2$ | Relative content $Y_1$ | $Y_2$ | Dissimilarity $Y_1$ | $Y_2$ | (a) Dissimilarity | (b) Normalized dissimilarity | (c) Euclidean distance |
|---|---|---|---|---|---|---|---|---|---|
| 1-2 | 0.333 | 0.500 | 0.333 | 0.500 | 0.667 | 1.000 | 1.667 | 0.467 | 1.202 |
| 1-3 | 0.333 | 0.000 | 0.333 | 0.000 | 0.667 | 0.000 | 0.667 | 0.133 | 0.667 |
| 1-4 | 0.000 | 0.667 | 0.667 | 0.667 | 0.667 | 1.333 | 2.000 | 0.578 | 1.491 |
| 1-5 | 0.000 | 0.333 | 1.333 | 1.000 | 1.333 | 1.333 | 2.667 | 0.711 | 1.667 |
| 2-3 | 0.000 | 0.500 | 0.000 | 0.500 | 0.000 | 1.000 | 1.000 | 0.333 | 1.000 |
| 2-4 | 0.250 | 0.333 | 0.750 | 0.333 | 1.000 | 0.667 | 1.667 | 0.422 | 1.202 |
| 2-5 | 0.200 | 0.000 | 1.000 | 0.667 | 1.200 | 0.667 | 1.867 | 0.462 | 1.373 |
| 3-4 | 0.250 | 0.667 | 0.750 | 0.667 | 1.000 | 1.333 | 2.333 | 0.644 | 1.667 |
| 3-5 | 0.200 | 0.333 | 1.000 | 1.000 | 1.200 | 1.333 | 2.533 | 0.684 | 1.794 |
| 4-5 | 0.000 | 0.333 | 0.667 | 0.333 | 0.667 | 0.667 | 1.333 | 0.356 | 0.943 |

From Eq. (3.2.6), the Gowda–Diday dissimilarity for $Y_j, j = 1, 2$, respectively, is

$$d_1(A_1, A_2) = d_{11}(A_1, A_2) + d_{21}(A_1, A_2) = 0.333 + 0.333 = 0.667, \quad (3.2.9)$$

$$d_2(A_1, A_2) = d_{12}(A_1, A_2) + d_{22}(A_1, A_2) = 0.500 + 0.500 = 1.000; \quad (3.2.10)$$

hence, the Gowda–Diday dissimilarity over all variables is

$$d(A_1, A_2) = d_1(A_1, A_2) + d_2(A_1, A_2) = 1.667. \quad (3.2.11)$$

Table 3.6(a), gives the complete set of Gowda–Diday dissimilarities for all household pairs.

(ii) Suppose now we want to calculate the normalized dissimilarities. In this case, the normalizing factors are $\mathcal{Y}_1 = 5$ and $\mathcal{Y}_2 = 3$ for $Y_j$, $j = 1, 2$, respectively. Hence, for example, the normalized Gowda–Diday dissimilarity between the first two households is

$$d(A_1, A_2) = d_1(A_1, A_2)/\mathcal{Y}_1 + d_2(A_1, A_2)/\mathcal{Y}_2 = 0.467. \quad (3.2.12)$$

The complete set of normalized Gowda–Diday dissimilarities for all household pairs is given in Table 3.6(b).

(iii) We can also calculate Euclidean distances based on these Gowda–Diday dissimilarities, that is, we substitute the Gowda–Diday dissimilarities into Eq. (3.1.8). Take the unweighted Euclidean distances. Then, for the first two households, we have, from Eqs. (3.1.8), (3.2.7), and (3.2.8), with $w_j = 1$,

$$d^{(2)}(A_1, A_2) = (0.667^2 + 1.000^2)^{1/2} = 1.202. \quad (3.2.13)$$

Likewise, we can find all these unweighted Euclidean weights. The resulting distance matrix is, from Table 3.6(c),

$$
\mathbf{D}^{(2)} = \begin{bmatrix}
0 & 1.202 & 0.667 & 1.491 & 1.667 \\
1.202 & 0 & 1.000 & 1.202 & 1.373 \\
0.667 & 1.000 & 0 & 1.667 & 1.794 \\
1.491 & 1.202 & 1.667 & 0 & 0.943 \\
1.667 & 1.373 & 1.794 & 0.943 & 0
\end{bmatrix}. \qquad (3.2.14)
$$

□

### 3.2.4 Ichino–Yaguchi Distance

**Definition 3.19** Let $A_{u_1}$ and $A_{u_2}$ be any two objects from the space $\Omega = \Omega_1 \times \cdots \times \Omega_p$ taking non-modal multi-valued list values of the form in Eq. (3.2.1) (i.e., $p_{ujk} = p_{uj} \equiv 1/s_{uj}$). Then, the **Ichino–Yaguchi dissimilarity distance** (Ichino and Yaguchi, 1994) between $A_{u_1}$ and $A_{u_2}$, for the variable $Y_j$, is

$$
d_j(A_{u_1}, A_{u_2}) = r_j - r_j^* + \gamma(2r_j^* - r_j^1 - r_j^2), \quad j = 1, \ldots, p, \qquad (3.2.15)
$$

where $0 \le \gamma \le 0.5$ is a pre-specified constant, and where $r_j$, $r_j^*$, $r_j^1$ and $r_j^2$ are as defined in Definition 3.18. □

**Example 3.8** (i) Consider the data displayed in Table 3.4. Let us find the distances between the second and fourth households, namely, $d_j(A_2, A_4)$, for each $j = 1, 2, 3$. Let us take $\gamma = 0.5$. Then, substituting into Eq. (3.2.15), we have, for each $Y_j, j = 1, 2$,

$$
d_1(A_2, A_4) = 4 - 1 + 0.5(2 \times 1 - 3 - 2) = 1.5, \quad \text{for } Y_1,
$$
$$
d_2(A_2, A_4) = 3 - 2 + 0.5(2 \times 2 - 2 - 3) = 0.5, \quad \text{for } Y_2.
$$

Likewise, the Ichino–Yaguchi distances for all household pairs can be found and are given in Table 3.7 (second and third columns). Distances which take into account the scales can also be found (see Eq. (3.2.16) below).

(ii) Let us consider some city block distances based on these Ichino–Yaguchi distances. An unweighted city block distance can be found from Eq. (3.2.7) with weights $w_j = 1$, and $q = 1$, to give the distance between the second and fourth households as $d^{(1)}(A_2, A_4) = 1.5 + 0.5 = 2$.

**Table 3.7** Ichino–Yaguchi distances[a] (Example 3.8)

| Object pair | $Y_1$ $d_1$ | $Y_2$ $d_2$ | City block distance Unweighted | Normalized | Distance | Normalized distance | Weighted distance |
|---|---|---|---|---|---|---|---|
| 1-2 | 0.5 | 0.5 | 1.0 | 0.267 | 0.707 | 0.500 | 0.258 |
| 1-3 | 0.5 | 0.0 | 0.5 | 0.100 | 0.500 | 0.354 | 0.158 |
| 1-4 | 1.0 | 1.0 | 2.0 | 0.533 | 1.414 | 1.000 | 0.516 |
| 1-5 | 2.0 | 1.5 | 3.5 | 0.900 | 2.500 | 1.768 | 0.880 |
| 2-3 | 0.0 | 0.5 | 0.5 | 0.167 | 0.500 | 0.354 | 0.204 |
| 2-4 | 1.5 | 0.5 | 2.0 | 0.467 | 1.581 | 1.118 | 0.516 |
| 2-5 | 2.5 | 1.0 | 3.5 | 0.833 | 2.693 | 1.904 | 0.890 |
| 3-4 | 1.5 | 1.0 | 2.5 | 0.633 | 1.803 | 1.275 | 0.626 |
| 3-5 | 2.5 | 1.5 | 4.0 | 1.000 | 2.915 | 2.062 | 1.000 |
| 4-5 | 1.0 | 0.5 | 1.5 | 0.367 | 1.118 | 0.791 | 0.376 |

The table has two top-level spanning headers: **Ichino–Yaguchi** (over $Y_1$ $d_1$, $Y_2$ $d_2$, and City block distance) and **Euclidean** (over Distance, Normalized distance, Weighted distance).

a)  $\gamma = 0.5$

If we want to account for the scale of the variables, we divide each $d_j(A_{u_1}, A_{u_2})$, $u_1, u_2 = 1, \ldots, 5$, by $|\mathcal{Y}_j|$, $j = 1, 2$, where $|\mathcal{Y}_1| = 5$ and $|\mathcal{Y}_2| = 3$. In this case, the city block distance becomes

$$d^{(1)}(A_2, A_4) = [4 - 1 + 0.5(2 \times 1 - 3 - 2)]/5$$
$$+ [3 - 2 + 0.5(2 \times 2 - 2 - 3)]/3 = 0.467.$$

The complete set of non-normalized and normalized distances is shown in Table 3.7 (fourth and fifth columns, respectively).

(iii) We can also find Euclidean distances based on the Ichino–Yaguchi distances. The unweighted non-normalized distances, the unweighted normalized distances, and the weighted normalized distances for each household pair are shown in Table 3.7 (sixth, seventh, and eighth columns, respectively). For example, the weighted normalized Euclidean distance between the second and fourth households, $d^{(2)}(A_2, A_4)$, is calculated from Eq. (3.1.9) with $q = 2$, $p = 2$, and $w_j = |\mathcal{Y}_j|$, $j = 1, 2$, i.e.,

$$d^{(2)}(A_2, A_4) = [(1/2)(1.5^2/5 + 0.5^2/3)]^{1/2} = 0.516.$$

Hence, the relevant distances matrices can be found. For example, the normalized Ichino–Yaguchi distance matrix is, for $j = 1$, 2, respectively,

$$
\mathbf{D}_1 = \begin{bmatrix}
0 & 0.100 & 0.100 & 0.200 & 0.400 \\
0.100 & 0 & 0.000 & 0.300 & 0.500 \\
0.100 & 0.100 & 0 & 0.300 & 0.500 \\
0.200 & 0.300 & 0.300 & 0 & 0.200 \\
0.400 & 0.500 & 0.500 & 0.200 & 0
\end{bmatrix},
$$

$$
\mathbf{D}_2 = \begin{bmatrix}
0 & 0.167 & 0.000 & 0.333 & 0.500 \\
0.167 & 0 & 0.167 & 0.167 & 0.333 \\
0.000 & 0.167 & 0 & 0.333 & 0.500 \\
0.500 & 0.167 & 0.333 & 0 & 0.167 \\
0.500 & 0.333 & 0.500 & 0.167 & 0
\end{bmatrix}. \tag{3.2.16}
$$

□

## 3.3 Distance Measures: Interval Data

Suppose our sample space $\Omega$ consists of objects $A_u$, $u = 1, \ldots, m$, and suppose each object is described by interval values for the $p$-dimensional random variable $\mathbf{Y} = (Y_1, \ldots, Y_p)$ in the form

$$
\xi(A_u) \equiv \mathbf{Y}_u = ([a_{uj}, \, b_{uj}), \, j = 1, \ldots, p), \quad u = 1, \ldots, m, \tag{3.3.1}
$$

where the intervals $[a, \, b)$ in Eq. (3.3.1) can be open or closed at either end. The intervals take values on $\mathcal{Y}_j \equiv \Omega_j \equiv \mathbb{R}_j$, $j = 1, \ldots, p$, where $\mathbb{R}_j$ is the real line. A classical observation $Y = a$ is a special case where now the interval becomes $Y = [a, a]$.

### 3.3.1 Join and Meet Operators for Interval Data

As for multi-valued list data, we will need the concepts of join and meet between two objects for interval data, from Ichino (1988).

**Definition 3.20** Let $A_1 = [a_1, \, b_1]$ and $A_2 = [a_2, \, b_2]$ be any two interval-valued objects in $\Omega$. Then, their **join**, $A_1 \oplus A_2$, is the interval

$$
A_1 \oplus A_2 = [\min\{a_1, \, a_2\}, \max\{b_1, \, b_2\}], \tag{3.3.2}
$$

i.e., the join consists of points in $\mathbb{R}$ that appear in $A_1$, or $A_2$, or both $A_1$ and $A_2$.

□

**Definition 3.21** Let $A_1 = [a_1, \, b_1]$ and $A_2 = [a_2, \, b_2]$ be any two interval-valued objects in $\Omega$. Then, their **meet**, $A_1 \otimes A_2$, is the interval

$$
\begin{aligned}
A_1 \otimes A_2 &= [\max\{a_1, \, a_2\}, \min\{b_1, \, b_2\}] \text{ when } A_1 \cap A_2 \neq \phi, \\
A_1 \otimes A_2 &= 0 \text{ when } A_1 \cap A_2 = \phi, \tag{3.3.3}
\end{aligned}
$$

where $\phi$ is the empty set, i.e., the meet is the interval that forms the intersection of intervals which overlap, but is zero when the intervals are disjoint. □

As for multi-valued list data, the join and meet of $p$-dimensional interval objects consist of the componentwise join and meet, respectively (see Eqs. (3.2.2) and (3.2.3)).

**Example 3.9**   Consider the interval-valued objects $A_1 = [3, 6]$, $A_2 = [4, 8]$, and $A_3 = [7, 9]$. Then, the join and meet of $A_1$ and $A_2$ are, from Eqs. (3.3.2) and (3.3.3), respectively,

$$A_1 \oplus A_2 = [\min\{3,4\}, \max\{6,8\}] = [3,8],$$

$$A_1 \otimes A_2 = [\max\{3,4\}, \min\{6,8\}] = [4,6].$$

However, the meet of $A_1$ and $A_3$ is $A_1 \otimes A_3 = 0$ since the intersection $A_1 \cap A_3 = [3, 6] \cap [7, 9] = \phi$ is an empty set. □

### 3.3.2   Hausdorff Distance

A frequently occurring distance measure for interval observations is the Hausdorff (1937) distance. After giving the definition of the basic Hausdorff distance, we define some other forms of this distance.

**Definition 3.22**   Let $A_{u_1} = ([a_{u_1 j}, b_{u_1 j}], j = 1, \ldots, p)$ and $A_{u_2} = ([a_{u_2 j}, b_{u_2 j}], j = 1, \ldots, p)$ be any two interval-valued objects in $\Omega = \Omega_1 \times \cdots \times \Omega_p$. Then, for each variable $Y_j$, $j = 1, \ldots, p$, the **Hausdorff distance** between $A_{u_1}$ and $A_{u_2}$, $d_j(A_{u_1}, A_{u_2})$, for $u_1, u_2 = 1, \ldots, m$, is

$$d_j(A_{u_1}, A_{u_2}) = \max\{|a_{u_1 j} - a_{u_2 j}|, |b_{u_1 j} - b_{u_2 j}|\}. \tag{3.3.4}$$

□

**Definition 3.23**   Let $A_{u_1} = ([a_{u_1 j}, b_{u_1 j}], j = 1, \ldots, p)$ and $A_{u_2} = ([a_{u_2 j}, b_{u_2 j}], j = 1, \ldots, p)$ be any two interval-valued objects in $\Omega = \Omega \times \cdots \times \Omega_p$. Then, the **Euclidean Hausdorff distance** between $A_{u_1}$ and $A_{u_2}$, $d(A_{u_1}, A_{u_2})$, for $u_1, u_2 = 1, \ldots, m$, is

$$d(A_{u_1}, A_{u_2}) = \left\{ \sum_{j=1}^p [d_j(A_{u_1}, A_{u_2})]^2 \right\}^{1/2} \tag{3.3.5}$$

where $d_j(A_{u_1}, A_{u_2})$ is the regular Hausdorff distance for the variable $Y_j$ defined in Eq. (3.3.4). □

**Definition 3.24** Let $A_u = ([a_{uj}, \quad b_{uj}], \quad j = 1, \ldots, p), \quad u = 1, \ldots, m$, be interval-valued objects in $\Omega = \Omega_1 \times \cdots \times \Omega_p$. Then, the **normalized Euclidean Hausdorff distance** between $A_{u_1}$ and $A_{u_2}$, $d(A_{u_1}, A_{u_2})$, for $u_1, u_2 = 1, \ldots, m$, is

$$d(A_{u_1}, A_{u_2}) = \left\{ \sum_{j=1}^{p} [H_j^{-1} d_j(A_{u_1}, A_{u_2})]^2 \right\}^{1/2} \tag{3.3.6}$$

where

$$H_j^2 = \frac{1}{2m^2} \sum_{u_1=1}^{m} \sum_{u_2=1}^{m} [d_j(A_{u_1}, A_{u_2})]^2 \tag{3.3.7}$$

and where $d_j(A_{u_1}, A_{u_2})$ is the regular Hausdorff distance for the variable $Y_j$ defined in Eq. (3.3.4). □

**Definition 3.25** Let $A_u = ([a_{uj}, \quad b_{uj}], \quad j = 1, \ldots, p), \quad u = 1, \ldots, m$, be interval-valued objects in $\Omega = \Omega_1 \times \cdots \times \Omega_p$. Then, the **span normalized Euclidean Hausdorff distance** between $A_{u_1}$ and $A_{u_2}$, $d(A_{u_1}, A_{u_2})$, is

$$d(A_{u_1}, A_{u_2}) = \left\{ \sum_{j=1}^{p} [|\mathcal{Y}_j|^{-1} d_j(A_{u_1}, A_{u_2})]^2 \right\}^{1/2} \tag{3.3.8}$$

where $|\mathcal{Y}_j|$ is the span of the observations, i.e.,

$$|\mathcal{Y}_j| = \max_{u \in \Omega_j} \{b_{uj}\} - \min_{u \in \Omega_j} \{a_{uj}\} \tag{3.3.9}$$

and where $d_j(A_{u_1}, A_{u_2})$ is the regular Hausdorff distance for the variable $Y_j$ defined in Eq. (3.3.4). □

Notice that the Euclidean Hausdorff distance of Eq. (3.3.5) is just an unweighted Minkowski distance (from Eq. (3.1.7)) with $w_j \equiv 1$ and $q = 2$. Also, the normalized Euclidean Hausdorff distance of Eq. (3.3.6) and the span normalized Euclidean Hausdorff distance of Eq. (3.3.8) are weighted Minkowski distances where the weights $w_j$ are, respectively, $H_j^{-2}$ and $|\mathcal{Y}_j|^{-2}$, and with $q = 2$. We can define a generalized Minkowski Hausdorff distance measure as follows.

**Definition 3.26** Let $A_{u_1} = ([a_{u_1 j}, \quad b_{u_1 j}], j = 1, \ldots, p)$ and $A_{u_2} = ([a_{u_2 j}, \quad b_{u_2 j}], j = 1, \ldots, p)$ be any two interval-valued objects in $\Omega = \Omega_1 \times \cdots \times \Omega_p$. Then, the **generalized Minkowski Hausdorff distance** between $A_{u_1}$ and $A_{u_2}$, $d^{(q)}(A_{u_1}, A_{u_2})$, is

$$d^{(q)}(A_{u_1}, A_{u_2}) = \left\{ \sum_{j=1}^{p} w_j [d_j(A_{u_1}, A_{u_2})]^q \right\}^{1/q} \tag{3.3.10}$$

where $d_j(A_{u_{i_1}}, A_{u_{i_2}})$ is the regular Hausdorff distance for the variable $Y_j$ defined in Eq. (3.3.4), and where the weights $w_j, j = 1, \ldots, p$, have the same interpretation as in Eq. (3.1.7). □

**Example 3.10** Table 3.8 contains interval observations (extracted from a study of DNA (deoxyribonucleic acid) sequences by Blaisdell et al., 1996) relating to symmetrized dinucleotide relative abundances for $m = 6$ classes of phage genomes for variables $Y_1 = CG$, $Y_2 = GC$, $Y_3 = TA$, and $Y_4 = AT$. The classes correspond to natural groupings of phage sequences. Most, but not all, of the sequences are *Escherichia coli* bacterial hosts. The organisms studied fell into six categories, specifically, enteric temperate double-strand DNA (ET-dsDNA), nonenteric temperate double-strand DNA (NT-dsDNA), lytic double-strand DNA (L-dsDNA), filamentous single-strand DNA (F-ssDNA), lytic single-strand DNA (L-ssDNA), and RNA (ribonucleic acid), corresponding to $u = 1, \ldots, 6$, respectively. Interval values are obtained when aggregating over the specific (classical) values for those organisms within each grouping. The quantity $n_u$ refers to the number of bacteria types grouped together to form each class.

Data for all $m = 14$ objects and all $p = 10$ variables are given in Table 3.16 in the appendix. In that table, the first six classes and the first four variables are as in Table 3.8. The last eight classes correspond to host bacteria, specifically Eco (for *E. coli*), Sty (*Salmonella typhimurium*), Bsu (*bacillus subtilis*), Pae (*Psuedomonas aeruginosa*), Cps (*Chlamydia psittaci*), Msm (*Mycobacterium smegmatis*), Ala (*Acholeplasma laidlawii*), and Psy (*Psuedomonas syringae*), and the last six variables are $Y_5 = CC(GG)$, $Y_6 = AA(TT)$, $Y_7 = AC(GT)$, $Y_8 = AG(CT)$, $Y_9 = CA(TG)$, and $Y_{10} = GA(TC)$. Notice that the values for these host bacteria are classically valued. See Blaisdell et al. (1996) for details of these classes, genome terms, and variables.

**Table 3.8** Relative abundances[a] (Example 3.10)

| Class $u$ | Genome name | $Y_1$ (CG) | $Y_2$ (GC) | $Y_3$ (TA) | $Y_4$ (AT) | $n_u$ |
|---|---|---|---|---|---|---|
| 1 | tmp enteric dsDNA | [98, 108] | [120, 126] | [69, 76] | [100, 109] | 5 |
| 2 | tmp nonenteric dsDNA | [59, 108] | [91, 107] | [51, 92] | [96, 99] | 2 |
| 3 | lyt dsDNA | [88, 125] | [88, 154] | [82, 91] | [85, 112] | 6 |
| 4 | fil ssDNA | [80, 101] | [117, 133] | [68, 89] | [92, 97] | 4 |
| 5 | lyt ssDNA | [81, 99] | [108, 129] | [76, 86] | [93, 95] | 3 |
| 6 | RNA | [104, 113] | [96, 102] | [67, 95] | [94, 99] | 3 |

a) Blaisdell et al. (1996)

(i) To calculate the regular Hausdorff distances, we substitute into Eq. (3.3.4). Thus, for example, the Hausdorff distance between objects $A_2$ and $A_4$ is, for $Y_1$,

$$d_1(A_2, A_4) = \max\{|a_{21} - a_{41}|, |b_{21} - b_{41}|\}$$
$$= \max\{|59 - 80|, |108 - 101|\} = 21.$$

Likewise, the Hausdorff distances for all object pairs for each variable are as shown in Table 3.9(a)–(d). Hence, the Hausdorff distance matrix for $Y_1$ is

$$\mathbf{D}_1 = \begin{bmatrix} 0 & 39 & 17 & 18 & 17 & 6 \\ 39 & 0 & 29 & 21 & 22 & 45 \\ 17 & 29 & 0 & 24 & 26 & 16 \\ 18 & 21 & 24 & 0 & 2 & 24 \\ 17 & 22 & 26 & 2 & 0 & 23 \\ 6 & 45 & 16 & 24 & 23 & 0 \end{bmatrix}. \tag{3.3.11}$$

(ii) City distances can be found by substituting the regular Hausdorff distances into Eq. (3.3.10) with $q = 1$. When the distances are unweighted, the distances are simply the sum of those over each variable and are given in Table 3.9(e).

**Table 3.9** Hausdorff distances (Example 3.10)

| Object pair | By variable $Y_j$ | | | | City | | Euclidean | | |
|---|---|---|---|---|---|---|---|---|---|
| | $d_1$ | $d_2$ | $d_3$ | $d_4$ | Unweighted | Weighted | Unweighted | Normalized | Span |
| | (a) | (b) | (c) | (d) | (e) | (f) | (g) | (h) | (i) |
| 1-2 | 39 | 29 | 18 | 10 | 96 | 24.00 | 52.783 | 3.716 | 6.845 |
| 1-3 | 17 | 32 | 15 | 15 | 79 | 19.75 | 41.988 | 3.337 | 5.774 |
| 1-4 | 18 | 7 | 13 | 12 | 50 | 12.50 | 26.192 | 2.469 | 3.850 |
| 1-5 | 17 | 12 | 10 | 14 | 53 | 13.25 | 27.000 | 2.593 | 4.012 |
| 1-6 | 6 | 24 | 19 | 10 | 59 | 14.75 | 32.757 | 2.705 | 4.602 |
| 2-3 | 29 | 47 | 31 | 13 | 120 | 30.00 | 64.653 | 4.738 | 8.620 |
| 2-4 | 21 | 26 | 17 | 4 | 68 | 17.00 | 37.709 | 2.608 | 4.908 |
| 2-5 | 22 | 22 | 25 | 4 | 73 | 18.25 | 40.112 | 3.076 | 5.428 |
| 2-6 | 45 | 5 | 16 | 2 | 68 | 17.00 | 48.063 | 3.274 | 6.085 |
| 3-4 | 24 | 29 | 14 | 15 | 82 | 20.50 | 42.872 | 3.389 | 5.853 |
| 3-5 | 26 | 25 | 6 | 17 | 74 | 18.50 | 40.324 | 3.340 | 5.589 |
| 3-6 | 16 | 52 | 15 | 13 | 96 | 24.00 | 57.914 | 3.854 | 7.498 |
| 4-5 | 2 | 9 | 8 | 2 | 21 | 5.25 | 12.369 | 0.964 | 1.700 |
| 4-6 | 24 | 31 | 6 | 2 | 63 | 15.75 | 39.712 | 2.376 | 4.925 |
| 5-6 | 23 | 27 | 9 | 4 | 63 | 15.75 | 36.810 | 2.329 | 4.636 |

If we weight the variables according to the city distance weight $c_j = 1/p$, then the distances of Table 3.9(f) emerge. For example, for the object pairs $A_2$ and $A_4$ we obtain, from Table 3.9(a)–(d), $d^{(1)}(A_2, A_4) = (21 + 26 + 17 + 4)/4 = 17.00$.

(iii) Consider now some Euclidean Hausdorff distances. In particular, consider the normalized Euclidean Hausdorff distance of Eq. (3.3.6). We need to calculate the weights $H_j$, $j = 1, \dots, 4$, from Eq. (3.3.7). Therefore, for $H_1$, i.e., $j = 1$, we first need to sum over all the pairwise distances in the matrix $\mathbf{D}_1$ in Eq. (3.3.11). Hence, we find that

$$H_1^2 = \frac{1}{2 \times 6^2} [0^2 + 39^2 + \cdots + 24^2 + 23^2 + 0^2] = 246.306.$$

Likewise, $H_2^2 = 332.472$, $H_3^2 = 109.667$, and $H_4^2 = 46.583$.

Notice that as written in Eq. (3.3.7), the distances $d_j(A, B)$ and $d_j(B, A)$ are both included. However, the distances are symmetrical so that $d_j(A, B) = d_j(B, A)$. Hence, by summing over distinct object pairs only, the formula for $H_j$ in Eq. (3.3.7) pertains but without the divisor "2".

Then, substituting the pairwise distances from Table 3.9(a)–(d), and using these $H_j$ values, we obtain the normalized Euclidean Hausdorff distances. For example,

$$d(A_2, A_4) = \left[ \frac{21^2}{246.306} + \frac{26^2}{332.472} + \frac{17^2}{109.667} + \frac{4^2}{46.583} \right]^{1/2} = 2.608.$$

The complete set of normalized Euclidean Hausdorff distances for all object pairs is displayed in Table 3.9(h).

To find the span normalized Euclidean Hausdorff distances, we use the span weights $|\mathcal{Y}_j|$ instead of the $H_j$ weights. For these data, we observe, from Table 3.8 and Eq. (3.3.9), that

$$|\mathcal{Y}_1| = \max\{108, 108, 125, 101, 99, 113\} - \min\{98, 59, 88, 80, 81, 104\} = 66$$

and, likewise, $|\mathcal{Y}_2| = 66$, $|\mathcal{Y}_3| = 44$, and $|\mathcal{Y}_4| = 27$. Hence, the span normalized Euclidean Hausdorff distance between the pair $A_2$ and $A_6$ (say) is, from Table 3.9(a)–(d) and Eq. (3.3.8),

$$d(A_2, A_4) = \left[ \left( \frac{45}{66} \right)^2 + \left( \frac{5}{66} \right)^2 + \left( \frac{16}{44} \right)^2 + \left( \frac{2}{27} \right)^2 \right]^{1/2} = 6.085. \quad (3.3.12)$$

In a similar manner, we can find the complete set of span normalized Euclidean Hausdorff distances for all object pairs as displayed in Table 3.9(i).

We can also obtain the unweighted (and unnormalized) Euclidean Hausdorff distances for each object pair, by using Table 3.9(a)–(d) and Eq. (3.3.5); these are provided in Table 3.9(g).

From these distances by object pairs, we can obtain the relevant distance matrices. Thus, for example, the normalized Euclidean Hausdorff distance matrix, $\mathbf{D}_n^{(2)}$ (say), and the span normalized Euclidean Hausdorff distance matrix, $\mathbf{D}_{sn}^{(2)}$ (say), are given, respectively, by

$$
\mathbf{D}_n^{(2)} =
\begin{bmatrix}
0 & 3.716 & 3.337 & 2.469 & 2.593 & 2.705 \\
3.716 & 0 & 4.738 & 2.608 & 3.076 & 3.274 \\
3.337 & 4.738 & 0 & 3.389 & 3.340 & 3.854 \\
2.469 & 2.608 & 3.389 & 0 & 0.964 & 2.376 \\
2.593 & 3.076 & 3.340 & 0.964 & 0 & 2.329 \\
2.705 & 3.274 & 3.864 & 2.376 & 2.329 & 0
\end{bmatrix},
\quad (3.3.13)
$$

$$
\mathbf{D}_{sn}^{(2)} =
\begin{bmatrix}
0 & 6.845 & 5.774 & 3.850 & 4.012 & 4.602 \\
6.845 & 0 & 8.620 & 4.908 & 5.428 & 6.085 \\
5.774 & 8.620 & 0 & 5.853 & 5.589 & 7.498 \\
3.850 & 4.908 & 5.853 & 0 & 1.700 & 4.925 \\
4.012 & 5.428 & 5.589 & 1.700 & 0 & 4.636 \\
4.602 & 6.085 & 7.498 & 4.925 & 4.636 & 0
\end{bmatrix}.
\quad (3.3.14)
$$

$\square$

### 3.3.3 Gowda–Diday Dissimilarity

**Definition 3.27**  Let $A_{u_1} = ([a_{u_1j}, b_{u_1j}], j = 1, \ldots, p)$, and $A_{u_2} = ([a_{u_2j}, b_{u_2j}], j = 1, \ldots, p)$ be any two interval-valued objects in $\Omega = \Omega_1 \times \cdots \times \Omega_p$. Then, from Gowda and Diday (1991, 1992), for each variable $Y_j$, $j = 1, \ldots, p$, the **Gowda–Diday dissimilarity** between $A_{u_1}$ and $A_{u_2}$, $d_j(A_{u_1}, A_{u_2})$, is

$$
d_j(A_{u_1}, A_{u_2}) = D_{1j}(A_{u_1}, A_{u_2}) + D_{2j}(A_{u_1}, A_{u_2}) + D_{3j}(A_{u_1}, A_{u_2}) \qquad (3.3.15)
$$

where

$$
D_{1j}(A_{u_1}, A_{u_2}) = (|(b_{u_1j} - a_{u_1j}) - (b_{u_2j} - a_{u_2j})|)/k_j, \qquad (3.3.16)
$$
$$
D_{2j}(A_{u_1}, A_{u_2}) = (b_{u_1j} - a_{u_1j} + b_{u_2j} - a_{u_2j} - 2I_j)/k_j, \qquad (3.3.17)
$$
$$
D_{3j}(A_{u_1}, A_{u_2}) = |a_{u_1j} - a_{u_2j}|/|\mathcal{Y}_j| \qquad (3.3.18)
$$

with

$$
k_j = |\max(b_{u_1j}, b_{u_2j}) - \min(a_{u_1j}, a_{u_2j})|, \qquad (3.3.19)
$$

$$
I_j =
\begin{cases}
|\max(a_{u_1j}, a_{u_2j}) - \min(b_{u_1j}, b_{u_2j})|, & \text{if} \quad A_{u_1} \cap A_{u_2} \neq \phi \\
0, & \text{if} \quad A_{u_1} \cap A_{u_2} = \phi
\end{cases},
$$
$$
\qquad (3.3.20)
$$

$$
|\mathcal{Y}_j| = \max_{u \in \Omega_j}\{b_{uj}\} - \min_{u \in \Omega_j}\{a_{uj}\}. \qquad (3.3.21)
$$

$\square$

Notice that the term $k_j$ in Eq. (3.3.19) is the length of the variable $Y_j$ that is covered by $A_{u_1}$ and $A_{u_2}$, in Eq. (3.3.20), $\phi$ is the empty set and $I_j$ is the length of the intersection of $A_{u_1}$ and $A_{u_2}$, and $|\mathcal{Y}_j|$ in Eq. (3.3.21) is the total length of all observed $Y_j$ values $A_{u_1}$ and $A_{u_2}$ in $\Omega_j$. Also, for variable $Y_j$, $D_{1j}(A_{u_1}, A_{u_2})$ and $D_{2j}(A_{u_1}, A_{u_2})$ of Eqs. (3.3.16) and (3.3.17) are span and content components of this distance (akin to the corresponding components for multi-valued data), while $D_{3j}(A_{u_1}, A_{u_2})$ in Eq. (3.3.18) is the relative measure component of $A_{u_1}$ and $A_{u_2}$ compared with all $A_u$ values in $\Omega$.

The Gowda–Diday dissimilarity between the two intervals $A_{u_1}$ and $A_{u_2}$ over all variables $Y_j$, $j = 1, \dots, p$, for all $u = 1, \dots, m$, follows readily from Eq. (3.3.15) as

$$d(A_{u_1}, A_{u_2}) = \sum_{j=1}^{p} d_j(A_{u_1}, A_{u_2})$$

$$= \sum_{j=1}^{p} [D_{1j}(A_{u_1}, A_{u_2}) + D_{2j}(A_{u_1}, A_{u_2}) + D_{3j}(A_{u_1}, A_{u_2})]. \quad (3.3.22)$$

**Example 3.11**   Let us obtain the Gowda–Diday dissimilarities for the relative abundances data of Table 3.8, described in Example 3.10. We illustrate by calculating the distances between the observation pair $A_2$ and $A_4$.

(i) To calculate the regular Gowda–Diday dissimilarities, we substitute into Eqs. (3.3.15)–(3.3.22). Hence, from Table 3.8, for $Y_1$, we first have, from Eqs. (3.3.19)–(3.3.21),

$$k_1 = |\max\{108, 101\} - \min\{59, 80\}| = 49,$$

$$I_1 = |\max\{59, 80\} - \min\{108, 101\}| = 21,$$

$$|\mathcal{Y}_1| = \max\{108, 108, 125, 101, 99, 113\} - \min\{98, 59, 88, 80, 81, 104\} = 66;$$

hence, substituting into Eqs. (3.3.16)–(3.3.18), we have

$$D_{11}(A_2, A_4) = (|(108 - 59) - (101 - 80)|)/49 = 0.571,$$

$$D_{21}(A_2, A_4) = (108 - 59 + 101 - 80 - 2 \times 21)/49 = 0.571,$$

$$D_{31}(A_2, A_4) = |59 - 80|/66 = 0.318.$$

Then, from Eq. (3.3.15), it follows that $d_1(A_2, A_4) = 1.461$. Similarly, we can obtain $d_2(A_2, A_4) = 1.156$, $d_3(A_2, A_4) = 1.362$, $d_4(A_2, A_4) = 1.291$. Hence, from Eq. (3.3.22), the Gowda–Diday dissimilarity over all variables $Y_j$, $j = 1, \dots, 4$, is $d(A_2, A_4) = 5.270$. Likewise, these distances can be calculated for all object pairs and all variables and are given in Table 3.10 in columns (a)–(d), respectively, and the dissimilarities over all variables are in column (e).

(ii) Minkowski distances of Eq. (3.1.7) can also be calculated from the basic Gowda–Diday dissimilarities derived thus far. The corresponding city block distances for $q = 1$ and city weights $c_j = 1/p = 1/4$ are as shown in Table 3.10(f); the details are left as an exercise for the reader.

**Table 3.10** Gowda–Diday dissimilarities (Example 3.11)

| Object pair | By variable $Y_j$ | | | | $\sum d_j$ | City weighted | Unweighted Euclidean | | |
|---|---|---|---|---|---|---|---|---|---|
| | $d_1$ | $d_2$ | $d_3$ | $d_4$ | $d$ | | Non-normalized | Normalized | Chebychev |
| | (a) | (b) | (c) | (d) | (e) | (f) | (g) | (h) | (i) |
| 1-2 | 2.183 | 1.354 | 2.068 | 1.533 | 7.137 | 1.784 | 3.636 | 1.818 | 2.183 |
| 1-3 | 1.611 | 2.303 | 1.114 | 1.889 | 6.917 | 1.729 | 3.565 | 1.782 | 2.303 |
| 1-4 | 1.558 | 1.295 | 1.356 | 1.355 | 5.565 | 1.391 | 2.790 | 1.395 | 1.558 |
| 1-5 | 1.517 | 1.610 | 1.336 | 1.385 | 5.847 | 1.462 | 2.932 | 1.466 | 1.610 |
| 1-6 | 0.891 | 0.764 | 1.545 | 1.422 | 4.622 | 1.556 | 2.406 | 1.203 | 1.545 |
| 2-3 | 1.318 | 1.561 | 2.266 | 2.185 | 7.329 | 1.832 | 3.752 | 1.876 | 2.266 |
| 2-4 | 1.461 | 1.156 | 1.362 | 1.291 | 5.270 | 1.317 | 2.644 | 1.322 | 1.461 |
| 2-5 | 1.599 | 1.363 | 2.080 | 1.111 | 6.153 | 1.538 | 3.158 | 1.579 | 2.080 |
| 2-6 | 2.348 | 1.326 | 1.091 | 0.874 | 5.639 | 1.410 | 3.038 | 1.519 | 2.348 |
| 3-4 | 1.188 | 1.955 | 1.536 | 1.889 | 6.567 | 1.642 | 3.340 | 1.670 | 1.955 |
| 3-5 | 1.288 | 1.667 | 0.936 | 2.148 | 6.039 | 1.510 | 3.151 | 1.575 | 2.148 |
| 3-6 | 1.756 | 1.939 | 1.698 | 1.963 | 7.357 | 1.839 | 3.685 | 1.843 | 1.963 |
| 4-5 | 0.301 | 0.856 | 1.229 | 1.237 | 3.624 | 0.906 | 1.966 | 0.983 | 1.237 |
| 4-6 | 1.636 | 1.183 | 0.523 | 0.646 | 3.988 | 0.997 | 2.183 | 1.092 | 1.636 |
| 5-6 | 1.473 | 1.455 | 1.490 | 1.370 | 5.789 | 1.447 | 2.896 | 1.448 | 1.490 |

Let us consider the unweighted non-normalized Euclidean distance derived from Eq. (3.1.8) with weights $w_j \equiv 1$, $j = 1, \ldots, p$, and $q = 2$. Then, we see that for objects $A_2$ and $A_4$, from Table 3.10, it follows that

$$d^{(2)}(A_2, A_4) = \left( \sum_{j=1}^{4} [d_j(A_2, A_4)]^2 \right)^{1/2}$$
$$= (1.461^2 + 1.156^2 + 1.362^2 + 1.291^2)^{1/2} = 2.644,$$

and likewise for all the object pairs. The complete set of unweighted non-normalized Euclidean Gowda–Diday dissimilarities is in Table 3.10(g).

Normalized Euclidean distances as given by Eq. (3.1.9) can also be calculated. Thus, for example, for objects $A_2$ and $A_4$, from Table 3.10, it follows that the unweighted normalized distance is

$$d^{(2)}(A_2, A_4) = \left( (1/p) \sum_{j=1}^{p=4} [d_j(A_2, A_4)]^2 \right)^{1/2}$$
$$= [(1.461^2 + 1.156^2 + 1.362^2 + 1.291^2)/4]^{1/2} = 1.322.$$

Table 3.10(h) provides all the unweighted normalized Euclidean Gowda–Diday dissimilarities for these data. Other variations of un/weighted non/normalized Euclidean distances based on the Gowda–Diday dissimilarities can be similarly calculated (see Exercise 3.8).

(iii) Finally, we obtain the Chebychev distances based on the Gowda–Diday dissimilarities from Eq. (3.1.11). These are shown in Table 3.10(i).

Hence, the distance matrices can be obtained. For example, the unweighted normailzed Euclidean Gowda–Diday dissimilarity matrix is

$$
\mathbf{D}^{(2)} =
\begin{bmatrix}
0 & 1.818 & 1.782 & 1.395 & 1.466 & 1.203 \\
1.818 & 0 & 1.876 & 1.322 & 1.579 & 1.519 \\
1.782 & 1.876 & 0 & 1.670 & 1.575 & 1.843 \\
1.395 & 1.322 & 1.670 & 0 & 0.983 & 1.092 \\
1.466 & 1.579 & 1.575 & 0.983 & 0 & 1.448 \\
1.203 & 1.519 & 1.843 & 1.092 & 1.448 & 0
\end{bmatrix}. \qquad (3.3.23)
$$

□

**Example 3.12**  The data of Table 3.11 consist of minimum and maximum temperatures for the months of January = $Y_1$, April = $Y_2$, July = $Y_3$, and October = $Y_4$, with elevation = $Y_5$ as a classical valued observation, for each of $m = 10$ weather stations in China. Let us take the first $m = 5$ stations only (see also Exercise 3.11). Let us base our Mahalanobis distance on the Gowda–Diday dissimilarities obtained from Eqs. (3.3.15)–(3.3.22). We can show that

Table 3.11  Temperatures and elevation at Chinese weather stations (Example 3.12)

| | | January | April | July | October | Elevation |
|---|---|---|---|---|---|---|
| $u$ | Station | $Y_1$ | $Y_2$ | $Y_3$ | $Y_4$ | $Y_5$ |
| 1 | MuDanJia | $[-18.4, -7.5]$ | $[-0.1, 13.2]$ | $[17.0, 26.5]$ | $[0.6, 13.1]$ | 4.82 |
| 2 | HaErBin | $[-20.0, -9.6]$ | $[0.2, 11.9]$ | $[17.8, 27.2]$ | $[-0.2, 12.5]$ | 3.44 |
| 3 | BoKeTu | $[-23.4, -15.5]$ | $[-4.5, 9.5]$ | $[12.9, 23.0]$ | $[-4.0, 8.9]$ | 14.78 |
| 4 | NenJiang | $[-27.9, -16.0]$ | $[-1.5, 12.0]$ | $[16.1, 25.0]$ | $[-2.6, 10.9]$ | 4.84 |
| 5 | LaSa | $[-8.4, 9.0]$ | $[1.7, 16.4]$ | $[10.8, 23.2]$ | $[1.4, 18.7]$ | 73.16 |
| 6 | TengChon | $[2.3, 16.9]$ | $[9.9, 24.3]$ | $[17.4, 22.8]$ | $[14.5, 23.5]$ | 32.96 |
| 7 | KunMing | $[2.8, 16.6]$ | $[10.4, 23.4]$ | $[16.9, 24.4]$ | $[12.4, 19.7]$ | 37.82 |
| 8 | WuZhou | $[10.0, 17.7]$ | $[15.8, 23.9]$ | $[24.2, 33.8]$ | $[19.2, 27.6]$ | 2.38 |
| 9 | NanNing | $[11.5, 17.7]$ | $[17.8, 24.2]$ | $[25.8, 33.5]$ | $[20.3, 26.9]$ | 1.44 |
| 10 | ShanTou | $[11.8, 19.2]$ | $[16.4, 22.7]$ | $[25.6, 32.6]$ | $[20.4, 27.3]$ | 0.02 |

**Table 3.12** Mahalanobis distances for Chinese weather stations based on Gowda–Diday dissimilarities (Example 3.12)

| Station | By variable $Y_j$ | | | | | $\sum d_j$ | Euclidean distance | Mahalanobis distance |
|---|---|---|---|---|---|---|---|---|
| | $d_1$ | $d_2$ | $d_3$ | $d_4$ | $d_5$ | $d$ | | |
| pair | (a) | (b) | (c) | (d) | (e) | (f) | (g) | (h) |
| 1-2 | 0.38 | 0.25 | 0.21 | 0.156 | 0.02 | 1.015 | 1.031 | 0.243 |
| 1-3 | 1.14 | 0.71 | 0.85 | 0.74 | 0.14 | 3.586 | 12.859 | 3.026 |
| 1-4 | 1.19 | 0.26 | 0.34 | 0.55 | 0.00 | 2.339 | 5.469 | 1.287 |
| 1-5 | 1.48 | 0.47 | 1.17 | 0.65 | 0.98 | 4.751 | 22.577 | 5.313 |
| 2-3 | 0.95 | 0.80 | 0.98 | 0.63 | 0.16 | 3.520 | 12.391 | 2.916 |
| 2-4 | 1.08 | 0.35 | 0.50 | 0.42 | 0.02 | 2.369 | 5.613 | 1.321 |
| 2-5 | 1.51 | 0.63 | 1.28 | 0.73 | 1.00 | 5.149 | 26.510 | 6.238 |
| 3-4 | 0.85 | 0.51 | 0.72 | 0.33 | 0.14 | 2.552 | 6.511 | 1.532 |
| 3-5 | 1.48 | 0.96 | 0.50 | 1.10 | 0.84 | 4.875 | 23.768 | 5.593 |
| 4-5 | 1.47 | 0.64 | 1.070 | 0.91 | 0.98 | 5.074 | 25.750 | 6.059 |

the covariance matrix, obtained from Eq. (2.4.15), is

$$\Sigma = \begin{bmatrix} 72.004 & 27.094 & 2.805 & 32.430 & 181.246 \\ 27.094 & 19.605 & 11.070 & 20.853 & 36.545 \\ 2.805 & 11.070 & 13.121 & 10.909 & -45.006 \\ 32.430 & 20.853 & 10.909 & 22.721 & 49.312 \\ 181.246 & 36.545 & -45.006 & 49.312 & 17.502 \end{bmatrix}. \quad (3.3.24)$$

Hence, the determinant $|\Sigma| = 1385.939$ and $|\Sigma|^{1/5} = 4.250$. The Gowda–Diday dissimilarities for each pair of stations (shown in the first column of Table 3.12) are given in column (f); the components $d_j$ of this distance by variable $Y_j$, $j = 1, \ldots, 5$, are also given, in columns (a)–(e). The Euclidean Gowda–Diday dissimilarities are shown in Table 3.12(g). The Mahalanobis distances are obtained from Definition 3.11, by multiplying the Euclidean Gowda–Diday dissimilarities by $\text{Det}(\Sigma)^{1/5}$, and are shown in column (h). Thence, the Mahalanobis distance matrix is

$$\mathbf{D}^{Mah} = \begin{bmatrix} 0 & 0.243 & 3.026 & 1.287 & 5.312 \\ 0.243 & 0 & 2.916 & 1.321 & 6.238 \\ 3.026 & 2.916 & 0 & 1.532 & 5.593 \\ 1.287 & 1.321 & 1.532 & 0 & 6.059 \\ 5.312 & 6.238 & 5.5593 & 6.059 & 0 \end{bmatrix}.$$

□

### 3.3.4   Ichino–Yaguchi Distance

**Definition 3.28**   Let $A_{u_1} = (A_{u_1j}, j = 1, \ldots, p) = ([a_{u_1j}, \quad b_{u_1j}], \quad j = 1, \ldots, p)$ and $A_{u_2} = (A_{u_2j}, j = 1, \ldots, p) = ([a_{u_2j}, \quad b_{u_2j}], \quad j = 1, \ldots, p)$ be any two interval-valued objects in $\Omega = \Omega_1 \times \cdots \times \Omega_p$. Then, from Ichino and Yaguchi (1994), for each variable $Y_j$, $j = 1, \ldots, p$, the **Ichino–Yaguchi distance** between $A_{u_1}$ and $A_{u_2}$, $d_j(A_{u_1}, A_{u_2})$, is

$$d_j(A_{u_1}, A_{u_2}) = |A_{u_1j} \oplus A_{u_2j}| - |A_{u_1j} \otimes A_{u_2j}| + \gamma(2|A_{u_1j} \otimes A_{u_2j}| - |A_{u_1j}| - |A_{u_2j}|),$$
$$(3.3.25)$$

where $A_{u_1j} \oplus A_{u_2j}$ and $A_{u_1j} \otimes A_{u_2j}$ are the meet and join operators defined in Eqs. (3.3.2) and (3.3.3), $|A| = (b - a)$ is the length of the interval $A = [a, b]$, and $0 \leq \gamma \leq 0.5$ is a pre-specified constant.   □

As for the preceding distances, so too can the Ichino–Yaguchi distances be used as the basis for the generalized Minkowski distances of Eq. (3.1.7) and likewise for the special cases of city block distances and the various unweighted and weighted, non-normalized and normalized, Euclidean distances of Eqs. (3.1.8) and (3.1.9) and the Chebychev distance of Eq. (3.1.11) (see Example 3.13).

**Example 3.13**   Let us obtain the Ichino–Yaguchi distances for the relative abundances data of Table 3.8, described in Example 3.10.

(i) The basic Ichino–Yaguchi distances are obtained by substituting into Eq. (3.3.25). Let us take $\gamma = 0.5$. We illustrate the derivation by finding the distances between objects $A_2$ and $A_4$. For the variable $Y_1$, we have, from Table 3.8 and Eqs. (3.3.2), (3.3.3), and (3.3.25),

$$\begin{aligned} d_1(A_2, A_4) &= |[\min\{59, 80\}, \max\{108, 101\}]| - |[\max\{59, 80\}, \min\{108, 101\}]| \\ &\quad + 0.5(2|[\max\{59, 80\}, \min\{108, 101\}]| - |[59, 108]| - |[80, 101]|) \\ &= |[59, 108]| - |[80, 101]| + 0.5(2|[80, 101]| - |[59, 108]| - |[80, 101]|) \\ &= 49 - 21 + 0.5(2 \times 21 - 49 - 21) = 14 \end{aligned}$$

and for $Y_2$,

$$\begin{aligned} d_2(A_2, A_4) &= |[\min\{91, 117\}, \max\{107, 133\}]| - |[91, 107] \otimes [117, 133]| \\ &\quad + 0.5(2\{|[91, 107] \otimes [117, 133]|\} - |[91, 107]| - |[117, 133]| \\ &= |[91, 133]| - 0 + 0.5(2 \times 0 - |[91, 107]| - |[117, 133]|) \\ &= 42 - 0 + 0.5(0 - 16 - 16) = 26, \end{aligned}$$

where we notice that the meet $|[91, 107] \otimes [117, 133]| = 0$ since the two intervals $[91, 107]$ and $[117, 133]$ do not overlap. Similarly, for $Y_3$ and $Y_4$, respectively, we have

$$d_3(A_2, A_4) = 10, \qquad d_4(A_2, A_4) = 3.$$

**Table 3.13** Ichino–Yaguchi distances[a] (Example 3.13)

| Object pair | By variable $Y_j$ | | | | $\sum d_j$ | City weighted | Euclidean | | Span Euclidean | | Chebychev |
| | $d_1$ | $d_2$ | $d_3$ | $d_4$ | $d$ | | Non-normalized | Normalized | Non-normalized | Normalized | |
| | (a) | (b) | (c) | (d) | (e) | (f) | (g) | (h) | (i) | (j) | (k) |
|---|---|---|---|---|---|---|---|---|---|---|---|
| 1-2 | 19.5 | 24.0 | 17.0 | 7.0 | 67.5 | 16.875 | 35.976 | 17.988 | 4.782 | 2.391 | 24 |
| 1-3 | 13.5 | 30.0 | 14.0 | 9.0 | 66.5 | 16.625 | 36.868 | 18.434 | 4.884 | 2.442 | 30 |
| 1-4 | 12.5 | 5.0 | 7.0 | 10.0 | 34.5 | 8.625 | 18.173 | 9.086 | 2.750 | 1.375 | 12.5 |
| 1-5 | 13.0 | 7.5 | 8.5 | 10.5 | 39.5 | 9.875 | 20.193 | 10.096 | 3.023 | 1.511 | 13 |
| 1-6 | 5.5 | 24.0 | 10.5 | 8.0 | 48.0 | 12.000 | 27.937 | 13.969 | 3.750 | 1.875 | 24 |
| 2-3 | 23.0 | 25.0 | 16.0 | 12.0 | 76.0 | 19.000 | 39.421 | 19.710 | 5.351 | 2.676 | 25 |
| 2-4 | 14.0 | 26.0 | 10.0 | 3.0 | 53.0 | 13.250 | 31.321 | 15.660 | 3.977 | 1.989 | 26 |
| 2-5 | 15.5 | 19.5 | 15.5 | 3.5 | 54.0 | 13.500 | 29.547 | 14.773 | 3.913 | 1.957 | 19.5 |
| 2-6 | 25.0 | 5.0 | 9.5 | 1.0 | 40.5 | 10.125 | 27.226 | 13.613 | 3.455 | 1.727 | 25 |
| 3-4 | 16.0 | 25.0 | 8.0 | 11.0 | 60.0 | 15.000 | 32.650 | 16.325 | 4.391 | 2.196 | 25 |
| 3-5 | 16.5 | 22.5 | 5.5 | 12.5 | 57.0 | 14.250 | 31.064 | 15.532 | 4.274 | 2.137 | 22.5 |
| 3-6 | 14.0 | 30.0 | 9.5 | 11.0 | 64.5 | 16.125 | 36.156 | 18.078 | 4.810 | 2.405 | 30 |
| 4-5 | 1.5 | 6.5 | 5.5 | 1.5 | 15.0 | 3.750 | 8.775 | 4.387 | 1.202 | 0.601 | 6.5 |
| 4-6 | 18.0 | 26.0 | 3.5 | 2.0 | 49.5 | 12.375 | 31.879 | 15.939 | 3.947 | 1.973 | 26 |
| 5-6 | 18.5 | 19.5 | 9.0 | 2.5 | 49.5 | 12.375 | 28.456 | 14.228 | 3.608 | 1.804 | 19.5 |

a) $\gamma = 0.5$

Hence, by summing over all $Y_j, j = 1, \ldots, p$, we obtain the Ichino–Yaguchi distance for $A_2$ and $A_4$ to be $d(A_2, A_4) = 53$.

The complete set of Ichino–Yaguchi distances for each variable is given in Table 3.13 in columns (a)–(d) and their sum over all variables is in column (e).

(ii) Let us now consider some Minkowski distances based on these regular Ichino–Yaguchi distances. First, city block distances obtained from Eq. (3.1.7) with $q = 1$ and weights $c_j = 1/p, j = 1, \ldots, p$, are shown in Table 3.13(f). The details are omitted.

Standard Euclidean distances are obtained from Eq. (3.3.8). Thus, for example, the unweighted non-normalized Euclidean distance between objects $A_2$ and $A_4$ is, from Table 3.13(a)–(d), with $w_j = 0, j = 1, \ldots, p$,

$$d^{(2)}(A_2, A_4) = (14^2 + 26^2 + 10^2 + 3^2)^{1/2} = 31.321.$$

The unweighted non-normalized Euclidean distances for all object pairs are shown in Table 3.13(g). The unweighted normalized Euclidean distances obtained from Eq. (3.1.9) are given in Table 3.13(h).

Suppose now we wish to calculate span weighted Euclidean distances. In this case, weights $w_j = 1/|\mathcal{Y}_j|, j = 1, \ldots, p$, are used. These span $|\mathcal{Y}_j|$ values are, from Eq. (3.3.9) and Table 3.8,

$$|\mathcal{Y}_1| = \max\{108, 108, 125, 101, 99, 113\} - \min\{98, 59, 88, 80, 81, 104\} = 66$$

and, likewise,

$$|\mathcal{Y}_2| = 66, \quad |\mathcal{Y}_3| = 44, \quad |\mathcal{Y}_4| = 27.$$

Thence, span weighted non/normalized Euclidean Ichino–Yaguchi distances can be found. Thus, for example, the span normalized Euclidean distance between objects $A_2$ and $A_4$ is obtained by substituting these weights into Eq. (3.1.9), as

$$d^{(2)}(A_2, A_4) = [(14^2/66 + 26^2/66 + 10^2/44 + 3^2/27)/4]^{1/2} = 1.989.$$

The complete set of span normalized Euclidean Ichino–Yaguchi distances is given in Table 3.13(j). Table 3.13, columns (i) and (k), respectively, give the span non-normalized Euclidean Ichino–Yaguchi distances and Chebychev Ichino–Yaguchi distances for all object pairs.

The relevant distance matrices can be easily obtained. For example, the span normalized Euclidean Ichino–Yaguchi distance matrix becomes

$$\mathbf{D}^{(2)} = \begin{bmatrix} 0 & 2.391 & 2.442 & 1.375 & 1.511 & 1.875 \\ 2.391 & 0 & 2.676 & 1.989 & 1.957 & 1.727 \\ 2.442 & 2.676 & 0 & 2.196 & 2.137 & 2.405 \\ 1.375 & 1.989 & 2.196 & 0 & 0.601 & 1.973 \\ 1.511 & 1.957 & 2.137 & 0.601 & 0 & 1.804 \\ 1.875 & 1.727 & 2.405 & 1.973 & 1.804 & 0 \end{bmatrix}. \tag{3.3.26}$$

□

### 3.3.5 de Carvalho Extensisons of Ichino–Yaguchi Distances

Proximity measures based on the Ichino (1988) meet function (of Definition 3.20) were introduced by de Carvalho (1994, 1998) with so-called agreement and disagreement indices. These were used to develop five different comparison functions which in turn gave five different extentions to the Ichino–Yaguchi distance measures.

**Definition 3.29**   Let $A$ and $B$ be two interval-valued objects. Then, **de Carvalho distances** between $A$ and $B$, for comparison functions $cf_k$, $k = 1, \ldots, 5$, are

$$d_{j,cf_k}(A, B) = 1 - cf_k, \tag{3.3.27}$$

where, for a given variable $Y_j$,

$$cf_1 = \alpha/(\alpha + \beta + \chi), \tag{3.3.28}$$

$$cf_2 = 2\alpha/(2\alpha + \beta + \chi), \tag{3.3.29}$$

$$cf_3 = \alpha/(\alpha + 2\beta + 2\chi), \tag{3.3.30}$$

$$cf_4 = [\alpha/(\alpha + \beta) + \alpha/(\alpha + \chi)]/2, \tag{3.3.31}$$

$$cf_4 = \alpha/[(\alpha + \beta)(\alpha + \chi)]^{1/2}, \tag{3.3.32}$$

where

$$\alpha = |A \otimes B|, \quad \beta = |A \otimes B^c|, \quad \chi = |A^c \otimes B|, \quad \delta = |A^c \otimes B^c|, \tag{3.3.33}$$

with $A^c$ being the complement of $A$, $|A|$ the length of $A$, and $\otimes$ the meet operator. The $\alpha, \beta, \chi$, and $\delta$ terms serve as an agreement–agreement index, agreement–disagreement index, disagreement–agreement index, and disagreement–disagreement index, respectively.                    □

Note that in Eqs. (3.3.27)–(3.3.33) each of the agreement–disagreement indices and hence each of the comparison functions varies with variable $Y_j$ and object pairs $A$ and $B$. De Carvalho (1994) then aggregates his distance functions as non-normalized and/or normalized versions of the Minkowski distances with $q = 2$. However, it is equally possible to calculate other weighted Minkowski distances along the lines already considered with previous distances in this chapter. We illustrate these distances using the format originally developed by de Carvalho.

**Example 3.14**  Let us obtain the de Carvalho distances for the relative abundances data of Table 3.8, described in Example 3.10. We first need to obtain the agreement–disagreement indices for each object pair. Thus, for example, for $A_2$ and $A_4$, for $Y_1$, we have

$$\alpha_1(A_2, A_4) = \|[59,108] \otimes [80,101]\| = \|[80,101]\| = 21,$$

$$\beta_1(A_2, A_4) = \|[59,108] \otimes [80,101]^c\| = \|[59,80] \cup [101,108]\| = 28,$$

$$\chi_1(A_2, A_4) = \|[59,108]^c \otimes [80,101]\| = |\phi| = 0,$$

$$\delta_1(A_2, A_4) = \|[59,108]^c \otimes [80,101]^c\| = |\phi| = 0.$$

Likewise, for $Y_j, j = 2, 3, 4$, we have, respectively,

$$\alpha_2(A_2, A_4) = 0, \quad \beta_2(A_2, A_4) = 16, \quad \chi_2(A_2, A_4) = 16, \quad \delta_2(A_2, A_4) = 10,$$

$$\alpha_3(A_2, A_4) = 21, \quad \beta_3(A_2, A_4) = 20, \quad \chi_3(A_2, A_4) = 0, \quad \delta_3(A_2, A_4) = 0,$$

$$\alpha_4(A_2, A_4) = 1, \quad \beta_4(A_2, A_4) = 26, \quad \chi_2(A_2, A_4) = 4, \quad \delta_2(A_2, A_4) = 0.$$

Then, by substituting into Eqs. (3.3.28)–(3.3.32), we can obtain the comparison functions and hence the corresponding distances. Thus, for example, the distance $d_{cf4}(A_2, B_4)$ becomes, for $Y_1$, from Eq. (3.3.27),

$$d_{1,cf_4}(A_2, A_4) = 1 - [21/(21 + 28) + 21/(21 + 0)]^{1/2} = 0.286.$$

Likewise, we obtain $d_{2,cf_4}(A_2, A_4) = 1$, $d_{3,cf_4}(A_2, A_4) = 0.244$ and $d_{4,cf_4}(A_2, A_4) = 0.733$. Hence, substituting into Eq. (3.1.8), we have, for $w_j = 1, j = 1, \ldots, p$, the unweighted de Carvalho Euclidean distance based on the comparison function $cf_4$,

$$d_{cf_4}^{(2)}(A_2, A_4) = [0.286^2 + 1^2 + 0.244^2 + 0.733^2]^{1/2} = 1.296,$$

and the corresponding normalized unweighted de Carvalho Euclidean distance based on the comparison function $cf_4$, obtained from Eq. (3.1.9), is

$$d_{cf_4}^{(2)}(A_2, A_4) = [(0.286^2 + 1^2 + 0.244^2 + 0.733^2)/4]^{1/2} = 0.648.$$

The complete set of non-normalized and normalized Euclidean de Carvalho distances based on each of the comparison functions is shown in Table 3.14.

Span weighted non-normalized and normalized Euclidean de Carvalho distances, where the weights are $w_j = 1/|\mathcal{Y}_j|$, are provided in Table 3.15. Their calculations are left as an exercise for the reader.  □

**Table 3.14** de Carvalho unweighted distances (Example 3.14)

| Object pair | Non-normalized Euclidean | | | | | Normalized Euclidean | | | | |
|---|---|---|---|---|---|---|---|---|---|---|
| | $cf_1$ | $cf_2$ | $cf_3$ | $cf_4$ | $cf_5$ | $cf_1$ | $cf_2$ | $cf_3$ | $cf_4$ | $cf_5$ |
| 1-2 | 1.822 | 1.714 | 1.899 | 1.527 | 1.626 | 0.911 | 0.857 | 0.950 | 0.763 | 0.813 |
| 1-3 | 1.674 | 1.508 | 1.805 | 1.205 | 1.377 | 0.837 | 0.754 | 0.903 | 0.602 | 0.689 |
| 1-4 | 1.622 | 1.452 | 1.767 | 1.347 | 1.399 | 0.811 | 0.726 | 0.883 | 0.674 | 0.700 |
| 1-5 | 1.854 | 1.781 | 1.912 | 1.726 | 1.753 | 0.927 | 0.890 | 0.956 | 0.863 | 0.877 |
| 1-6 | 1.761 | 1.642 | 1.858 | 1.573 | 1.608 | 0.880 | 0.821 | 0.929 | 0.787 | 0.804 |
| 2-3 | 1.568 | 1.307 | 1.753 | 0.877 | 1.125 | 0.784 | 0.653 | 0.876 | 0.439 | 0.563 |
| 2-4 | 1.516 | 1.352 | 1.677 | 1.296 | 1.323 | 0.758 | 0.676 | 0.838 | 0.648 | 0.662 |
| 2-5 | 1.724 | 1.607 | 1.828 | 1.498 | 1.553 | 0.862 | 0.804 | 0.914 | 0.749 | 0.776 |
| 2-6 | 1.263 | 1.043 | 1.485 | 0.862 | 0.962 | 0.631 | 0.522 | 0.743 | 0.431 | 0.481 |
| 3-4 | 1.492 | 1.197 | 1.707 | 0.879 | 1.053 | 0.746 | 0.599 | 0.853 | 0.439 | 0.526 |
| 3-5 | 1.556 | 1.306 | 1.741 | 0.981 | 1.176 | 0.778 | 0.653 | 0.871 | 0.490 | 0.588 |
| 3-6 | 1.589 | 1.342 | 1.763 | 0.794 | 1.121 | 0.794 | 0.671 | 0.882 | 0.397 | 0.561 |
| 4-5 | 0.962 | 0.663 | 1.251 | 0.528 | 0.597 | 0.481 | 0.331 | 0.626 | 0.264 | 0.298 |
| 4-6 | 1.546 | 1.477 | 1.640 | 1.475 | 1.476 | 0.773 | 0.738 | 0.820 | 0.738 | 0.738 |
| 5-6 | 1.763 | 1.654 | 1.854 | 1.589 | 1.622 | 0.881 | 0.827 | 0.927 | 0.795 | 0.811 |

**Table 3.15** de Carvalho span weighted distances (Example 3.14)

| Object pair | Non-normalized Euclidean | | | | | Normalized Euclidean | | | | |
|---|---|---|---|---|---|---|---|---|---|---|
| | $cf_1$ | $cf_2$ | $cf_3$ | $cf_4$ | $cf_5$ | $cf_1$ | $cf_2$ | $cf_3$ | $cf_4$ | $cf_5$ |
| 1-2 | 0.278 | 0.265 | 0.288 | 0.242 | 0.254 | 0.139 | 0.132 | 0.144 | 0.121 | 0.127 |
| 1-3 | 0.244 | 0.218 | 0.266 | 0.179 | 0.201 | 0.122 | 0.109 | 0.133 | 0.089 | 0.100 |
| 1-4 | 0.255 | 0.236 | 0.272 | 0.224 | 0.230 | 0.128 | 0.118 | 0.136 | 0.112 | 0.115 |
| 1-5 | 0.286 | 0.278 | 0.291 | 0.273 | 0.276 | 0.143 | 0.139 | 0.146 | 0.137 | 0.138 |
| 1-6 | 0.270 | 0.256 | 0.282 | 0.246 | 0.251 | 0.135 | 0.128 | 0.141 | 0.123 | 0.125 |
| 2-3 | 0.243 | 0.207 | 0.268 | 0.131 | 0.176 | 0.122 | 0.104 | 0.134 | 0.065 | 0.088 |
| 2-4 | 0.230 | 0.202 | 0.254 | 0.194 | 0.198 | 0.115 | 0.101 | 0.127 | 0.097 | 0.099 |
| 2-5 | 0.267 | 0.253 | 0.280 | 0.239 | 0.246 | 0.133 | 0.126 | 0.140 | 0.119 | 0.123 |
| 2-6 | 0.171 | 0.136 | 0.208 | 0.112 | 0.125 | 0.085 | 0.068 | 0.104 | 0.056 | 0.063 |
| 3-4 | 0.228 | 0.185 | 0.259 | 0.130 | 0.160 | 0.114 | 0.092 | 0.129 | 0.065 | 0.080 |
| 3-5 | 0.244 | 0.211 | 0.268 | 0.148 | 0.187 | 0.122 | 0.106 | 0.134 | 0.074 | 0.094 |
| 3-6 | 0.237 | 0.199 | 0.264 | 0.119 | 0.166 | 0.119 | 0.100 | 0.132 | 0.059 | 0.083 |
| 4-5 | 0.155 | 0.108 | 0.199 | 0.082 | 0.095 | 0.077 | 0.054 | 0.100 | 0.041 | 0.048 |
| 4-6 | 0.209 | 0.192 | 0.231 | 0.191 | 0.191 | 0.105 | 0.096 | 0.116 | 0.096 | 0.096 |
| 5-6 | 0.256 | 0.233 | 0.274 | 0.220 | 0.226 | 0.128 | 0.117 | 0.137 | 0.110 | 0.113 |

## 3.4 Other Measures

Working in the field of ecological taxonomy, many authors developed similarity/ dissimilarity measures to compare two objects $A$ and $B$ by considering the number of common categories ($a$, say), and the number of categories in $A$ but not in $B$, and the number of categories in $B$ but not in $A$ ($b$, and $c$, respectively, say), and number of not-common categories ($d$, say) between $A$ and $B$. For example, one of the earliest such measures is the Jaccard (1907) dissimilarity between two objects $A$ and $B$ with $d(A, B) = (b + c)/(a + b + c)$. See, for example, Sneath and Sokal (1973), Batagelj and Bren (1995), Gordon (1999), and Consonni and Todeschini (2012) for extended summaries of these measures. These coefficients tend to relate to binary coded values. In a different direction, some researchers have used Pearson's product-moment correlation coefficient as their measure of dissimilarity between taxonomies, e.g., Sokal (1963) in taxonomy and Parks (1966) in geology. How these might be extended to multi-valued data in $\mathcal{Y}$ is by and large an open problem; likewise, for multivariate $p \geq 1$ variables.

## Exercises

**3.1** Refer to the objects ($A$, $B$, $C$) of Examples 3.3 and 3.4. Calculate all the distances displayed in Tables 3.2 and 3.3.

**3.2** Refer to Example 3.7 and the data of Table 3.4.
(a) Verify that the weights displayed in Table 3.6 are correct.
(b) Find the weighted Euclidean distances based on the Gowda–Diday dissimilarities where the weights are selected to account for scale for each variable, i.e., $w_j = 1/|\mathcal{Y}_j|, j = 1, 2$.

**3.3** Refer to Example 3.7 and the data of Table 3.4.
(a) Do the Gowda–Diday-based Euclidean distances of Table 3.6 give Robinson matrices?
(b) For those distance matrices in (a) that were not Robinson matrices, can you find a new ordering of the households that do produce Robinson matrices?

**3.4** Refer to Example 3.8 and the data of Table 3.4.
(a) Do the Ichino–Yaguchi-based distances of Table 3.7 give Robinson distance matrices?

(b) For those distance matrices in (a) that were not Robinson matrices, can you find a new ordering of the households that do produce Robinson matrices?

**3.5** Refer to Example 3.8 and the data of Table 3.4. Verify that the Inchino–Yaguci-based weights displayed in Table 3.7 are correct.

**3.6** Refer to Example 3.10 and the data of Table 3.8.
(a) Verify the Hausdorff distances displayed in Table 3.9(a)–(d).
(b) Do the Hausdorff distances for $Y_1$ (in Table 3.9(a)) give a Robinson matrix? If not, is there a reordered sequence of the objects which does produce a Robinson matrix?

**3.7** Refer to Example 3.10, the data of Table 3.8, and the results of Table 3.9(a)–(d).
(a) Verify the three Euclidean Hausdorff distances given in Table 3.9(h) and (i).
(b) Show that weighted and unweighted the city block distances based on the Hausdorff distances (in Table 3.9(a)–(d)) are as displayed in Table 3.9(e)–(f).
(c) Suppose now you added a span normalization factor to your weighted city block distances (i.e., when you use the weights $w_j = 1/|\mathcal{Y}_j|$ and include the city weight $c_j = 1/p, j = 1, \ldots, p$). What is the resulting distance matrix?

**3.8** Refer to Example 3.11, the data of Table 3.8, and the results of Table 3.10.
(a) Verify the basic Gowda–Diday dissimilarities of columns (a)–(e).
(b) Verify the Gowda–Diday-based Euclidean distances of columns (g)–(h).
(c) Obtain the span weighted non-normalized and span weighted normalized Euclidean distances based on the Gowda–Diday dissimilarities. Compare these results with the analogous distances based on Ichino–Yaguchi distances (shown in Table 3.13(i)–(j)).

**3.9** Refer to Example 3.13, the data of Table 3.8, and the results of Table 3.13.
(a) Verify the basic Ichino–Yaguchi distances of columns (a)–(e).
(b) Verify the Ichino–Yaguchi-based Euclidean distances of columns (g)–(j).
(c) Obtain span weighted city block distances based on the basic Ichino–Yaguchi distances (of columns (a) and (b)), and compare your results with the city block distances in column (f).

**3.10** The unweighted normalized Euclidean Gowda–Diday distances in Table 3.10(h) and the corresponding unweighted normalized Euclidean Ichino–Yaguchi distances in Table 3.13(h) were obtained from the normalization of $1/p$ in Eq. (3.1.8) with effective weights $w_j \equiv 1$, $j = 1, \ldots, p$.

   (a) Find the weighted Euclidean Gowda–Diday distances using weights $w_j = H_j^{-1}, j = 1, \ldots, p$, with $H_j$ given in Eq. (3.3.7).

   (b) Repeat (a) for Ichino–Yaguchi-based distances.

   (c) Compare the results you obtained in (a) and (b) with the normalized Euclidean Hausdorff distances of Table 3.9(h).

**3.11** Refer to Example 3.12 and the data of Table 3.11. Using all $m = 10$ observations, show that the Mahalanobis distance matrix based on the Euclidean Gowda–Diday dissimilarities is

$\mathbf{D}^{Mah}$

$$
=\begin{bmatrix}
0 & 0.055 & 0.648 & 0.279 & 1.162 & 1.621 & 1.431 & 1.794 & 2.078 & 2.107 \\
0.055 & 0 & 0.623 & 0.287 & 1.359 & 1.724 & 1.609 & 1.647 & 1.917 & 1.942 \\
0.648 & 0.623 & 0 & 0.331 & 1.192 & 1.840 & 1.770 & 2.035 & 2.203 & 2.236 \\
0.279 & 0.287 & 0.331 & 0 & 1.281 & 1.698 & 1.505 & 2.030 & 2.208 & 2.254 \\
1.162 & 1.359 & 1.192 & 1.281 & 0 & 1.827 & 1.690 & 3.661 & 3.946 & 3.962 \\
1.621 & 1.724 & 1.840 & 1.698 & 1.827 & 0 & 0.172 & 1.428 & 1.826 & 1.740 \\
1.431 & 1.609 & 1.770 & 1.505 & 1.690 & 0.172 & 0 & 1.594 & 1.894 & 1.803 \\
1.794 & 1.647 & 2.035 & 2.030 & 3.661 & 1.428 & 1.594 & 0 & 0.209 & 0.212 \\
2.078 & 1.917 & 2.203 & 2.208 & 3.946 & 1.826 & 1.894 & 0.209 & 0 & 0.083 \\
2.107 & 1.942 & 2.236 & 2.254 & 3.962 & 1.740 & 1.803 & 0.212 & 0.083 & 0
\end{bmatrix}.
$$

# Appendix

Table 3.16 Relative abundances[a] (all data)

| Class | CG $Y_1$ | GC $Y_2$ | TA $Y_3$ | AT $Y_4$ | CC $Y_5$ | AA $Y_6$ | AC $Y_7$ | AG $Y_8$ | CA $Y_9$ | GA $Y_{10}$ | $n_u$ |
|---|---|---|---|---|---|---|---|---|---|---|---|
| 1 | [98, 108] | [120, 126] | [69, 76] | [100, 109] | [92, 101] | [112, 125] | [83, 91] | [81, 96] | [110, 116] | [90, 100] | 5 |
| 2 | [59, 108] | [91, 107] | [51, 92] | [96, 99] | [86, 108] | [90, 103] | [98, 105] | [98, 104] | [104, 118] | [95, 132] | 2 |
| 3 | [88, 125] | [88, 154] | [82, 91] | [85, 112] | [91, 116] | [97, 146] | [71, 108] | [66, 112] | [96, 112] | [64, 107] | 6 |
| 4 | [80, 101] | [117, 133] | [68, 89] | [92, 97] | [97, 112] | [115, 123] | [85, 89] | [93, 98] | [98, 118] | [90, 95] | 4 |
| 5 | [81, 99] | [108, 129] | [76, 86] | [93, 95] | [93, 97] | [108, 118] | [80, 95] | [98, 118] | [94, 111] | [97, 116] | 3 |
| 6 | [104, 113] | [96, 102] | [67, 95] | [94, 99] | [83, 103] | [104, 111] | [91, 103] | [98, 100] | [92, 106] | [105, 117] | 3 |
| 7 | [116, 116] | [127, 127] | [76, 76] | [110, 110] | [90, 90] | [121, 121] | [89, 89] | [83, 83] | [111, 111] | [93, 93] | 1 |
| 8 | [123, 123] | [129, 129] | [82, 82] | [114, 114] | [91, 91] | [122, 122] | [85, 85] | [82, 82] | [103, 103] | [93, 93] | 1 |
| 9 | [104, 104] | [126, 126] | [65, 65] | [101, 101] | [96, 96] | [123, 123] | [76, 76] | [92, 92] | [108, 108] | [107, 107] | 1 |
| 10 | [109, 109] | [117, 117] | [58, 58] | [111, 111] | [87, 87] | [114, 114] | [87, 87] | [99, 99] | [109, 109] | [108, 108] | 1 |
| 11 | [81, 81] | [120, 120] | [79, 79] | [83, 83] | [99, 99] | [119, 119] | [83, 83] | [115, 115] | [98, 98] | [105, 105] | 1 |
| 12 | [124, 124] | [100, 100] | [45, 45] | [116, 116] | [81, 81] | [95, 95] | [111, 111] | [83, 83] | [108, 108] | [124, 124] | 1 |
| 13 | [76, 76] | [108, 108] | [89, 89] | [88, 88] | [109, 109] | [112, 112] | [100, 100] | [101, 101] | [109, 109] | [89, 89] | 1 |
| 14 | [105, 105] | [121, 121] | [64, 64] | [108, 108] | [88, 88] | [115, 115] | [88, 88] | [93, 93] | [115, 115] | [101, 101] | 1 |

a) Blaisdell et al. (1996)

# 4

# Dissimilarity, Similarity, and Distance Measures: Modal Data

In Chapter 3, the basic principles and properties of dissimilarity/similarity and distance measures were introduced. Detailed measures for multi-valued list observations and interval-valued observations were described and illustrated. In this chapter, we discuss dissimilarity/similarity and distance measures for modal data, i.e., modal multi-valued data in section 4.1 and histogram-valued data in section 4.2.

## 4.1 Dissimilarity/Distance Measures: Modal Multi-valued List Data

As in Chapter 3, we suppose our sample space $\Omega$ consists of objects $A_u$, $u = 1, \ldots, m$, and suppose the realization of each object contains modal multi-valued list values for the $p$-dimensional random variable $\mathbf{Y} = (Y_1, \ldots, Y_p)$ in the form

$$\xi(A_u) \equiv \mathbf{Y}_u = (\{Y_{ujk_j}, p_{ujk_j}; \ k_j = 1, \ldots, s_{uj}\}, j = 1, \ldots, p), \quad u = 1, \ldots, m,$$

$$(4.1.1)$$

where $Y_{ujk}$ is the $k$th $\equiv k_j$th category observed from the set of possible categories $\mathcal{Y}_j = \{Y_{j1}, \ldots, Y_{js_j}\}$ for the variable $Y_j$ and for the object $A_u$, and where $p_{ujk_j}$ is the probability or relative frequency associated with $Y_{ujk_j}$, with $\sum_{k_j} p_{ujk_j} = 1$. Unlike the non-modal multi-valued list values of Chapter 3, where it was assumed that for each observed category $p_{ujk_j} \equiv 1/s_{uj}$ for all $(u, j, k_j)$, for modal multi-valued observations these $p_{ujk_j}$ are in general unequal. Furthermore, for the modal case, without loss of generality, we will assume that $s_{uj} = s_j$ for all $u = 1, \ldots, m$, that is, when a given category $Y_{jk_j}$ in $\mathcal{Y}_j$ does not occur, it has probability $p_{jk_j'} = 0$.

*Clustering Methodology for Symbolic Data*, First Edition. Lynne Billard and Edwin Diday.
© 2020 John Wiley & Sons Ltd. Published 2020 by John Wiley & Sons Ltd.

### 4.1.1 Union and Intersection Operators for Modal Multi-valued List Data

We will need the concepts of union of two objects and intersection of two objects. These are analogous to the Ichino (1988) concepts of join and meet operators for non-modal multi-valued list objects described in section 3.2.1. Recall that the description of each object can be written as containing all possible categories. Hence, objects, for a given variable $Y_j$, are distinguished by their respective probabilities or relative frequencies $p_{ujk_j}$ as $k_j$ changes across $k_j = 1, \dots, s_j$. Thus, Eq. (4.1.1) can be re-written as

$$\xi(A_u) \equiv \mathbf{Y}_u = (\{Y_{jk_j}, p_{ujk_j}; k_j = 1, \dots, s_j\}, j = 1, \dots, p), u = 1, \dots, m. \tag{4.1.2}$$

**Definition 4.1**  Let $A_{u_1}$ and $A_{u_2}$ be any two modal list multi-valued objects in $\Omega = \Omega_1 \times \cdots \times \Omega_p$ with realizations taking the general form of Eq. (4.1.2). Then, the **union** of $A_{u_1}$ and $A_{u_2}$, $A_{u_1} \cup A_{u_2}$, $u_1, u_2 = 1, \dots, m$, is the object described by

$$\xi(A_{u_1} \cup A_{u_2}) = (\{Y_{jk_j}, p_{(u_1 \cup u_2)jk_j}; k_j = 1, \dots, s_j\}, j = 1, \dots, p) \tag{4.1.3}$$

where $p_{(u_1 \cup u_2)jk_j} = \max\{p_{u_1jk_j}, p_{u_2jk_j}\}$. □

**Definition 4.2**  Let $A_{u_1}$ and $A_{u_2}$ be any two modal list multi-valued objects in $\Omega = \Omega_1 \times \cdots \times \Omega_p$ with realizations taking the form of Eq. (4.1.2). Then, the **intersection** of $A_{u_1}$ and $A_{u_2}$, $A_{u_1} \cap A_{u_2}$, $u_1, u_2 = 1, \dots, m$, is the object described by

$$\xi(A_{u_1} \cap A_{u_2}) = (\{Y_{jk_j}, p_{(u_1 \cap u_2)jk_j}; k_j = 1, \dots, s_j\}, j = 1, \dots, p) \tag{4.1.4}$$

where $p_{(u_1 \cap u_2)jk_j} = \min\{p_{u_1jk_j}, p_{u_2jk_j}\}$. □

It is observed that category probabilities for the union and/or intersection of two objects do not necessarily sum to one. This leads to Choquet capacities (Diday and Emilion, 2003). While the direct impact of this is of no consequence for dissimilarity/distance measures, it will be necessary to standardize these values for other entities such as descriptive statistics. This will be covered in section 4.2.3. It is also noted that the justification for the choices of "max" and "min" for the union and intersection probabilities, respectively, comes from conceptual lattices for symbolic data as they are the only choice which gives Galois closed sets; see Diday (1995) and Diday and Emilion (2003).

**Example 4.1**  Let $\mathcal{Y} = \mathcal{Y}_1 \times \mathcal{Y}_2 \times \mathcal{Y}_3$ be the set of possible multi-valued list values describing household characteristics obtained by aggregating individual household values over a region of interest, with $\mathcal{Y}_1 = \{$gas, electricity,

**Table 4.1** Modal list multi-valued data (Example 4.1)

| Region | Gas | Electricity | Coal | Wood | Other | 1–3 | ≥ 4 | Brick | Timber | Stone | Other |
|---|---|---|---|---|---|---|---|---|---|---|---|
| | | Fuel usage | | | | Number of bedrooms | | House material | | | |
| $A$ | 0.4 | 0.4 | 0.1 | 0.0 | 0.1 | 0.7 | 0.3 | 0.5 | 0.2 | 0.2 | 0.1 |
| $B$ | 0.3 | 0.3 | 0.2 | 0.1 | 0.1 | 0.8 | 0.2 | 0.4 | 0.3 | 0.1 | 0.2 |
| $A \cup B$ | 0.4 | 0.4 | 0.2 | 0.1 | 0.1 | 0.8 | 0.3 | 0.5 | 0.3 | 0.2 | 0.2 |
| $A \cap B$ | 0.3 | 0.3 | 0.1 | 0.0 | 0.1 | 0.7 | 0.2 | 0.4 | 0.2 | 0.1 | 0.1 |

coal, wood, other} being the set of possible fuel types used, $\mathcal{Y}_2 = \{1 - 3, \geq 4\}$ being the number of bedrooms, and $\mathcal{Y}_3 = \{$brick, timber, stone, other$\}$ being the type of material used to build the house. Table 4.1 shows the aggregated proportions of each possible categorical value for two regions $A$ and $B$. Thus, for example, in Region A, 70% of the households have one to three bedrooms while 30% have more than three bedrooms.

Then, taking variable $Y_2$, we see that for the union of regions $A$ and $B$, we have

$$p_{(A \cup B)21} = \max\{0.7, 0.8\} = 0.8, \quad p_{(A \cup B)22} = \max\{0.3, 0.2\} = 0.3$$

and, for the intersection of $A$ and $B$, we have

$$p_{(A \cap B)21} = \min\{0.7, 0.8\} = 0.7, \quad p_{(A \cap B)22} = \min\{0.3, 0.2\} = 0.2.$$

Likewise, for $Y_1$ and $Y_3$, we have, respectively,

$$p_{(A \cup B)11} = 0.4, \quad p_{(A \cup B)12} = 0.4, \quad p_{(A \cup B)13} = 0.2, \quad p_{(A \cup B)14} = 0.1, \quad p_{(A \cup B)15} = 0.1,$$

$$p_{(A \cap B)11} = 0.3, \quad p_{(A \cap B)12} = 0.3, \quad p_{(A \cap B)13} = 0.1, \quad p_{(A \cap B)14} = 0.0, \quad p_{(A \cap B)15} = 0.1;$$

$$p_{(A \cup B)31} = 0.5, \quad p_{(A \cup B)32} = 0.3, \quad p_{(A \cup B)33} = 0.2, \quad p_{(A \cup B)34} = 0.2,$$

$$p_{(A \cap B)31} = 0.4, \quad p_{(A \cap B)32} = 0.2, \quad p_{(A \cap B)33} = 0.1, \quad p_{(A \cap B)34} = 0.1.$$

These results are summarized in Table 4.1. □

### 4.1.2 A Simple Modal Multi-valued List Distance

Possibly the simplest distance measure for modal list multi-valued data is that given by Chavent (2000). This extends the non-modal list dissimilarity of Definition 3.17 to the case of modal list multi-valued observations.

**Definition 4.3** Let $A_{u_1}$ and $A_{u_2}$ be any two modal list valued objects in $\Omega = \Omega_1 \times \cdots \times \Omega_p$, $u_1, u_2 = 1, \ldots, m$, with realizations taking values of the

**Table 4.2** List multi-valued data (Example 4.2)

| Region | $Y_1$ = Number of bedrooms | | | | $Y_2$ = Number of bathrooms | | | $Y_3$ = College | |
|---|---|---|---|---|---|---|---|---|---|
| $u$ | 1 | 2 | 3 | 4+ | 1 | 2 | 3+ | No | Yes |
| 1 | 0.1 | 0.3 | 0.4 | 0.2 | 0.3 | 0.6 | 0.1 | 0.4 | 0.6 |
| 2 | 0.1 | 0.2 | 0.5 | 0.2 | 0.1 | 0.4 | 0.5 | 0.1 | 0.9 |
| 3 | 0.2 | 0.5 | 0.3 | 0.0 | 0.4 | 0.6 | 0.0 | 0.8 | 0.2 |
| $\sum_u p_{ujk_j}$ | 0.4 | 1.0 | 1.2 | 0.4 | 0.8 | 1.6 | 0.6 | 1.3 | 1.7 |

form in Eq. (4.1.2). Then, a **modal list multi-valued distance measure**, $d(A_{u_1}, A_{u_2})$, between $A_{u_1}$ and $A_{u_2}$ satisfies

$$d^2(A_{u_1}, A_{u_2}) = \sum_{j=1}^{p} \sum_{k_j=1}^{s_j} \left( p \sum_{u=1}^{m} p_{ujk_j} \right)^{-1} (p_{u_1 jk_j} - p_{u_2 jk_j})^2. \qquad (4.1.5)$$

□

**Example 4.2** Table 4.2 gives a data set of modal multi-valued observations for $Y_1$ = number of bedrooms, $Y_2$ = number of bathrooms, and $Y_3$ = highest educational level attained for households aggregated by region (such as a census tract). The complete categorical spaces were assumed to be $\mathcal{Y}_1 = \{1, 2, 3, 4+\}$, $\mathcal{Y}_2 = \{1, 2, 3+\}$, and $\mathcal{Y}_3 = \{\text{college (no), college (yes)}\}$. The table shows the proportion of households for each region taking the various categories in $\mathcal{Y}_j$, $j = 1, 2, 3$, respectively. Thus, for example, in Region 1, 40% of the households had three bedrooms and 10% had three or more bathrooms. Also shown are the summation terms $\sum_u p_{ujk_j}$ in the denominator of Eq. (4.1.5).

Consider the last two regions $A_2$ and $A_3$. Then, from Eq. (4.1.5) and Table 4.2, we have

$$d^2(A_2, A_3) = \frac{1}{3} \left[ \frac{(0.1 - 0.2)^2}{0.4} + \frac{(0.2 - 0.5)^2}{1.0} + \frac{0.5 - 0.3)^2}{1.2} + \frac{(0.2 - 0.0)^2}{0.4} \right.$$
$$+ \frac{(0.1 - 0.4)^2}{0.8} + \frac{(0.4 - 0.6)^2}{1.6} + \frac{(0.5 - 0.0)^2}{0.6}$$
$$\left. + \frac{(0.1 - 0.8)^2}{1.3} + \frac{(0.9 - 0.2)^2}{1.6} \right] = 0.4892$$

and so $d(A_2, A_3) = 0.699$. Similarly, we can find all the relevant distances; the distance matrix is

$$\mathbf{D} = \begin{bmatrix} 0 & 0.401 & 0.374 \\ 0.401 & 0 & 0.699 \\ 0.374 & 0.699 & 0 \end{bmatrix}. \qquad (4.1.6)$$

□

### 4.1.3   Extended Multi-valued List Gowda–Diday Dissimilarity

Gowda and Diday (1991, 1992) obtained dissimilarity measures for non-modal list multi-valued observations, as described in section 3.2.3. Kim and Billard (2012) extended those measures to the modal multi-valued list case as follows.

**Definition 4.4**   Let $A_{u_1}$ and $A_{u_2}$ be any two modal list multi-valued objects in $\Omega = \Omega_1 \times \cdots \times \Omega_p$ with realizations taking the form Eq. (4.1.2). Then, the **extended multi-valued list Gowda–Diday dissimilarity** between $A_{u_1}$ and $A_{u_2}$, $d(A_{u_1}, A_{u_2})$, $u_1, u_2 = 1, \ldots, m$, is given by

$$d(A_{u_1}, A_{u_2}) = \sum_{j=1}^{p} [d_{1j}(A_{u_1}, A_{u_2}) + d_{2j}(A_{u_1}, A_{u_2})] \tag{4.1.7}$$

where, for each variable $Y_j$,

$$d_{1j}(A_{u_1}, A_{u_2}) = \sum_{k_j=1}^{s_j} |p_{u_1 j k_j} - p_{u_2 j k_j}| / \sum_{k_j=1}^{s_j} p_{(u_1 \cup u_2) j k_j}, \tag{4.1.8}$$

$$d_{2j}(A_{u_1}, A_{u_2}) = \sum_{k_j=1}^{s_j} (p_{u_1 j k_j} + p_{u_2 j k_j} - 2 p_{(u_1 \cap u_2) j k_j}) / \sum_{k_j=1}^{s_j} (p_{u_1 j k_j} + p_{u_2 j k_j}) \tag{4.1.9}$$

with the union and intersection operators as defined in Eqs. (4.1.3) and (4.1.4), respectively.                                                                       □

As for the Gowda–Diday measure for non-modal list multi-valued observations (see Definition 3.18), there are two components: one, $d_{1j}(A_{u_1}, A_{u_2})$, represents the relative size of $A_{u_1}$ and $A_{u_2}$, while the other, $d_{2j}(A_{u_1}, A_{u_2})$, represents their relative content. The divisor term in Eqs. (4.1.8) and (4.1.9) takes a value between one and two, and is greater than or equal to the numerator in each of these equations. Furthermore, it follows that if two objects are perfectly matched (i.e., they have the same probabilities $p_{u_1 j k_j} = p_{u_2 j k_j}$ for all $k_j = 1, \ldots, s_j$), then $d_{1j}(A_{u_1}, A_{u_2}) = d_{2j}(A_{u_1}, A_{u_2}) = 0$. Also, if two objects are completely different and so do not overlap (i.e., $p_{(u_1 \cap u_2) j k_j} = 0$ for all $k_j = 1, \ldots, s_j$), then $d_{1j}(A_{u_1}, A_{u_2}) = d_{2j}(A_{u_1}, A_{u_2}) = 1$. Hence, both $0 \leq d_{1j}(A_{u_1}, A_{u_2}) \leq 1$ and $0 \leq d_{2j}(A_{u_1}, A_{u_2}) \leq 1$ hold.

**Example 4.3**   The data of Table 4.3 (extracted from US Census, 2002, p. 174f) show the relative frequencies for categories related to two random variables, $Y_1$ = household makeup with $\mathcal{Y}_1$ = {married couple with children under 18, married couple without children under 18, female head with children under 18, female head without children under 18, householder living alone aged under 65,

**Table 4.3** Household characteristics data (Example 4.3)

| | $Y_1$ = Household makeup | | | | | | $Y_2$ = Occupancy and tenure | | |
| | Married couple | | Female head | | Living alone | | | | |
| County | Children aged | | Children aged | | Aged | | Owner | Renter | |
| $u$ | < 18 | ≥ 18 | < 18 | ≥ 18 | < 65 | ≥ 65 | occupied | occupied | Vacant |
|---|---|---|---|---|---|---|---|---|---|
| 1 | 0.274 | 0.273 | 0.082 | 0.069 | 0.217 | 0.085 | 0.530 | 0.439 | 0.031 |
| 2 | 0.319 | 0.275 | 0.109 | 0.064 | 0.144 | 0.089 | 0.527 | 0.407 | 0.066 |
| 3 | 0.211 | 0.304 | 0.092 | 0.048 | 0.236 | 0.109 | 0.528 | 0.388 | 0.084 |
| 4 | 0.368 | 0.281 | 0.110 | 0.051 | 0.114 | 0.076 | 0.527 | 0.414 | 0.059 |
| 5 | 0.263 | 0.335 | 0.064 | 0.048 | 0.160 | 0.130 | 0.609 | 0.326 | 0.065 |
| 6 | 0.332 | 0.306 | 0.065 | 0.056 | 0.159 | 0.082 | 0.587 | 0.378 | 0.035 |

householder living alone aged 65 or over}, and $Y_2$ = household occupancy and tenure with $\mathcal{Y}_2$ = {owner occupied, renter occupied, vacant}. Data are given for six counties; let us focus on just the first four counties.

Consider the object pair $A_2$ and $A_3$. First, we see that their union and intersection probabilities are, from Eqs. (4.1.3) and (4.1.4), respectively,

$$A_2 \cup A_3 = \{0.319, 0.304, 0.109, 0.064, 0.236, 0.109; 0.528, 0.407, 0.084\}, \tag{4.1.10}$$

$$A_2 \cap A_3 = \{0.211, 0.275, 0.092, 0.048, 0.144, 0.089; 0.527, 0.388, 0.066\}. \tag{4.1.11}$$

Then, from Eq. (4.1.8), for $Y_j, j = 1, 2$, we have the relative size components

$$
\begin{aligned}
d_{11}(A_2, A_3) = (&|0.319 - 0.211| + |0.275 - 0.304| + |0.109 - 0.092| \\
&+ |0.064 - 0.048| + |0.144 - 0.236| + |0.089 - 0.109|)/ \\
&(0.319 + 0.304 + 0.109 + 0.064 + 0.236 + 0.109) \\
= &\, 0.247, \\
d_{12}(A_2, A_3) = (&|0.527 - 0.528| + |0.407 - 0.388| + |0.066 - 0.084|)/ \\
&(0.528 + 0.407 + 0.084) \\
= &\, 0.037;
\end{aligned}
$$

and from Eq. (4.1.9), for $Y_j, j = 1, 2$, we have the relative contents components

$$d_{21}(A_2, A_3) = 0.131, \quad d_{22}(A_2, A_3) = 0.018.$$

**Table 4.4** Extended Gowda–Diday dissimilarities (Example 4.3)

| Object | $Y_1$ | | | $Y_2$ | | | |
|---|---|---|---|---|---|---|---|
| pair | $d_{11}$ | $d_{21}$ | $d_1$ | $d_{12}$ | $d_{22}$ | $d_2$ | $\sum d_j$ |
| 1-2 | 0.145 | 0.074 | 0.219 | 0.068 | 0.033 | 0.101 | 0.319 |
| 1-3 | 0.155 | 0.079 | 0.234 | 0.101 | 0.050 | 0.150 | 0.384 |
| 1-4 | 0.230 | 0.120 | 0.350 | 0.055 | 0.026 | 0.081 | 0.431 |
| 2-3 | 0.247 | 0.131 | 0.378 | 0.037 | 0.018 | 0.055 | 0.433 |
| 2-4 | 0.106 | 0.054 | 0.160 | 0.014 | 0.007 | 0.021 | 0.181 |
| 3-4 | 0.302 | 0.162 | 0.464 | 0.051 | 0.024 | 0.074 | 0.538 |

Hence, from Eq. (4.1.7), the extended Gowda–Diday dissimilarity between $A_2$ and $A_3$ is

$$d(A_2, A_3) = \sum_{j=1}^{2} [d_{1j}(A_2, A_3) + d_{1j}(A_2, A_3)]$$

$$= 0.247 + 0.131 + 0.037 + 0.018 = 0.433. \tag{4.1.12}$$

The complete set of extended Gowda–Diday dissimilarities for all object pairs is shown in Table 4.4. □

As for the distances considered in Chapter 3, so too can we obtain Minkowski distances based on the these modal multi-valued list distances/dissimilarities. We illustrate this for some Euclidean distances in Example 4.4.

**Example 4.4** Let us obtain the unweighted non-normalized Euclidean distances based on the extended Gowda–Diday dissimilarities calculated in Example 4.3. Then, from Eq. (3.1.8) with $w_j = 1, j = 1, 2$, and Table 4.4, we have for the pair $A_2$ and $A_3$,

$$d^{(2)}(A_2, A_3) = (0.378^2 + 0.055^2)^{1/2} = 0.382. \tag{4.1.13}$$

The weighted (by $w_j = 1/|\mathcal{Y}_j|$ where here $\mathcal{Y}_1 = 6$ and $\mathcal{Y}_2 = 3$) and normalized (by $1/p = 1/2$) Euclidean distance between $A_2$ and $A_3$ becomes, from Eq. (3.1.9),

$$d^{(2)}(A_2, A_3) = ((0.378^2/6 + 0.0.55^2/3)/2)^{1/2} = 0.111. \tag{4.1.14}$$

The complete set of weighted and unweighted non-normalized Euclidean distances and the weighted and unweighted normalized (by $1/p$) Euclidean distances are displayed in Table 4.5. □

**Table 4.5** Euclidean distances based on extended Gowda–Diday dissimilarities (Example 4.4)

| Object | Non-normalized | | Normalized | |
|---|---|---|---|---|
| pair | Unweighted | Weighted | Unweighted | Weighted |
| 1-2 | 0.241 | 0.107 | 0.170 | 0.075 |
| 1-3 | 0.278 | 0.129 | 0.196 | 0.091 |
| 1-4 | 0.360 | 0.150 | 0.254 | 0.106 |
| 2-3 | 0.382 | 0.157 | 0.270 | 0.111 |
| 2-4 | 0.162 | 0.067 | 0.114 | 0.047 |
| 3-4 | 0.470 | 0.194 | 0.332 | 0.137 |

### 4.1.4 Extended Multi-valued List Ichino–Yaguchi Dissimilarity

As for the Gowda and Diday dissimilarity measures for list multi-valued observations, Kim and Billard (2012) extended the Ichino and Yaguchi (1994) distances described in section 3.2.4 for non-modal list multi-valued observations to the modal list multi-valued case as follows.

**Definition 4.5**   Let $A_{u_1}$ and $A_{u_2}$ be any two modal multi-valued list objects in $\Omega = \Omega_1 \times \cdots \times \Omega_p$ with realizations taking the form Eq. (4.1.2). Then, the **extended multi-valued list Ichino–Yaguchi dissimilarity measure** between $A_{u_1}$ and $A_{u_2}$, $d_j(A_{u_1}, A_{u_2})$, $u_1, u_2 = 1, \ldots, m$, is given by, for each variable $Y_j$, $j = 1, \ldots, p$,

$$d_j(A_{u_1}, A_{u_2}) = \sum_{k_j=1}^{s_j} [p_{(u_1 \cup u_2)jk_j} - p_{(u_1 \cap u_2)jk_j} + \gamma(2p_{(u_1 \cap u_2)jk_j} - p_{u_1jk_j} - p_{u_2jk_j})],$$

$$(4.1.15)$$

where the union $p_{(u_1 \cup u_2)jk_j}$ and intersection $p_{(u_1 \cap u_2)jk_j}$ probabilities are as given in Eqs. (4.1.3) and (4.1.4), respectively, and where $0 \le \gamma \le 0.5$ is a pre-specified constant. □

The parameter $\gamma$ controls the effect of the minimum and maximum probabilities between observations. For example, for a given $j$, suppose there are two objects $A_1$ and $A_2$. Then, from Eq. (4.1.15), dropping the $j$ subscript,

$$\text{when } \gamma = 0, \quad d(A_1, A_2) = \sum_{k=1}^{s} [p_{(1 \cup 2)k} - p_{(1 \cap 2)k}],$$

$$\text{when } \gamma = 0.5, \quad d(A_1, A_2) = \sum_{k=1}^{s} [p_{(1 \cup 2)k} - (p_{1k} - p_{2k})/2].$$

Therefore, when $\gamma = 0$ and the observations $A_1$ and $A_2$ do not overlap (so that their intersection probabilities are $p_{(A_1 \cap A_2)k} = 0$, $k = 1, \ldots, s$), the extended Ichino–Yaguchi dissimilarity can be the same for many pairs of such observations. When $\gamma = 0.5$, the overlapping portions of any pair of observations is excluded. Malerba et al. (2001) recommended using intermediate values for $\gamma$.

**Example 4.5**  Let us take the household characteristics data set of Table 4.3 (see Example 4.3 for variable descriptions). Consider the pair of objects $A_2$ and $A_3$. Then, using the probabilities for the union and intersection of $A_2$ and $A_3$ found in Eqs. (4.1.10) and (4.1.11) and Table 4.3, and substituting into Eq. (4.1.15), for the variable $Y_2$, we have, for $\gamma = 0.25$,

$$
\begin{aligned}
d_2(A_2, A_3) &= [0.528 - 0.527 + 0.25(2 \times 0.527 - 0.527 - 0.528)] \\
&\quad + [0.407 - 0.388 + 0.25(2 \times 0.388 - 0.407 - 0.388)] \\
&\quad + [0.084 - 0.066 + 0.25(2 \times 0.066 - 0.066 - 0.084)] \\
&= 0.0285;
\end{aligned}
$$

likewise, for $Y_1$, we have $d_1(A_2, A_3) = 0.212$. The distance over all variables is simply the sum over $j = 1, \ldots, p$. Thus, here, the overall distance between $A_2$ and $A_3$ is $d(A_2, A_3) = 0.0285 + 0.212 = 0.2405$. The complete set of distances for all object pairs is given in Table 4.6, for the two cases $\gamma = 0.25$ and $\gamma = 0.5$.

Then, the extended Ichino–Yaguchi distance matrices when $\gamma = 0.25$ for $Y_1$ and $Y_2$ are, respectively,

$$
\mathbf{D}_1 = \begin{bmatrix}
0 & 0.117 & 0.126 & 0.195 \\
0.117 & 0 & 0.212 & 0.084 \\
0.126 & 0.212 & 0 & 0.267 \\
0.195 & 0.084 & 0.267 & 0
\end{bmatrix}, \quad
\mathbf{D}_2 = \begin{bmatrix}
0 & 0.053 & 0.080 & 0.042 \\
0.053 & 0 & 0.029 & 0.011 \\
0.080 & 0.029 & 0 & 0.039 \\
0.042 & 0.011 & 0.039 & 0
\end{bmatrix}.
$$

$$(4.1.16)$$

□

**Example 4.6**  Let us now consider some Euclidean distances based on the extended Ichino–Yaguchi distances of Table 4.6 (see Example 4.5). In particular, take the object pair $A_1$ and $A_3$, and let $\gamma = 0.25$.

(i) The non-normalized Euclidean distance is the Minkowski distance of order $q = 2$ given in Eq. (3.1.8) and Definition 3.9. Consider the unweighted distance with $w_j \equiv 1$. Then, taking the basic extended Ichino–Yaguchi distances given in Table 4.6, we have, from Eq. (3.1.8),

$$
d^{(2)}(A_1, A_3) = [0.126^2 + 0.080^2]^{1/2} = 0.149.
$$

Suppose now we use weights $w_j = 1/|\mathcal{Y}_j|$ where $|\mathcal{Y}_j| = s_j$, $j = 1, 2$, is the size of each category space. Thus, in this case, $|\mathcal{Y}_1| = 6$ and $|\mathcal{Y}_2| = 3$. Substituting into

**Table 4.6** Extended Ichino–Yaguchi distances (Example 4.5)

| Object | $\gamma = 0.25$ | | | $\gamma = 0.5$ | | |
|---|---|---|---|---|---|---|
| pair | $d_1$ | $d_2$ | $d$ | $d_1$ | $d_2$ | $d$ |
| 1-2 | 0.117 | 0.053 | 0.170 | 0.078 | 0.035 | 0.113 |
| 1-3 | 0.126 | 0.080 | 0.206 | 0.084 | 0.053 | 0.137 |
| 1-4 | 0.195 | 0.042 | 0.237 | 0.130 | 0.028 | 0.158 |
| 1-5 | 0.161 | 0.170 | 0.330 | 0.107 | 0.113 | 0.220 |
| 1-6 | 0.137 | 0.092 | 0.228 | 0.091 | 0.061 | 0.152 |
| 2-3 | 0.212 | 0.029 | 0.241 | 0.141 | 0.019 | 0.160 |
| 2-4 | 0.084 | 0.011 | 0.094 | 0.056 | 0.007 | 0.063 |
| 2-5 | 0.176 | 0.123 | 0.299 | 0.117 | 0.082 | 0.199 |
| 2-6 | 0.089 | 0.090 | 0.179 | 0.059 | 0.060 | 0.119 |
| 3-4 | 0.267 | 0.039 | 0.306 | 0.178 | 0.026 | 0.204 |
| 3-5 | 0.156 | 0.122 | 0.278 | 0.104 | 0.081 | 0.185 |
| 3-6 | 0.197 | 0.089 | 0.285 | 0.131 | 0.059 | 0.190 |
| 4-5 | 0.231 | 0.132 | 0.363 | 0.154 | 0.088 | 0.242 |
| 4-6 | 0.122 | 0.090 | 0.212 | 0.081 | 0.060 | 0.141 |
| 5-6 | 0.117 | 0.078 | 0.195 | 0.078 | 0.052 | 0.130 |

Eq. (3.1.8) gives the weighted non-normalized Euclidean distance as

$$d^{(2)}(A_1, A_3) = [0.126^2/6 + 0.080^2/3]^{1/2} = 0.069.$$

(ii) Normalized Euclidean distances can also be found from Eq. (3.1.9) where in this case $p = 2$. Thus, for example, the normalized weighted (by $|\mathcal{Y}_j|$ for each variable $Y_j, j = 1, 2$) Euclidean distance between $A_1$ and $A_3$ is

$$d^{(2)}(A_1, A_3) = [(0.126^2/6 + 0.080^2/3)/2]^{1/2} = 0.049.$$

The complete set of unweighted and weighted non-normalized and normalized Euclidean distances based on the extended Ichino–Yaguchi distances is displayed in Table 4.7 for both $\gamma = 0.25$ and $\gamma = 0.5$. From these results, we can obtain the corresponding distance matrices. For example, the weighted normalized Euclidean matrix with $\gamma = 0.25$ becomes

$$\mathbf{D}^{(2)} = \begin{bmatrix} 0 & 0.040 & 0.049 & 0.059 \\ 0.040 & 0 & 0.062 & 0.025 \\ 0.049 & 0.062 & 0 & 0.079 \\ 0.059 & 0.025 & 0.079 & 0 \end{bmatrix}. \tag{4.1.17}$$

□

**Table 4.7** Euclidean distances based on extended Ichino–Yaguchi distances (Example 4.6)

| Object pair | Non-normalized | | | | Normalized | | | |
|---|---|---|---|---|---|---|---|---|
| | Unweighted | | Weighted | | Unweighted | | Weighted | |
| | $\gamma = 0.25$ | $\gamma = 0.5$ | $\gamma = 0.25$ | $\gamma = 0.5$ | $\gamma = 0.25$ | $\gamma = 0.5$ | $\gamma = 0.25$ | $\gamma = 0.5$ |
| 1-2 | 0.128 | 0.085 | 0.057 | 0.038 | 0.091 | 0.060 | 0.040 | 0.027 |
| 1-3 | 0.149 | 0.099 | 0.069 | 0.046 | 0.105 | 0.070 | 0.049 | 0.032 |
| 1-4 | 0.199 | 0.133 | 0.083 | 0.055 | 0.141 | 0.094 | 0.059 | 0.039 |
| 1-5 | 0.233 | 0.156 | 0.118 | 0.079 | 0.165 | 0.110 | 0.083 | 0.056 |
| 1-6 | 0.164 | 0.110 | 0.077 | 0.051 | 0.116 | 0.077 | 0.054 | 0.036 |
| 2-3 | 0.213 | 0.142 | 0.088 | 0.059 | 0.151 | 0.101 | 0.062 | 0.041 |
| 2-4 | 0.085 | 0.056 | 0.035 | 0.023 | 0.060 | 0.040 | 0.025 | 0.016 |
| 2-5 | 0.214 | 0.143 | 0.101 | 0.067 | 0.152 | 0.101 | 0.071 | 0.048 |
| 2-6 | 0.126 | 0.084 | 0.063 | 0.042 | 0.089 | 0.060 | 0.045 | 0.030 |
| 3-4 | 0.270 | 0.180 | 0.111 | 0.074 | 0.191 | 0.127 | 0.079 | 0.052 |
| 3-5 | 0.198 | 0.132 | 0.095 | 0.063 | 0.140 | 0.093 | 0.067 | 0.045 |
| 3-6 | 0.216 | 0.144 | 0.095 | 0.063 | 0.152 | 0.102 | 0.067 | 0.045 |
| 4-5 | 0.266 | 0.177 | 0.121 | 0.081 | 0.188 | 0.125 | 0.086 | 0.057 |
| 4-6 | 0.151 | 0.101 | 0.072 | 0.048 | 0.107 | 0.071 | 0.051 | 0.034 |
| 5-6 | 0.141 | 0.094 | 0.066 | 0.044 | 0.099 | 0.066 | 0.046 | 0.031 |

## 4.2 Dissimilarity/Distance Measures: Histogram Data

Suppose our sample space $\Omega$ consists of objects $A_u$, $u = 1, \ldots, m$, and suppose the realizations of each object contain histogram values for the $p$-dimensional random variable $\mathbf{Y} = (Y_1, \ldots, Y_p)$ in the form

$$\xi(A_u) \equiv \mathbf{Y}_u = (\{[a_{ujk_j}, b_{ujk_j}), p_{ujk_j}; k_j = 1, \ldots, s_{uj}\}, j = 1, \ldots, p), \quad u = 1, \ldots, m,$$
$$(4.2.1)$$

where $p_{ujk_j}$ are the probabilities or relative frequencies associated with the sub-interval $[a_{ujk_j}, b_{ujk_j})$, and where $s_{uj}$ is the number of sub-intervals for the histogram observation $u$ for variable $Y_j$, $u = 1, \ldots, m$, $j = 1, \ldots, p$. The sub-intervals $[a_{(\cdot)}, b_{(\cdot)})$ in Eq. (4.2.1) can be open or closed at either end, though typically we take them as written in Eq. (4.2.1) except that the last sub-interval $[a_{ujs_j}, b_{ujs_j}]$ would be closed at both ends. The sub-intervals take values on $\mathcal{Y}_j \equiv \Omega_j \equiv \mathbb{R}_j$, $j = 1, \ldots, p$, where $\mathbb{R}_j$ is the real line. It is not necessary that sub-intervals for a given observation all have the same length. When the number of sub-intervals $s_{uj} = 1$ and hence $p_{ujk_j} = 1$ for all $u,j$, the observation is interval-valued as a special case.

The terms $p_{ujk_j}$ in Eq. (4.2.1) are more correctly non-negative measures associated with the sub-intervals $[a_{ujk_j}, b_{ujk_j})$, and are not necessarily probability measures. They can be credibility measures, possibility measures, or necessity measures (see Billard and Diday, 2006a, for detailed descriptions and examples of these alternative measures; see also Chapter 2). However, for concreteness, throughout this text they are assumed to be probabilities or relative frequencies.

In general, the number $s_{uj}$, the beginning value $a_{ujk_j}$, and the length ($b_{ujk_j} - a_{ujk_j}$) of the histogram sub-intervals will vary across $u$ and $j$. Conceptually, this is not an issue; however, computationally, this can be a problem in that the details of the relevant computations can become complicated. Therefore, in order to ease the computational burden, the observations are transformed so that all observations have common sub-intervals for a given variable $Y_j$. This transformation is described in section 4.2.1. Thence, it is the transformed histogram observations that are used in subsequent sections where different dissimilarity/distance measures are presented for histogram data.

Furthermore, since some histogram distance measures involve descriptive statistics (specifically, mean and variance) of the union and/or intersection of two histogram observations, these entities are also described in sections 4.2.2–4.2.3 below (see Kim (2009) and Kim and Billard (2013)).

## 4.2.1 Transformation of Histograms

We have histogram observations in the form of Eq. (4.2.1). For a given variable $Y_j$, $j = 1, \ldots, p$, we want to transform the differing sub-intervals across the observations $u = 1, \ldots, m$ into a set of common sub-intervals $[c_{jk_j}, d_{jk_j})$, $k_j = 1, \ldots, t_j$. Then, we can define

$$c_{j1} = \min_{u=1,\ldots,m} \{a_{uj1}\}, \quad d_{jt_j} = \max_{u=1,\ldots,m} \{b_{ujs_{uj}}\}, \tag{4.2.2}$$

$$d_{jk_j} = c_{j1} + k_j \Psi_j / t_j, \quad k_j = 1, \ldots, t_j - 1, \tag{4.2.3}$$

$$c_{jk_j} = d_{j,k_j-1}, \quad k_j = 2, \ldots, t_j, \tag{4.2.4}$$

where

$$\Psi_j = d_{jt_j} - c_{j1}, \tag{4.2.5}$$

$$t_j = \left\lceil \frac{\Psi_j}{\min_{u=1,\ldots,m; k_j=1,\ldots,s_{uj}} \{b_{ujk_j} - a_{ujk_j}\}} \right\rceil \tag{4.2.6}$$

where $\lceil a \rceil$ represents rounding up the value $a$ to its nearest integer.

The entity $\Psi_j$ in Eq. (4.2.5) corresponds to the $|\mathcal{Y}_j|$ of Eq. (3.3.21) for intervals, and equates to the span of values in $\mathbb{R}_j$ covered by the histogram itself. Notice further that the length $\Psi_j/t_j$ of the transformed sub-intervals equals the length of the smallest sub-interval in the original formulation (of Eq. (4.2.1)). If for a given data set of $m$ observations, most observations have similar sub-intervals but with one or two (say) observations having much smaller sub-interval lengths relatively, then the number of transformed sub-intervals $t_j$ can be quite large, adding to the computational burden perhaps unnecessarily. In such cases, instead of the minimum length $\Psi_j/t_j$ from Eqs. (4.2.5) and (4.2.6), a transformed sub-interval length $\Psi_j'$ equal to the average of all sub-intervals or the mode of sub-intervals lengths could be used instead.

The relative frequency for the respective transformed sub-interval corresponds to the proportion of overlap between the original sub-interval and the transformed sub-interval. In such calculations, in the absence of any additional information, it is assumed that observations within sub-intervals are uniformly distributed across the respective sub-intervals (as in Example 4.7). However, if, for example, the original classical observations which were aggregated to give the histogram values are available, it may be possible to determine the relative frequencies for the transformed sub-intervals directly (as in Example 4.8).

**Example 4.7**  Table 4.8 shows histogram observations for the variable $Y$ = cholesterol level, for $m = 4$ regional areas of a country. It is clear that sub-intervals vary across the four observations. To transform these into a set of common sub-intervals, we first apply Eq. (4.2.2) and Eqs. (4.2.5) and (4.2.6),

$$c_1 = \min_u\{80, 70, 55, 75\} = 55, \quad d_t = \max_u\{280, 270, 280, 275\} = 280,$$
$$\Psi = d_t - c_1 = 280 - 55 = 225,$$

**Table 4.8**  Cholesterol histogram data (Example 4.7)

| Region $u$ | $Y$ = Cholesterol |
|---|---|
| 1 | {[80, 120), 0.030; [120, 160), 0.265; [160, 200), 0.583; [200, 240), 0.107; [240, 280], 0.015} |
| 2 | {[70, 120), 0.031; [120, 170), 0.376; [170, 220), 0.519; [220, 270], 0.074} |
| 3 | {[55, 100), 0.008; [100, 145), 0.138; [145, 190), 0.516; [190, 235), 0.309; [235, 280], 0.029} |
| 4 | {[75, 125), 0.049; [125, 175), 0.432; [175, 225), 0.473; [225, 275], 0.046} |

$$t = \left\lceil \frac{225}{\min_{u,k}\{40, \ldots, 40; 50, \ldots, 50; 45, \ldots, 45; 50, \ldots, 50\}} \right\rceil$$

$$= \left\lceil \frac{225}{40} \right\rceil = 6.$$

Hence, from Eqs. (4.2.3) and (4.2.4), the transformed sub-intervals are

$$[c_1, d_1) = [55, 95), \qquad [c_2, d_2) = [95, 135), \qquad [c_3, d_3) = [135, 175),$$
$$[c_4, d_4) = [175, 215), \qquad [c_5, d_5] = [215, 255), \qquad [c_6, d_6] = [255, 295].$$
$$\tag{4.2.7}$$

To obtain the respective relative frequencies $p_{uk}$, $k = 1, \ldots, t = 6$, for each observation $u = 1, \ldots, 4$ we calculate the proportions of the transformed sub-intervals that overlap the original sub-intervals. Thus, for example, for region 1, we have

$$p_{11} = \frac{95 - 80}{120 - 80}(0.030) = 0.011,$$

$$p_{12} = \frac{120 - 95}{120 - 80}(0.030) + \frac{135 - 120}{160 - 120}(0.265) = 0.118,$$

$$p_{13} = \frac{160 - 135}{160 - 120}(0.265) + \frac{175 - 160}{200 - 160}(0.583) = 0.384,$$

$$p_{14} = \frac{200 - 175}{200 - 160}(0.583) + \frac{215 - 200}{240 - 200}(0.107) = 0.405,$$

$$p_{15} = \frac{240 - 215}{240 - 200}(0.107) + \frac{255 - 240}{280 - 240}(0.015) = 0.073,$$

$$p_{16} = \frac{280 - 255}{280 - 240}(0.015) = 0.009;$$

likewise, for the other regions. The full transformed data set is given in Table 4.9. □

**Example 4.8** Table 4.10 gives histogram values for the Fisher (1936) Iris data set aggregated over values for 50 individual observations for each species for variables $Y_1$ = sepal width, $Y_2$ = sepal length, $Y_3$ = petal width, and $Y_4$ = petal length. Thus, observation $A_u$, $u = 1, 2, 3$, corresponds to the species *Versicolor*, *Viriginica*, and *Setosa*, respectively. For example, in the species *Setosa*, 74% have a petal length between 1.0 and 1.5 units, and the remaining 26% have a length in the range 1.5 to 2.0 units.

Suppose we obtain the transformed histogram for sepal length ($Y_1$). Then, from Eq. (4.2.2) and Eqs. (4.2.5) and (4.2.6), we have

$$c_{11} = \min_{u}\{1.8, 2.1, 3.0\} = 1.8, \qquad d_{1t_1} = \max_{u}\{3.6, 3.9, 4.4\} = 4.4,$$

$$\Psi_1 = d_{1t_1} - c_{11} = 4.4 - 1.8 = 2.6,$$

$$t_1 = \left\lceil \frac{2.6}{\min_{u,k_1}\{0.6, \ldots, 0.6; 0.8, 0.8, 0.8\}} \right\rceil = \left\lceil \frac{2.6}{0.6} \right\rceil = 5.$$

**Table 4.9** Cholesterol transformed histogram data (Example 4.7)

| Region $u$ | $Y$ = Cholesterol |
|---|---|
| 1 | {[55, 95), 0.011; [95, 135), 0.118; [135, 175), 0.384; [175, 215), 0.405; [215, 255), 0.073; [255, 295], 0.009} |
| 2 | {[55, 95), 0.016; [95, 135), 0.128; [135, 175), 0.315; [175, 215), 0.415; [215, 255), 0.104; [255, 295], 0.022} |
| 3 | {[55, 95), 0.007; [95, 135), 0.108; [135, 175), 0.375; [175, 215), 0.344; [215, 255), 0.150; [255, 295], 0.016} |
| 4 | {[55, 95), 0.020; [95, 135), 0.116; [135, 175), 0.346; [175, 215), 0.378; [215, 255), 0.122; [255, 295], 0.018} |

**Table 4.10** Fisher iris species histogram data (Example 4.8)

| $A_u$ | Species | |
|---|---|---|
| | | $Y_1$ = **Sepal width** |
| $A_1$ | *Versicolor* | {[1.8, 2.4), 0.12; [2.4, 3.0), 0.56; [3.0, 3.6], 0.32} |
| $A_2$ | *Virginica* | {[2.1, 2.7), 0.14; [2.7, 3.3), 0.70; [3.3, 3.9], 0.16} |
| $A_3$ | *Setosa* | {[2.0, 2.8), 0.02; [2.8, 3.6), 0.66; [3.6, 4.4], 0.32} |
| | | $Y_2$ = **Sepal length** |
| $A_1$ | *Versicolor* | {[4.8, 5.6), 0.22; [5.6, 6.4), 0.56; [6.4, 7.2], 0.22} |
| $A_2$ | *Virginica* | {[4.8, 6.0), 0.14; [6.0, 7.2), 0.64; [7.2, 8.4], 0.22} |
| $A_3$ | *Setosa* | {[4.2, 5.0), 0.40; [5.0, 5.8], 0.60} |
| | | $Y_3$ = **Petal width** |
| $A_1$ | *Versicolor* | {[1.0, 1.5), 0.90; [1.5, 2.0], 0.10} |
| $A_2$ | *Virginica* | {[1.2, 1.6), 0.08; [1.6, 2.0), 0.46; [2.0, 2.4), 0.40; [2.4, 2.8], 0.06} |
| $A_3$ | *Setosa* | {[0.0, 0.4), 0.96; [0.4, 0.8], 0.04} |
| | | $Y_4$ = **Petal length** |
| $A_1$ | *Versicolor* | {[3.0, 4.0), 0.32; [4.0, 5.0), 0.66; [5.0, 6.0), 0.02} |
| $A_2$ | *Virginica* | {[4.5, 5.5), 0.50; [5.5, 6.5), 0.42; [6.5, 7.5], 0.08} |
| $A_3$ | *Setosa* | {[1.0, 1.5), 0.74; [1.5, 2.0], 0.26} |

Hence, from Eqs. (4.2.3) and (4.2.4), the transformed sub-intervals are

$$[c_{11}, d_{11}) = [1.8, 2.4), \quad [c_{12}, d_{12}) = [2.4, 3.0), \quad [c_{13}, d_{13}) = [3.0, 3.6),$$
$$[c_{14}, d_{14}) = [3.6, 4.2), \quad [c_{15}, d_{15}] = [4.2, 4.8].$$

To obtain the relative frequencies for the transformed sub-intervals, we first notice that in this example, the first three transformed sub-intervals for the first species *Versicolor* are the same as in the original formulation with correspondingly the same relative frequencies; the remaining two transformed sub-intervals clearly have relative frequency zero. For the second species *Virginica*, we could obtain proportional values (along the lines used in Example 4.7). Or, since in this case the individual classical observations are available, we can easily obtain the relative frequencies for the transformed sub-intervals directly. Hence, we have relative frequencies $p_{ujk_j}$, for $u = 2, j = 1, k_j = 1, \ldots, 5$,

$$p_{211} = 0.02, \quad p_{212} = 0.40, \quad p_{213} = 0.52, \quad p_{214} = 0.06, \quad p_{215} = 0.0.$$

Similarly, for the third species *Setosa*, relative frequencies $p_{ujk_j}$, for $u = 3$, $j = 1, k_j = 1, \ldots, 5$, are

$$p_{311} = 0.02, \quad p_{312} = 0.04, \quad p_{313} = 0.62, \quad p_{314} = 0.28, \quad p_{315} = 0.04.$$

Likewise, transformed sub-intervals for all variables can be obtained. These are shown in Table 4.11.                                                             □

### 4.2.2 Union and Intersection Operators for Histograms

We will need the concepts of union and intersection of two histogram observations in order to construct dissimilarity and distance measures. These are analogous to the union and intersection of modal list multi-valued objects of section 4.1.1. While not conceptually necessary, as explained in section 4.2.1, it is computationally easier if the histogram observations have common sub-intervals across all histograms in the data set. Therefore, the histogram formulation of Eq. (4.2.1) is rewritten, for the $p$-dimensional random variable $\mathbf{Y} = (Y_1, \ldots, Y_p)$, in the form

$$\xi(A_u) \equiv \mathbf{Y}_u = (\{[a_{jk_j}, b_{jk_j}), p_{ujk_j}; \ k_j = 1, \ldots, s_j\}, \ j = 1, \ldots, p), \quad u = 1, \ldots, m, \tag{4.2.8}$$

where $p_{ujk_j}$ are the probabilities or relative frequencies associated with the common sub-interval $[a_{jk_j}, b_{jk_j})$ for variable $Y_j$ for object $A_u$, and where $s_j$ is the common number of sub-intervals, $u = 1, \ldots, m$, $j = 1, \ldots, p$, and sample space $\Omega$ with observations $A_1, \ldots, A_m$. As in Eq. (4.2.1), typically, we take the sub-intervals to be as written here except that the last sub-interval $[a_{js_j}, b_{js_j}]$ is closed at both ends.

**Definition 4.6** Let $A_{u_1}$ and $A_{u_2}$ be any two histogram-valued objects in $\Omega = \Omega_1 \times \cdots \times \Omega_p$ with realizations taking the form Eq. (4.2.8). Then, the **union** of $A_{u_1}$ and $A_{u_2}$, $A_{u_1} \cup A_{u_2}$, $u_1, u_2 = 1, \ldots, m$, is the object described by

$$\xi(A_{u_1} \cup A_{u_2}) = (\{[a_{jk_j}, b_{jk_j}), p_{(u_1 \cup u_2)jk_j}; \ k_j = 1, \ldots, s_j\}, \ j = 1, \ldots, p) \tag{4.2.9}$$

**Table 4.11** Transformed iris species histogram data, all variables (Example 4.8)

| $A_u$ | Species | |
|---|---|---|
| | | $Y_1$ = **Sepal width** |
| $A_1$ | *Versicolor* | $\{[1.8, 2.4), 0.12; [2.4, 3.0), 0.56; [3.0, 3.6), 0.32; [3.6, 4.2), 0.0;$ $[4.2, 4.8], 0.0\}$ |
| $A_2$ | *Virginica* | $\{[1.8, 2.4), 0.02; [2.4, 3.0), 0.40; [3.0, 3.6), 0.52; [3.6, 4.2), 0.06;$ $[4.2, 4.8], 0.0\}$ |
| $A_3$ | *Setosa* | $\{[1.8, 2.4), 0.02; [2.4, 3.0), 0.04; [3.0, 3.6), 0.62; [3.6, 4.2), 0.28;$ $[4.2, 4.8], 0.04\}$ |
| | | $Y_2$ = **Sepal length** |
| $A_1$ | *Versicolor* | $\{[4.2, 5.0), 0.02; [5.0, 5.8), 0.40; [5.8, 6.6), 0.42; [6.6, 7.4), 0.16;$ $[7.4, 8.2], 0.0\}$ |
| $A_2$ | *Virginica* | $\{[4.2, 5.0), 0.02; [5.0, 5.8), 0.04; [5.8, 6.6), 0.50; [6.6, 7.4), 0.30;$ $[7.4, 8.2], 0.14;\}$ |
| $A_3$ | *Setosa* | $\{[4.2, 5.0), 0.40; [5.0, 5.8), 0.60; [5.8, 6.6), 0.0; [6.6, 7.4), 0.0;$ $[7.4, 8.2], 0.0\}$ |
| | | $Y_3$ = **Petal width** |
| $A_1$ | *Versicolor* | $\{[0.0, 0.4), 0.0; [0.4, 0.8), 0.0; [0.8, 1.2), 0.20; [1.2, 1.6), 0.70;$ $[1.6, 2.0), 0.10; [2.0, 2.4), 0.0; [2.4, 2.8], 0.0\}$ |
| $A_2$ | *Virginica* | $\{[0.0, 0.4), 0.0; [0.4, 0.8), 0.0; [0.8, 1.2), 0.0; [1.2, 1.6), 0.06;$ $[1.6, 2.0), 0.36; [2.0, 2.4), 0.46; [2.4, 2.8], 0.12\}$ |
| $A_3$ | *Setosa* | $\{[0.0, 0.4), 0.82; [0.4, 0.8), 0.18; [0.8, 1.2), 0.0; [1.2, 1.6), 0.0;$ $[1.6, 2.0), 0.0; [2.0, 2.4), 0.0; [2.4, 2.8], 0.0\}$ |
| | | $Y_4$ = **Petal length** |
| $A_1$ | *Versicolor* | $\{[1.0, 1.5), 0.0; [1.5, 2.0), 0.0; [2.0, 2.5), 0.0; [2.5, 3.0), 0.0; [3.0,$ $3.5), 0.06; [3.5, 4.0), 0.16; [4.0, 4.5), 0.36;$ $[4.5, 5.0), 0.38; [5.0, 5.5), 0.04; [5.5, 6.0), 0.0; [6.0, 6.5), 0.0;$ $[6.5, 7.0), 0.0; [7.0, 7.5], 0.0\}$ |
| $A_2$ | *Virginica* | $\{[1.0, 1.5), 0.0; [1.5, 2.0), 0.0; [2.0, 2.5), 0.0; [2.5, 3.0), 0.0; [3.0,$ $3.5), 0.0; [3.5, 4.0), 0.0; [4.0, 4.5), 0.0;$ $[4.5, 5.0), 0.12; [5.0, 5.5), 0.32; [5.5, 6.0), 0.34; [6.0, 6.5), 0.14;$ $[6.5, 7.0), 0.08;[7.0, 7.5], 0.0\}$ |
| $A_3$ | *Setosa* | $\{[1.0, 1.5), 0.48; [1.5, 2.0), 0.52; [2.0, 2.5), 0.0; [2.5, 3.0), 0.0;$ $[3.0, 3.5), 0.0; [3.5, 4.0), 0.0; [4.0, 4.5), 0.0;$ $[4.5, 5.0), 0.0; [5.0, 5.5), 0.0; [5.5, 6.0), 0.0; [6.0, 6.5), 0.0; [6.5,$ $7.0), 0.0; [7.0, 7.5], 0.0\}$ |

where

$$p_{(u_1 \cup u_2)jk_j} = \max\{p_{u_1jk_j}, p_{u_2jk_j}\}, \ k_j = 1, \ldots, s_j, \ j = 1, \ldots, p. \qquad (4.2.10)$$

□

**Definition 4.7** Let $A_{u_1}$ and $A_{u_2}$ be any two histogram-valued objects in $\Omega = \Omega_1 \times \cdots \times \Omega_p$ with realizations taking the form Eq. (4.2.8). Then, the **intersection** of $A_{u_1}$ and $A_{u_2}$, $A_{u_1} \cap A_{u_2}$, $u_1, u_2 = 1, \ldots, m$, is the object described by

$$\xi(A_{u_1} \cap A_{u_2}) = (\{[a_{jk_j}, \ b_{jk_j}), p_{(u_1 \cap u_2)jk_j}; \ k_j = 1, \ldots, s_j\}, j = 1, \ldots, p) \tag{4.2.11}$$

where

$$p_{(u_1 \cap u_2)jk_j} = \min\{p_{u_1jk_j}, p_{u_2jk_j}\}, \ k_j = 1, \ldots, s_j, j = 1, \ldots, p. \tag{4.2.12}$$

□

The union and intersection probabilities in Eqs. (4.2.10) and (4.2.12) do not sum to one. Instead, it is easily shown that, for $u_1, u_2 = 1, \ldots, m$,

$$\sum_{k_j=1}^{s_j} p_{(u_1 \cup u_2)jk_j} \geq 1, \quad \sum_{k_j=1}^{s_j} p_{(u_1 \cap u_2)jk_j} \leq 1. \tag{4.2.13}$$

**Example 4.9** Consider the transformed histogram data of Table 4.9 for $Y$ = cholesterol levels for $m = 4$ regions.

(i) Suppose we want the union of the first two regions. First, notice, from Eq. (4.2.9), that the union object has the same sub-intervals as the individual objects (since we are using the common sub-intervals formulation of Eq. (4.2.8)). We only need to determine the union probabilities $p_{(1 \cup 2)k}$, $k = 1, \ldots, s$. Thus, applying Eq. (4.2.10), we have

$$p_{(1 \cup 2)1} = \max\{0.011, 0.016\} = 0.016, \quad p_{(1 \cup 2)2} = \max\{0.018, 0.128\} = 0.128,$$

$$p_{(1 \cup 2)3} = \max\{0.384, 0.315\} = 0.384, \quad p_{(1 \cup 2)4} = \max\{0.405, 0.415\} = 0.415,$$

$$p_{(1 \cup 2)5} = \max\{0.073, 0.104\} = 0.104, \quad p_{(1 \cup 2)6} = \max\{0.009, 0.022\} = 0.022;$$

notice $\sum_k p_{(1 \cup 2)k} = 1.069 > 1$. Hence, the union histogram becomes

$$Y_{(1 \cup 2)} = \{[55, 95), 0.016; [95, 135), 0.128; [135, 175), 0.384;$$

$$[175, 215), 0.415; [215, 255), 0.104; [255, 295), 0.022\}.$$

(ii) Suppose we want the intersection of the first two regions. Then, from Eq. (4.2.12), we have the intersection probabilities

$$p_{(1 \cap 2)1} = \min\{0.011, 0.016\} = 0.011, \quad p_{(1 \cap 2)2} = \min\{0.018, 0.128\} = 0.018,$$

$$p_{(1 \cap 2)3} = \min\{0.384, 0.315\} = 0.315, \quad p_{(1 \cap 2)4} = \min\{0.405, 0.415\} = 0.405,$$

$$p_{(1 \cap 2)5} = \min\{0.073, 0.104\} = 0.073, \quad p_{(1 \cap 2)6} = \min\{0.009, 0.022\} = 0.009,$$

and $\sum_k p_{(1 \cap 2)k} = 0.931 < 1$. Hence, the intersection histogram becomes

$$Y_{(1 \cap 2)} = \{[55, 95), 0.011; [95, 135), 0.018; [135, 175), 0.315;$$

$$[175, 215), 0.405; [215, 255), 0.073; [255, 295], 0.009\}.$$

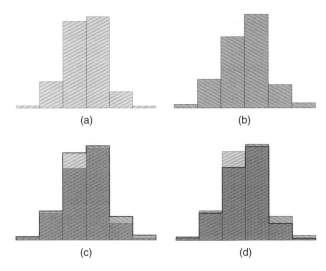

**Figure 4.1** Histograms (a) Region 1, (b) Region 2, (c) Region 1 ∪ Region 2, (d) Region 1 ∩ Region 2 (Example 4.9).

The union and intersection of these two regions are illustrated in Figure 4.1. Figure 4.1(a) and (b) correspond to the histograms for region 1 and region 2, respectively. Superimposing the two histograms upon each other, the histogram of the union of both histograms is the heavy line tracing the outline of this superposition (Figure 4.1(c)), while in Figure 4.1(d) the heavy line tracing the parts contained in both histograms represents the intersection histogram. □

### 4.2.3 Descriptive Statistics for Unions and Intersections

Suppose we have the sample space $\Omega = \Omega_1 \times \cdots \times \Omega_p$ consisting of objects $A_u$, $u = 1, \ldots, m$, and suppose each object has realization consisting of histogram values for the $p$-dimensional random variable $\mathbf{Y} = (Y_1, \ldots, Y_p)$ in the form of Eq. (4.2.8). Then, from Eqs. (2.4.3) and (2.4.11), it follows that the mean $\bar{Y}_{uj}$ and variance $S^2_{uj}$ for the variable $Y_j$, $j = 1, \ldots, p$, for a specific observation $A_u$, $u = 1, \ldots, m$, are, respectively,

$$\bar{Y}_{uj} = \frac{1}{2} \sum_{k_j=1}^{s_j} p_{ujk_j}(a_{jk_j} + b_{jk_j}), \tag{4.2.14}$$

$$S^2_{uj} = \frac{1}{3} \sum_{k_j=1}^{s_j} p_{ujk_j}[(a_{jk_j} - \bar{Y}_{uj})^2 + (a_{jk_j} - \bar{Y}_{uj})(b_{jk_j} - \bar{Y}_{uj}) + (b_{jk_j} - \bar{Y}_{uj})^2].$$

$$\tag{4.2.15}$$

Use of Eq. (2.4.3) directly to obtain expressions for the sample mean of the union, and likewise the intersection, of two histogram observations is not appropriate because of the inequalities in Eq. (4.2.13). Therefore, standardization will be necessary (see the following Definitions 4.8 and 4.9). However, these inequalities are not at issue for the sample variances since in the distances measured it is required that the relationship Eq. (4.2.23) below holds.

**Definition 4.8**    Let $A_{u_1}$ and $A_{u_2}$ be any two histogram-valued objects in $\Omega = \Omega_1 \times \cdots \times \Omega_p$ taking realizations of the form Eq. (4.2.8). Then, the **mean of the union** of $A_{u_1}$ and $A_{u_2}$, $\bar{Y}_{(u_1 \cup u_2)j}$, $u_1, u_2 = 1, \ldots, m$, for the variable $Y_j, j = 1, \ldots, p$, is given by

$$\bar{Y}_{(u_1 \cup u_2)j} = \frac{1}{2} \sum_{k_j=1}^{s_j} p^*_{(u_1 \cup u_2)jk_j} (a_{jk_j} + b_{jk_j}) \tag{4.2.16}$$

where

$$p^*_{(u_1 \cup u_2)jk_j} = p_{(u_1 \cup u_2)jk_j} / \sum_{k_j=1}^{s_j} p_{(u_1 \cup u_2)jk_j} \tag{4.2.17}$$

with the union probabilities $p_{(u_1 \cup u_2)jk_j}$ as defined in Eq. (4.2.10). □

**Definition 4.9**    Let $A_{u_1}$ and $A_{u_2}$ be any two histogram-valued objects in $\Omega = \Omega_1 \times \cdots \times \Omega_p$ taking realizations of the form Eq. (4.2.8). Then, the **mean of the intersection** of $A_{u_1}$ and $A_{u_2}$, $\bar{Y}_{(u_1 \cap u_2)j}$, $u_1, u_2 = 1, \ldots, m$, for the variable $Y_j$, $j = 1, \ldots, p$, is given by

$$\bar{Y}_{(u_1 \cap u_2)j} = \frac{1}{2} \sum_{k_j=1}^{s_j} p^*_{(u_1 \cap u_2)jk_j} (a_{jk_j} + b_{jk_j}) \tag{4.2.18}$$

where

$$p^*_{(u_1 \cap u_2)jk_j} = p_{(u_1 \cap u_2)jk_j} / \sum_{k_j=1}^{s_j} p_{(u_1 \cap u_2)jk_j} \tag{4.2.19}$$

with the intersection probabilities $p_{(u_1 \cap u_2)jk_j}$ as defined in Eq. (4.2.12). □

The standardized union and intersection probabilities of Eqs. (4.2.17) and (4.2.19) now, respectively, satisfy

$$\sum_{k_j=1}^{s_j} p^*_{(u_1 \cup u_2)jk_j} = 1, \quad \sum_{k_j=1}^{s_j} p^*_{(u_1 \cap u_2)jk_j} = 1. \tag{4.2.20}$$

**Definition 4.10**    Let $A_{u_1}$ and $A_{u_2}$ be any two histogram-valued objects in $\Omega = \Omega_1 \times \cdots \times \Omega_p$ taking realizations of the form Eq. (4.2.8). Then, the **variance of the union** of $A_{u_1}$ and $A_{u_2}$, $S^2_{(u_1 \cup u_2)j}$, $u_1, u_2 = 1, \ldots, m$, for the variable $Y_j, j = 1, \ldots, p$, is given by

$$S^2_{(u_1 \cup u_2)j} = \frac{1}{3} \sum_{k_j=1}^{s_j} p_{(u_1 \cup u_2)jk_j} [(a_{jk_j} - \bar{Y}_{Uj})^2 + (a_{jk_j} - \bar{Y}_{Uj})(b_{jk_j} - \bar{Y}_{Uj}) + (b_{jk_j} - \bar{Y}_{Uj})^2]$$

$$(4.2.21)$$

where $\bar{Y}_{Uj} \equiv \bar{Y}_{(u_1 \cup u_2)j}$ is as given in Eq. (4.2.16). $\qquad\square$

**Definition 4.11** Let $A_{u_1}$ and $A_{u_2}$ be any two histogram-valued objects in $\Omega = \Omega_1 \times \cdots \times \Omega_p$ taking realizations of the form Eq. (4.2.8). Then, the **variance of the intersection** of $A_{u_1}$ and $A_{u_2}$, $S^2_{(u_1 \cap u_2)j}$, $u_1, u_2 = 1, \ldots, m$, for the variable $Y_j$, $j = 1, \ldots, p$, is given by

$$S^2_{(u_1 \cap u_2)j} = \frac{1}{3} \sum_{k_j=1}^{s_j} p_{(u_1 \cap u_2)jk_j} [(a_{jk_j} - \bar{Y}_{\cap j})^2 + (a_{jk_j} - \bar{Y}_{\cap j})(b_{jk_j} - \bar{Y}_{\cap j}) + (b_{jk_j} - \bar{Y}_{\cap j})^2]$$

$$(4.2.22)$$

where $\bar{Y}_{\cap j} \equiv \bar{Y}_{(u_1 \cap u_2)j}$ is as given in Eq. (4.2.18). $\qquad\square$

It is easily verified that the union and intersection standard deviations $S_{(\cdot)}$, respectively, satisfy the relations

$$S_{(u_1 \cup u_2)j} \geq \max\{S_{u_1j}, S_{u_2j}\}, \quad S_{(u_1 \cap u_2)j} \leq \min\{S_{u_1j}, S_{u_2j}\}. \qquad (4.2.23)$$

Notice that these descriptive statistics are simply the respective internal means and variances (and hence standard deviations) for a single observation (i.e., the "within" observation values).

**Example 4.10** Consider the transformed histogram data of Table 4.9 for $Y =$ cholesterol levels for $m = 4$ regions. Consider the first two regions.

(i) Let us take the union of these two regions. The union probabilities $p_{(1 \cup 2)k}$, $k = 1, \ldots, s = 6$, and $\sum_k p_{(1 \cup 2)k}$, were calculated in Example 4.9(i). Substituting into Eq. (4.2.16), we obtain the mean of the union, $\bar{Y}_{(1 \cup 2)}$, as

$$\bar{Y}_{(1 \cup 2)} = \frac{1}{2 \times 1.069} [(95 + 55)(0.016) + \cdots + (295 + 255)(0.022)] = 174.794.$$

Substituting into Eq. (4.2.21), we obtain the variance, $S^2_{(1 \cup 2)}$, as

$$S^2_{(1 \cup 2)} = \frac{1}{3} \{[(55 - 174.794)^2 + (55 - 174.794)(95 - 174.794)$$

$$+ (95 - 174.794)^2](0.016) + \ldots$$

$$+ [(255 - 174.794)^2 + (255 - 174.794)(295 - 174.794)$$

$$+ (295 - 174.794)^2](0.022)\}$$

$$= 1677.288,$$

and hence the standard deviation $S_{(1 \cup 2)} = 40.955$.

(ii) Let us take the intersection of these two regions. The intersection probabilities $p_{(1\cap2)k}$, $k = 1, \ldots, s = 6$, were calculated in Example 4.9(ii), as was also $\sum_k p_{(1\cap2)k}$. Substituting into Eq. (4.2.18), we obtain the mean of the intersection, $\bar{Y}_{(1\cap2)}$, as

$$\bar{Y}_{(1\cap2)} = \frac{1}{2 \times 0.931}[(95 - 55)(0.011) + \cdots + (295 - 255)(0.009)]$$

$$= 173.818.$$

Substituting into Eq. (4.2.22), we obtain the variance, $S^2_{(1\cap2)}$, as

$$S^2_{(1\cap2)} = \frac{1}{3}\{[(55 - 173.818)^2 + (55 - 173.818)(95 - 173.818)$$

$$+ (95 - 173.818)^2](0.011) + \ldots$$

$$+ [(255 - 173.818)^2 + (255 - 173.818)(295 - 173.818)$$

$$+ (295 - 173.818)^2](0.009)\}$$

$$= 1298.434,$$

and hence the standard deviation $S_{(1\cap2)} = 36.034$.

The corresponding statistics for the remaining pairs of regions are left for the reader (see Exercise 4.1). A summary of the means, variances, and standard deviations for all regions and region pairs is provided in Table 4.12(a). □

## 4.2.4 Extended Gowda–Diday Dissimilarity

The dissimilarities for interval data developed by Gowda and Diday (1991, 1992) (see section 3.3.3) have been extended to histogram observations by Kim and Billard (2013) as follows.

**Definition 4.12** Let $A_{u_1}$ and $A_{u_2}$ be any two histogram-valued objects in $\Omega = \Omega_1 \times \cdots \times \Omega_p$ with realizations taking the form Eq. (4.2.8). Then, the **extended Gowda–Diday dissimilarity measure** between $A_{u_1}$ and $A_{u_2}$, $d(A_{u_1}, A_{u_2})$, is given by

$$d(A_{u_1}, A_{u_2}) = \sum_{j=1}^{p}[D_{1j}(A_{u_1}, A_{u_2}) + D_{2j}(A_{u_1}, A_{u_2}) + D_{3j}(A_{u_1}, A_{u_2})] \quad (4.2.24)$$

where, for each variable $Y_j$, $j = 1, \ldots, p$,

$$D_{1j}(A_{u_1}, A_{u_2}) = |S_{u_1j} - S_{u_2j}|/(S_{u_1j} + S_{u_2j}), \quad (4.2.25)$$

$$D_{2j}(A_{u_1}, A_{u_2}) = (S_{u_1j} + S_{u_2j} - 2S_{(u_1 \cap u_2)j})/(S_{u_1j} + S_{u_2j}), \quad (4.2.26)$$

$$D_{3j}(A_{u_1}, A_{u_2}) = |\bar{Y}_{u_1j} - \bar{Y}_{u_2j}|/\Psi_j, \quad (4.2.27)$$

**Table 4.12** Cholesterol–glucose histogram descriptive statistics

| Object | (a) Cholesterol | | | (b) Glucose | | |
|---|---|---|---|---|---|---|
| | Mean | Variance | Standard deviation | Mean | Variance | Standard deviation |
| | $\bar{Y}$ | $S^2$ | $S$ | $\bar{Y}$ | $S^2$ | $S$ |
| $A_1$ | 172.520 | 1330.383 | 36.474 | 106.780 | 529.365 | 23.008 |
| $A_2$ | 176.160 | 1639.188 | 40.487 | 106.360 | 516.084 | 22.717 |
| $A_3$ | 177.800 | 1571.893 | 39.647 | 104.400 | 395.573 | 19.889 |
| $A_4$ | 175.800 | 1659.093 | 40.732 | 107.220 | 368.805 | 19.204 |
| $A_1 \cup A_2$ | 174.794 | 1677.288 | 40.955 | 106.248 | 634.757 | 25.194 |
| $A_1 \cap A_2$ | 173.818 | 1298.434 | 36.034 | 106.989 | 410.510 | 20.261 |
| $A_1 \cup A_3$ | 177.620 | 1687.493 | 41.079 | 107.811 | 576.952 | 24.020 |
| $A_1 \cap A_3$ | 172.249 | 1214.401 | 34.848 | 102.491 | 337.052 | 18.359 |
| $A_1 \cup A_4$ | 175.431 | 1701.668 | 41.251 | 106.995 | 556.828 | 23.597 |
| $A_1 \cap A_4$ | 172.706 | 1289.492 | 35.910 | 107.005 | 341.438 | 18.478 |
| $A_2 \cup A_3$ | 177.459 | 1837.577 | 42.867 | 105.081 | 602.851 | 24.553 |
| $A_2 \cap A_3$ | 176.387 | 1374.280 | 37.071 | 105.808 | 310.471 | 17.620 |
| $A_2 \cup A_4$ | 176.159 | 1763.387 | 41.993 | 106.621 | 660.053 | 25.691 |
| $A_2 \cap A_4$ | 175.781 | 1534.888 | 39.178 | 107.030 | 225.124 | 15.004 |
| $A_3 \cup A_4$ | 176.798 | 1776.318 | 42.146 | 106.545 | 509.360 | 22.569 |
| $A_3 \cap A_4$ | 176.803 | 1456.669 | 38.166 | 104.698 | 257.361 | 16.042 |

where for the variable $Y_j$, $j = 1, \ldots, p$, $\bar{Y}_{uj}$ and $S_{uj}$ are the mean and standard deviation, respectively, for a single object $(A_u, u = u_1, u_2)$ obtained from Eqs. (4.2.14) and (4.2.15), $S_{(u_1 \cap u_2)j}$ is the standard deviation of the intersection of $A_{u_1}$ and $A_{u_2}$ obtained from Eq. (4.2.22), and $\Psi_j = (b_{js_j} - a_{j1})$ is the span of values across $A_u$ as given in Eq. (4.2.5). □

As for the Gowda–Diday dissimilarity measure between interval observations, so too are there three components representing measures of relative size $(D_{1j}(A_{u_1}, A_{u_2}))$, relative content $(D_{2j}(A_{u_1}, A_{u_2}))$, and relative locations $(D_{3j}(A_{u_1}, A_{u_2}))$ between $A_{u_1}$ and $A_{u_2}$, respectively. However, in contrast to the corresponding components for interval observations, for histograms these terms are now based on the means and standard deviations which provide measures of the location and dispersion of the two objects. It is easy to show that each component satisfies $0 \leq D_{rj}(A_{u_1}, A_{u_2}) \leq 1$, $r = 1, 2, 3$.

**Example 4.11** To illustrate, consider the transformed histogram data of Table 4.9 for $Y$ = cholesterol levels for $m = 4$ regions. Let us calculate the extended Gowda–Diday dissimilarity between the observations for the second and third regions. Substituting from Table 4.12 and Eq. (4.2.5) into Eqs. (4.2.25)–(4.2.27), respectively, we obtain

$$d_1(A_2, A_3) = |40.487 - 39.647|/(40.487 + 39.647) = 0.010,$$

$$d_2(A_2, A_3) = (40.487 + 39.647 - 2 \times 36.034)/(40.487 + 39.647) = 0.075,$$

$$d_3(A_2, A_3) = |176.160 - 177.800|/(295 - 55) = 0.007;$$

hence, substituting into Eq. (4.2.24), we obtain the extended Gowda–Diday dissimilarity measure as

$$d(A_2, A_3) = \sum_{r=1}^{3} d_r(A_2, A_3) = 0.092.$$

Table 4.13(a) displays the extended Gowda–Diday dissimilarity measures for all regions pairs. □

**Example 4.12** Let us now calculate some Euclidean distances based on the extended Gowda–Diday dissimilarity. Suppose there are two variables $Y_1$ = cholesterol with histogram values and $Y_2$ = glucose with histogram values as in Tables 4.8 and 4.14, respectively. Transformation of the observed sub-intervals into a common set of intervals, calculation of the basic descriptive statistics, and hence calculation of the extended Gowda–Diday dissimilarities for cholesterol were given in Tables 4.9, 4.12(a), and 4.13(a), respectively (see Examples 4.7, 4.9, and 4.10). Corresponding results for glucose are given in Tables 4.15, 4.12(b), and 4.13(b); the details are left to the reader as an exercise.

(i) First, let us obtain city block distances from Eq. (3.1.7) with $q = 1$, and suppose we take weights $c_j = 1/p = 1/2$. Then, it follows readily that,

**Table 4.13** Extended Gowda–Diday dissimilarities for cholesterol–glucose histograms

| Object | (a) Cholesterol | | | | (b) Glucose | | | |
|--------|------|------|------|------|------|------|------|------|
| pair | $D_1$ | $D_2$ | $D_3$ | $\sum D_r$ | $D_1$ | $D_2$ | $D_3$ | $\sum D_r$ |
| 1-2 | 0.052 | 0.064 | 0.015 | 0.131 | 0.006 | 0.114 | 0.004 | 0.124 |
| 1-3 | 0.042 | 0.084 | 0.022 | 0.148 | 0.073 | 0.144 | 0.020 | 0.237 |
| 1-4 | 0.055 | 0.070 | 0.014 | 0.139 | 0.090 | 0.124 | 0.004 | 0.218 |
| 2-3 | 0.010 | 0.075 | 0.007 | 0.092 | 0.066 | 0.173 | 0.016 | 0.255 |
| 2-4 | 0.003 | 0.035 | 0.002 | 0.040 | 0.084 | 0.284 | 0.007 | 0.375 |
| 3-4 | 0.013 | 0.050 | 0.008 | 0.072 | 0.018 | 0.179 | 0.023 | 0.220 |

**Table 4.14** Glucose histogram data (Example 4.12)

| Region $u$ | $Y$ = Glucose |
|---|---|
| 1 | $\{[55, 85), 0.083; [85, 115), 0.651; [115, 155], 0.266\}$ |
| 2 | $\{[50, 90), 0.150; [90, 120), .694; [120, 160], 0.156\}$ |
| 3 | $\{[65, 95), 0.259; [95, 125), 0.654; [125, 145], 0.087\}$ |
| 4 | $\{[70, 90), 0.152; [90, 110), 0.487; [110, 130), 0.209; [130, 150], 0.152\}$ |

**Table 4.15** Glucose transformed histogram data (Example 4.12)

| Region $u$ | $Y$ = Glucose |
|---|---|
| 1 | $\{[50, 70), 0.042; [70, 90), 0.150; [90, 110), 0.434;$ $[110, 130), 0.208; [130, 150), 0.133; [150, 170), 0.033\}$ |
| 2 | $\{[50, 70), 0.075; [70, 90), 0.075; [90, 110), 0.463;$ $[110, 130), 0.270; [130, 150), 0.078; [150, 170), 0.039\}$ |
| 3 | $\{[50, 70), 0.043; [70, 90), 0.173; [90, 110), 0.370;$ $[110, 130), 0.349; [130, 150), 0.065; [150, 170), 0.000\}$ |
| 4 | $\{[50, 70), 0.000; [70, 90), 0.152; [90, 110), 0.487;$ $[110, 130), 0.209; [130, 150), 0.152; [150, 170), 0.000\}$ |

for example, the city block distance between the first two regions is, from Table 4.13,

$$d^{(1)}(A_1, A_2) = \sum_{j=1}^{p} d_j(A_1, A_2) = \frac{1}{2}0.131 + \frac{1}{2}0.124 = 0.127.$$

Likewise, unweighted distances can be found from Eq. (3.1.7) with $w_j \equiv 1$. Weighted and unweighted city distances for all regional pairs are displayed in Table 4.16.

(ii) When $q = 2$ in Eq. (3.1.7), we have Euclidean distances. Consider the unweighted non-normalized Euclidean distance for the second and third regions. Hence, from Table 4.13 and Eq. (3.1.8) with $w_j = 1, j = 1, 2$, we have

$$d^{(2)}(A_2, A_3) = \left[\sum_{j=1}^{p} d_j^2(A_2, A_3)\right]^{1/2} = [0.092^2 + 0.255^2]^{1/2} = 0.590.$$

The unweighted normalized Euclidean distance is found from Eq. (3.1.9) as

$$d^{(2)}(A_2, A_3) = \left[\frac{1}{p}\sum_{j=1}^{p} d_j^2(A_2, A_3)\right]^{1/2} = [(0.092^2 + 0.255^2)/2]^{1/2} = 0.192.$$

**Table 4.16** Minkowski distances based on extended Gowda–Diday dissimilarities for cholesterol–glucose histograms (Example 4.12)

| Object pair | City Unweighted | City Weighted | Euclidean Non-normalized Unweighted | Euclidean Non-normalized Weighted | Euclidean Normalized Unweighted | Euclidean Normalized Weighted |
|---|---|---|---|---|---|---|
| 1-2 | 0.255 | 0.127 | 0.505 | 0.129 | 0.127 | 0.091 |
| 1-3 | 0.385 | 0.192 | 0.620 | 0.174 | 0.197 | 0.123 |
| 1-4 | 0.357 | 0.178 | 0.597 | 0.162 | 0.183 | 0.115 |
| 2-3 | 0.348 | 0.174 | 0.590 | 0.151 | 0.192 | 0.107 |
| 2-4 | 0.415 | 0.207 | 0.644 | 0.191 | 0.267 | 0.135 |
| 3-4 | 0.292 | 0.146 | 0.541 | 0.127 | 0.164 | 0.090 |

Let us suppose there are weights $w_1 = 0.75$ and $w_2 = 0.25$ associated with the variables $Y_1$ and $Y_2$, respectively. Then, to calculate the weighted non-normalized Euclidean distance between the second and third regions, we substitute from Table 4.13 into Eq. (3.1.9) to give

$$d^{(2)}(A_2, A_3) = \left[ \frac{1}{p} \sum_{j=1}^{p} w_j d_j^2(A_2, A_3) \right]^{1/2} = [0.75 \times 0.092^2 + 0.25 \times 0.255^2]^{1/2}$$

$$= 0.151.$$

The complete set of weighted and unweighted non-normalized and normalized Euclidean distances based on the extended Gowda–Diday dissimilarities for all regional pairs is provided in Table 4.16. Thence, distance matrices can be found. For example, the weighted normalized Euclidean extended Gowda–Diday dissimilarity matrix becomes

$$\mathbf{D}^{(2)} = \begin{bmatrix} 0 & 0.091 & 0.123 & 0.115 \\ 0.091 & 0 & 0.107 & 0.135 \\ 0.123 & 0.107 & 0 & 0.090 \\ 0.115 & 0.135 & 0.090 & 0 \end{bmatrix}. \tag{4.2.28}$$

□

### 4.2.5 Extended Ichino–Yaguchi Distance

The distances for interval data developed by Ichino and Yaguchi (1994) (see section 3.3.4) have been extended to histogram observations by Kim and Billard (2013) as follows.

**Definition 4.13** Let $A_{u_1}$ and $A_{u_2}$ be any two histogram-valued objects in $\Omega = \Omega_1 \times \cdots \times \Omega_p$ taking realizations of the form Eq. (4.2.8). Then, the **extended**

**Ichino–Yaguchi distance measure** between $A_{u_1}$ and $A_{u_2}$, $d_j(A_{u_1}, A_{u_2})$, for each variable $Y_j$, $j = 1, \ldots, p$, for $u_1, u_2 = 1, \ldots, m$, is given by

$$d_j(A_{u_1}, A_{u_2}) = S_{(u_1 \cup u_2)j} - S_{(u_1 \cap u_2)j} + \gamma(2S_{(u_1 \cap u_2)j} - S_{u_1 j} - S_{u_2 j}) \qquad (4.2.29)$$

where $S_{uj}$, $u = u_1, u_2$, is the standard deviation of the histogram $A_u$ defined in Eq. (4.2.15), and $S_{(u_1 \cup u_2)j}$ and $S_{(u_1 \cap u_2)j}$ are the standard deviations of the union and intersection of the $A_{u_1}$ and $A_{u_2}$ histograms defined in Eqs. (4.2.21) and (4.2.22), respectively, and where $0 \leq \gamma \leq 0.5$ is a pre-specified constant.  □

These distances in Eq. (4.2.29) vary depending on the units of measurement for each variable. The following definition gives a normalized extended distance.

**Definition 4.14** Let $A_{u_1}$ and $A_{u_2}$ be any two histogram-valued objects in $\Omega = \Omega_1 \times \cdots \times \Omega_p$ with realizations taking the form Eq. (4.2.8), and let $d_j(A_{u_1}, A_{u_2})$ be the extended Ichino–Yaguchi distance measure defined in Eq. (4.2.29) in Definition 4.13. Then, the **normalized extended Ichino–Yaguchi distance measure** between $A_{u_1}$ and $A_{u_2}$, $d_j^*(A_{u_1}, A_{u_2})$, for each variable $Y_j$, $j = 1, \ldots, p$, for $u_1, u_2 = 1, \ldots, m$, is

$$d_j^*(A_{u_1}, A_{u_2}) = d_j(A_{u_1}, A_{u_2})/N_j \qquad (4.2.30)$$

where the normalization factor $N_j$ is given by

$$N_j^2 = (5V_{1j} + 2V_{2j} - 6V_{3j})/24 \qquad (4.2.31)$$

where

$$V_{1j} = a_{j1}^2 + b_{j1}^2 + a_{js_j}^2 + b_{js_j}^2, \qquad (4.2.32)$$

$$V_{2j} = a_{j1}b_{j1} + a_{js_j}b_{js_j}, \qquad (4.2.33)$$

$$V_{3j} = (a_{j1} + b_{j1})(a_{js_j} + b_{js_j}). \qquad (4.2.34)$$

□

Notice that this normalizing factor is a function of the first sub-interval $[a_{1j}, b_{1j})$ and the last sub-interval $[a_{s_j}, b_{s_j}]$. Furthermore, $N_j$ is the maximum standard deviation value across the unions of histogram pairs. It can be shown that, when the union probabilities (of Eq. (4.2.10)) satisfy $p_{(u_1 \cup u_2)j1} = p_{(u_1 \cup u_2)js_j} = 1$ and $p_{(u_1 \cup u_2)jk} = 0$ when $k_j = 2, \ldots, (s_j - 1)$, then

$$\max_{u_1, u_2 = 1, \ldots, m} \{S_{(u_1 \cup u_2)j}\} = N_j, \qquad (4.2.35)$$

i.e., $N_j$ maximizes the span of the standard deviation values across unions of histogram pairs (akin to the normalizing distance span $|\mathcal{Y}_j|$ used for interval values; see, e.g., Eq. (3.3.9)).

**Example 4.13**   Take the transformed $Y_1$ = cholesterol level and $Y_2$ = glucose level data of Tables 4.9 and 4.15, respectively. Consider the first two regions.

(i) The basic extended Ichino–Yaguchi distance is found by substituting the descriptive statistics values from Table 4.12 into Eq. (4.2.29), for $Y_j, j = 1, 2$, respectively, i.e., for $\gamma = 0.25$,

$$d_1(A_1, A_2) = 40.955 - 36.034 + 0.25(2 \times 36.034 - 36.474 - 40.487) = 3.697,$$
$$d_2(A_1, A_2) = 25.194 - 20.261 + 0.25(2 \times 20.261 - 23.008 - 22.717) = 3.633;$$

hence, the extended Ichino–Yaguchi distance for both variables is

$$d(A_1, A_2) = d_1(A_1, A_2) + d_2(A_1, A_2) = 7.330.$$

To calculate the span-normed distance that normalizes across these different scales, we adjust by using the normalizing factor $N_j$ of Eq. (4.2.31). Therefore, for $Y_1$ = cholesterol, we find

$$V_{11} = 55^2 + 95^2 + 255^2 + 295^2 = 164100,$$
$$V_{21} = 55 \times 95 + 255 \times 295 = 80450,$$
$$V_{31} = (55 + 95)(255 + 295) = 82500;$$

thus the span-normalizing factor for cholesterol is

$$N_1 = [(5 \times 16400 + 2 \times 80450 - 6 \times 82500)/24]^{1/2} = 142.361.$$

Similarly, the span-normalizing factor for $Y_2$ = glucose is $N_2 = 64.936$. Therefore, the span-normalizing (span-normed) extended Ichino–Yaguchi distances, for each $j = 1, 2$, between the first two regions become, from Eq. (4.2.30),

$$d_1^*(A_1, A_2) = 3.697/142.361 = 0.026, \qquad d_2^*(A_1, A_2) = 7.330/64.936 = 0.056,$$

and hence, over all variables, $d^*(A_1, A_2) = 0.082$. The corresponding distances $d_j$ for each variable $Y_j, j = 1, 2$, and their sum $d = d_1 + d_2$, between all pairs of regions are shown in Table 4.17, for the non-normalized distances (obtained from Eq. (4.2.29)) and the span-normalized distances (obtained from Eq. (4.2.30)), for the two cases $\gamma = 0.25, 0.5$.

(ii) Again, various Euclidean distances based on these extended Ichino–Yaguchi distance measures can be obtained. For example, substituting from Table 4.17 into Eq. (3.1.8) with $w_j = 1$, we can find the (non-normalized) Euclidean distance based on the span-normalized extended Ichino–Yaguchi distance, $d^{(2)*}(A_1, A_2)$ say, between the first two regions as, for $\gamma = 0.5$,

$$d^{(2)*}(A_1, A_2) = [0.017^2 + 0.036^2]^{1/2} = 0.040.$$

If we want the normalized (by $1/p$) Euclidean distance, then we use Eq. (3.1.9). For example, the (unweighted, i.e., $w_j = 1$) span-normed normalized Euclidean distance for the first two regions becomes, for $\gamma = 0.5$,

$$d^{(2)*}(A_1, A_2) = [(0.017^2 + 0.036^2)/2]^{1/2} = 0.028.$$

4.2 Dissimilarity/Distance Measures: Histogram Data

**Table 4.17** Extended Ichino–Yaguchi distances for cholesterol–glucose histograms (Example 4.13)

| | $\gamma = 0.5$ | | | | | | $\gamma = 0.25$ | | | | | |
| | Unnormalized | | | Span-normalized | | | Unnormalized | | | Span-normalized | | |
| Object pair | $d_1$ | $d_2$ | $d$ | $d_1$ | $d_2$ | $d$ | $d_1$ | $d_2$ | $d$ | $d_1$ | $d_2$ | $d$ |
|---|---|---|---|---|---|---|---|---|---|---|---|---|
| 1-2 | 2.474 | 2.332 | 4.806 | 0.017 | 0.036 | 0.053 | 3.697 | 3.633 | 7.330 | 0.026 | 0.056 | 0.082 |
| 1-3 | 3.018 | 2.571 | 5.590 | 0.021 | 0.040 | 0.061 | 4.625 | 4.116 | 8.741 | 0.032 | 0.063 | 0.095 |
| 1-4 | 2.648 | 2.491 | 5.139 | 0.019 | 0.038 | 0.057 | 3.995 | 3.805 | 7.800 | 0.028 | 0.059 | 0.087 |
| 2-3 | 2.800 | 3.250 | 6.050 | 0.020 | 0.050 | 0.070 | 4.298 | 5.091 | 9.389 | 0.030 | 0.078 | 0.108 |
| 2-4 | 1.383 | 4.731 | 6.114 | 0.010 | 0.073 | 0.083 | 2.099 | 7.709 | 9.808 | 0.015 | 0.119 | 0.134 |
| 3-4 | 1.957 | 3.022 | 4.979 | 0.014 | 0.047 | 0.061 | 2.968 | 4.774 | 7.743 | 0.021 | 0.073 | 0.094 |

**Table 4.18** Euclidean distances based on extended Ichino–Yaguchi distances cholesterol–glucose histograms (Example 4.13)

| | $\gamma = 0.5$ | | | | $\gamma = 0.25$ | | | |
| | Euclidean | | Normalized Euclidean | | Euclidean | | Normalized Euclidean | |
| Object pair | Standard | Span-normed | Standard | Span-normed | Standard | Span-normed | Standard | Span-normed |
|---|---|---|---|---|---|---|---|---|
| 1-2 | 3.400 | 0.040 | 2.404 | 0.028 | 5.183 | 0.062 | 3.665 | 0.044 |
| 1-3 | 3.965 | 0.045 | 2.804 | 0.032 | 6.191 | 0.071 | 4.378 | 0.050 |
| 1-4 | 3.636 | 0.043 | 2.571 | 0.030 | 5.517 | 0.065 | 3.901 | 0.046 |
| 2-3 | 4.290 | 0.054 | 3.033 | 0.038 | 6.663 | 0.084 | 4.711 | 0.059 |
| 2-4 | 4.929 | 0.073 | 3.485 | 0.052 | 7.990 | 0.120 | 5.650 | 0.085 |
| 3-4 | 3.600 | 0.049 | 2.546 | 0.034 | 5.622 | 0.076 | 3.975 | 0.054 |

Table 4.18 displays the (unweighted, i.e., $w_j = 1$, $j = 1, 2$, in all cases) normalized and non-normalized Euclidean distances based on the standard and the span-normed extended Ichino–Yaguchi distances for all region pairs, for $\gamma = 0.25$ and $0.5$. From these, we can build the respective distance matrices. For example, the normalized Euclidean span-normed extended Ichino–Yaguchi distance matrix is, when $\gamma = 0.25$,

$$
\mathbf{D}^{(2)*} = \begin{bmatrix}
0 & 0.044 & 0.050 & 0.046 \\
0.044 & 0 & 0.059 & 0.085 \\
0.050 & 0.059 & 0 & 0.054 \\
0.046 & 0.085 & 0.054 & 0
\end{bmatrix}. \tag{4.2.36}
$$

□

### 4.2.6 Extended de Carvalho Distances

Kim and Billard (2013) adapted the de Carvalho (1994) extensions of the Ichino–Yaguchi distances for interval-valued data (see section 3.3.5) to the extended Ichino–Yaguchi distances for histogram-valued observations. Those extensions, as do the Ichino–Yaguchi distances for intervals, used the meet and join operators. For histogram data, these adaptations use the union and intersections operators (of section 4.2.2).

**Definition 4.15** Let $A_{u_1}$ and $A_{u_2}$ be any two histogram-valued objects in $\Omega = \Omega_1 \times \cdots \times \Omega_p$ taking realizations of the form Eq. (4.2.8). Then, the **extended de Carvalho distances** between $A_{u_1}$ and $A_{u_2}$, for comparison functions $cf_k$, $k = 1, \ldots, 5$, are

$$d_{j,cf_k}(A_{u_1}, A_{u_2}) = 1 - cf_k \qquad (4.2.37)$$

where, for a given variable $Y_j$, $j = 1, \ldots, p$,

$$cf_1 = (\alpha - \delta)/(\alpha + \beta + \chi + \delta), \qquad (4.2.38)$$

$$cf_2 = (2\alpha - \delta)/(2\alpha + \beta + \chi + \delta), \qquad (4.2.39)$$

$$cf_3 = (\alpha - \delta)/(\alpha + 2\beta + 2\chi + \delta), \qquad (4.2.40)$$

$$cf_4 = [(\alpha - \delta)/(\alpha + \beta + \delta) + (\alpha - \delta)/(\alpha + \chi + \delta)]/2, \qquad (4.2.41)$$

$$cf_5 = (\alpha - \delta)/[(\alpha + \beta + \delta)(\alpha + \chi + \delta)]^{1/2}, \qquad (4.2.42)$$

where

$$\alpha = S_{(u_1 \cap u_2)j}, \qquad \beta = S_{u_1} - S_{(u_1 \cap u_2)j}, \qquad \chi = S_{u_2} - S_{(u_1 \cap u_2)j},$$
$$\delta = S_{(u_1 \cup u_2)j} + S_{(u_1 \cap u_2)j} - S_{u_1} - S_{u_2}. \qquad (4.2.43)$$

The $\alpha$, $\beta$, $\chi$, and $\delta$ terms are equivalent to an agreement–agreement index, agreement–disagreement index, disagreement–agreement index, and disagreement–disagreement index, respectively.                                    □

**Example 4.14** Take the transformed $Y_1$ = sepal width, $Y_2$ = sepal length, $Y_3$ = petal width, and $Y_4$ = petal length data of Table 4.11. Consider the species pair *Versicolor* and *Setosa*. We first need to find the standard deviations for each species, their union, and their intersection. These together with the respective means are shown in Table 4.19. Then, substituting into Eq. (4.2.43), we have, for sepal width ($Y_1$),

$$\alpha_1 = 0.289, \qquad \beta_1 = 0.415 - 0.289 = 0.126, \qquad \chi_1 = 0.451 - 0.289 = 0.162,$$
$$\delta_1 = 0.609 + 0.289 - 0.415 - 0.451 = 0.032.$$

**Table 4.19** Mean and standard deviations for transformed iris histograms (Example 4.14)

| Object | Sepal width $\bar{Y}_1$ | $S_1$ | Sepal length $\bar{Y}_2$ | $S_2$ | Petal width $\bar{Y}_3$ | $S_3$ | Petal length $\bar{Y}_4$ | $S_4$ |
|---|---|---|---|---|---|---|---|---|
| 1 *Versicolor* | 2.820 | 0.417 | 5.976 | 0.642 | 1.360 | 0.244 | 4.340 | 0.498 |
| 2 *Virginicia* | 3.072 | 0.415 | 6.600 | 0.721 | 2.056 | 0.328 | 5.725 | 0.649 |
| 3 *Setosa* | 3.468 | 0.451 | 5.080 | 0.455 | 0.272 | 0.192 | 1.510 | 0.289 |
| $A_1 \cup A_2$ | 2.948 | 0.527 | 6.282 | 0.960 | 1.713 | 0.633 | 5.046 | 1.270 |
| $A_1 \cap A_2$ | 2.943 | 0.318 | 6.300 | 0.455 | 1.650 | 0.090 | 4.875 | 0.104 |
| $A_1 \cup A_3$ | 3.137 | 0.734 | 5.572 | 0.986 | 0.816 | 0.830 | 2.925 | 2.082 |
| $A_1 \cap A_3$ | 3.174 | 0.220 | 5.362 | 0.186 | 0.000 | 0.000 | 0.000 | 0.000 |
| $A_2 \cup A_3$ | 3.265 | 0.609 | 5.862 | 1.356 | 1.164 | 1.318 | 3.618 | 3.064 |
| $A_2 \cap A_3$ | 3.281 | 0.289 | 5.133 | 0.108 | 0.000 | 0.000 | 0.000 | 0.000 |

Therefore, from Eq. (4.2.38), the first extended de Carvalho comparison function distance for $Y_1$, $cf_{11}$, is

$$cf_{11} = (0.289 - 0.032)/(0.289 + 0.126 + 0.162 + 0.032) = 0.577.$$

Consider now the distance between *Versicolor* and *Setosa* for petal width ($Y_3$). In this case, the agreement–disagreement indices of Eq. (4.2.43) are, respectively,

$$\alpha_3 = 0, \quad \beta_3 = 0.328, \quad \chi_3 = 0.192, \quad \delta_3 = 0.797.$$

In particular, notice that the agreement index here is $\alpha = 0$; notice also that this is consistent with the data where we observe that relative to this variable, there is no overlap (i.e., intersection $A_2 \cap A_3 = \phi$) with the intersection mean and standard deviation therefore zero (see Table 4.19). Figure 4.2 illustrates the differences between the variables clearly. In particular, in the plot of sepal width ($Y_1$) versus sepal length ($Y_2$) (in Figure 4.2(a)), from either a $Y_1$ or $Y_2$ perspective there is clearly overlap between the species. On the other hand, the plots of petal width ($Y_3$) versus petal length ($Y_4$) (in Figure 4.2(b)) show that for these variables there is no overlap between the species *Setosa* and *Versicolor* (nor with *Virginica*). This is reflected in the agreement–disagreement values calculated here.

From Eq. (4.2.38), the first extended de Carvalho comparison function distance for $Y_3$ is $cf_{13} = 1.605$. Likewise, we can show that $cf_{12} = 1.133$ and $cf_{14} = 1.694$. These distances by variable for each comparison function $cf_{kj}$, $k = 1, \ldots, 5, j = 1 \ldots, 4$, for each pair of species are given in Table 4.20(a).

Euclidean distances can also be calculated. These are shown in Table 4.20(b) for unweighted city distances (i.e., $\sum_j cf_{kj}$ and weighted by $c_j = 1/p$, $p = 4$, city distances with $q = 1$ in Eq. (3.1.7)), for each $k = 1, \ldots, 5$. Standard (unweighted non-normalized) Euclidean distances (from Eq. (3.1.8)) and normalized

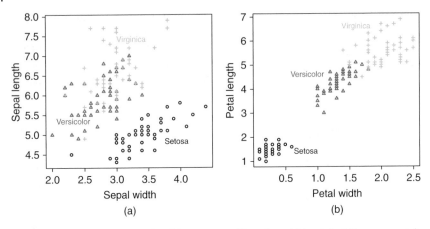

**Figure 4.2** Iris species: (a) sepal width versus sepal length and (b) petal width versus petal length (Example 4.14).

Euclidean distances (from Eq. (3.1.9)) with $w_j = 1$ in each case are given in Table 4.20(c). The details are left as an exercise (see Exercise 4.4). □

**Table 4.20** Extended de Carvalho Minkowski distances for iris histograms (Example 4.14)

| $cf_r$ | Object pair | (a) By variable $d_1$ | $d_2$ | $d_3$ | $d_4$ | (b) City Unweighted | Weighted | (c) Euclidean Non-normalized | Normalized |
|---|---|---|---|---|---|---|---|---|---|
| $cf_1$ | 1-2 | 0.419 | 0.579 | 1.096 | 1.097 | 3.191 | 0.798 | 1.707 | 0.854 |
| | 1-3 | 0.817 | 0.887 | 1.474 | 1.622 | 4.800 | 1.200 | 2.502 | 1.251 |
| | 2-3 | 0.577 | 1.133 | 1.605 | 1.694 | 5.008 | 1.252 | 2.657 | 1.329 |
| $cf_2$ | 1-2 | 0.262 | 0.393 | 0.959 | 1.014 | 2.627 | 0.657 | 1.473 | 0.737 |
| | 1-3 | 0.628 | 0.746 | 1.474 | 1.622 | 4.471 | 1.118 | 2.399 | 1.200 |
| | 2-3 | 0.391 | 1.049 | 1.605 | 1.694 | 4.739 | 1.185 | 2.588 | 1.294 |
| $cf_3$ | 1-2 | 0.578 | 0.714 | 1.059 | 1.056 | 3.407 | 0.852 | 1.755 | 0.877 |
| | 1-3 | 0.884 | 0.935 | 1.310 | 1.452 | 4.581 | 1.145 | 2.341 | 1.171 |
| | 2-3 | 0.712 | 1.078 | 1.434 | 1.531 | 4.755 | 1.189 | 2.464 | 1.232 |
| $cf_4$ | 1-2 | 0.286 | 0.446 | 1.140 | 1.155 | 3.027 | 0.757 | 1.707 | 0.854 |
| | 1-3 | 0.741 | 0.817 | 1.644 | 1.770 | 4.973 | 1.243 | 2.656 | 1.328 |
| | 2-3 | 0.445 | 1.210 | 1.757 | 1.823 | 5.236 | 1.309 | 2.842 | 1.421 |
| $cf_5$ | 1-2 | 0.286 | 0.447 | 0.861 | 0.845 | 2.439 | 0.610 | 1.318 | 0.659 |
| | 1-3 | 0.741 | 0.819 | 0.356 | 0.231 | 2.148 | 0.537 | 1.184 | 0.592 |
| | 2-3 | 0.445 | 0.792 | 0.245 | 0.179 | 1.661 | 0.415 | 0.958 | 0.489 |

### 4.2.7 Cumulative Density Function Dissimilarities

A dissimilarity measure based on cumulative density functions was developed in Kim (2009) and Kim and Billard (2013) for histogram-valued observations. Again, this is described for transformed histograms (see section 4.2.1).

**Definition 4.16** Let $A_{u_1}$ and $A_{u_2}$ be any two histogram-valued objects in $\Omega = \Omega_1 \times \cdots \times \Omega_p$ with realizations taking the form Eq. (4.2.8). Then, the **cumulative density function dissimilarity**, $d_{cdf}(A_{u_1}, A_{u_2})$, between $A_{u_1}$ and $A_{u_2}$ is given by

$$d^{cdf}(A_{u_1}, A_{u_2}) = \sum_{j=1}^{p} d_j^{cdf}(A_{u_1}, A_{u_2}) \qquad (4.2.44)$$

where

$$d_j^{cdf}(A_{u_1}, A_{u_2}) = \sum_{k_j=1}^{s_j}(L_{jk_j}|F_{u_1jk_j} - F_{u_2jk_j}|), \ j = 1, \ldots, p, \qquad (4.2.45)$$

where $L_{jk_j} = b_{jk_j} - a_{jk_j}$ is the length of the $k_j$th, $k_j = 1, \ldots, s_j$, sub-interval, and

$$F_{ujt} = \sum_{k_j=1}^{t} p_{ujk_j}, \ t = 1, \ldots, s_j, \qquad (4.2.46)$$

is the cumulative relative frequency of $Y_j$ up to the sub-interval $[a_{jt}, b_{jt}]$. □

**Definition 4.17** Let $A_{u_1}$ and $A_{u_2}$ be any two histogram-valued objects in $\Omega = \Omega_1 \times \cdots \times \Omega_p$ with realizations taking the form Eq. (4.2.8). Then, the **normalized cumulative density function dissimilarity**, $d_{ncdf}(A_{u_1}, A_{u_2})$, between $A_{u_1}$ and $A_{u_2}$ is given by

$$d^{ncdf}(A_{u_1}, A_{u_2}) = \sum_{j=1}^{p} \Psi_j^{-1} d_j^{cdf}(A_{u_1}, A_{u_2}) \qquad (4.2.47)$$

where $\Psi_j = b_{js_j} - a_{j1}$ is the span of the observed values for $Y_j$ and $d_j^{cdf}(A_{u_1}, A_{u_2})$ is the (non-normalized) cumulative density function dissimilarity of Eq. (4.2.45). □

**Example 4.15** Let us take the transformed histogram observations for cholesterol and glucose for the four regions, given in Tables 4.9 and 4.15, respectively. Suppose we want to find the cumulative density function dissimilarity for the first two regions. Take $Y_1$ = cholesterol. The cumulative relative frequencies $F_{ujt}, t = 1, \ldots, 6 = s_1$, are displayed in Table 4.21. Each sub-interval length is $L_{1k_1} = b_{1k_1} - a_{1k_1} = 40$, for all $k$.

**Table 4.21** Cholesterol relative frequencies: regions 1 and 2 (Example 4.15)

| Region | Histogram sub-interval $k$ | | | | | |
|---|---|---|---|---|---|---|
| | 1 | 2 | 3 | 4 | 5 | 6 |
| $u$ | $p_{u11}$ | $p_{u12}$ | $p_{u13}$ | $p_{u14}$ | $p_{u15}$ | $p_{u16}$ |
| 1 | 0.011 | 0.118 | 0.384 | 0.405 | 0.073 | 0.009 |
| 2 | 0.016 | 0.128 | 0.315 | 0.415 | 0.104 | 0.022 |
| | $F_{u11}$ | $F_{u12}$ | $F_{u13}$ | $F_{u14}$ | $F_{u15}$ | $F_{u16}$ |
| 1 | 0.011 | 0.129 | 0.513 | 0.918 | 0.991 | 1 |
| 2 | 0.016 | 0.144 | 0.459 | 0.874 | 0.978 | 1 |

**Table 4.22** Cumulative density function dissimilarities: regions histograms (Example 4.15)

| Object pair | Non-normalized | | | Normalized | | |
|---|---|---|---|---|---|---|
| | $d_1^{cdf}$ | $d_2^{cdf}$ | $d^{cdf}$ | $d_1^{ncdf}$ | $d_2^{ncdf}$ | $d^{ncdf}$ |
| 1-2 | 5.24 | 2.86 | 8.10 | 0.022 | 0.024 | 0.046 |
| 1-3 | 5.28 | 3.98 | 9.26 | 0.022 | 0.033 | 0.055 |
| 1-4 | 4.56 | 2.84 | 7.40 | 0.019 | 0.024 | 0.043 |
| 2-3 | 4.60 | 4.32 | 8.92 | 0.019 | 0.036 | 0.055 |
| 2-4 | 2.12 | 3.54 | 5.66 | 0.009 | 0.029 | 0.038 |
| 3-4 | 2.80 | 4.94 | 7.74 | 0.012 | 0.041 | 0.053 |

Then, from Eq. (4.2.45), we have

$$d_1^{cdf}(A_1, A_2) = 40 \times (|0.011 - 0.016| + |0.129 - 0.144| + |0.513 - 0.459| + |0.918 - 0.874| + |0.991 - 0.978| + |1 - 1|) = 5.24.$$

Likewise, for $Y_2$ = glucose, we have $d_2^{cdf}(A_1, A_2) = 2.86$. Hence, summing over all $j = 1, 2$, the cumulative density function dissimilarity for the first two regions is $d^{cdf}(A_1, A_2) = 5.24 + 2.86 = 8.10$. The corresponding cumulative density function dissimilarities for all pairs of regions are shown in Table 4.22.

To find the normalized cumulative density function dissimilarity, we first need the spans $\Psi_j$, $j = 1, 2$. From Tables 4.9 and 4.15, respectively, we have $\Psi_1 = 240$ and $\Psi_2 = 120$. Hence, from Eq. (4.2.47), e.g.,

$$d_1^{ncdf}(A_1, A_2) = 5.24/240 = 0.022, \quad d_2^{ncdf}(A_1, A_2) = 2.86/120 = 0.024;$$

therefore $d^{ncdf}(A_1, A_2) = 0.022 + 0.024 = 0.046$. The normalized cumulative density function dissimilarities for all regional pairs are given in Table 4.22. □

### 4.2.8 Mallows' Distance

Another distance for comparing two distributions is the Mallows' (1972) distance.

**Definition 4.18** Let $A_{u_1}$ and $A_{u_2}$ be any two histogram-valued objects in $\Omega = \Omega_1 \times \cdots \times \Omega_p$ with realizations being the distribution functions $F_u(\cdot)$, $u = u_1, u_2$, respectively, with $u = 1, \ldots, m$. Then, the **Mallows' distance** of order $q$ between $A_{u_1}$ and $A_{u_2}$, $d_M^q(u_1, u_2)$, is

$$d_M^q(u_1, u_2) = d_M^q(F_{u_1}, F_{u_2}) = \int_0^1 |F_{u_1}^{-1}(t) - F_{u_2}^{-1}(t)|^q dt \qquad (4.2.48)$$

where $F_u^{-1}(t)$ is the inverse distribution function, or quantile function, of $F_u(\cdot)$, $u = u_1, u_2$. ☐

When $q = 2$, Eq. (4.2.48) becomes

$$d_M^2(u_1, u_2) = d_M^2(F_{u_1}, F_{u_2}) = \int_0^1 (F_{u_1}^{-1}(t) - F_{u_2}^{-1}(t))^2 dt. \qquad (4.2.49)$$

Eq. (4.2.49) is a natural extension of classical point data to distributional data. Later, Irpino and Romano ( 2007) showed that

$$d_M^2(F_{u_1}, F_{u_2}) = (\mu_{u_1} - \mu_{u_2})^2 + (\sigma_{u_1} - \sigma_{u_2})^2 + 2\sigma_{u_1}\sigma_{u_2} - 2Cov_Q(u_1, u_2) \qquad (4.2.50)$$

where the quantile covariance is

$$Cov_Q(u_1, u_2) = \int_0^1 (F_{u_1}^{-1}(t) - \mu_{u_1})(F_{u_2}^{-1}(t) - \mu_{u_2})dt, \qquad (4.2.51)$$

and where $\mu_v$ and $\sigma_v$ are the mean and standard deviation, respectively, for the distribution $F_v(\cdot)$, $v = u_1, u_2$. A simple proof of the decomposition result of Eq. (4.2.50) is given in Košmelj and Billard (2012). The terms of the right-hand side of Eq. (4.2.50) refer to a location component, a spread component, and a shape component, respectively.

**Example 4.16** The data of Table 4.23 show the histogram distribution of salaries for two observations, with histogram sub-intervals $[0, 25), [25, 50), [50, 100), [100, 125]$ (in 1000s US\$). Thus, for example, for $u = 2$, the probability that the salary is between \$ 50 and\$ 100 is 0.148. Also shown are the respective cumulative distributions. Then, applying Eq. (4.2.49), we obtain the Mallows' distance $d_M^2(1, 2) = 206.981$. The details are omitted (see Exercise 4.2.5). ☐

Mallows' distance was also used by Irpino and Verde (2006) under the name Wasserstein distance. Computer scientists call this the "earth movers' distance";

**Table 4.23** Income distribution (Example 4.16)

|  | Probability distribution | | | |
|---|---|---|---|---|
|  | [0, 25) | [25, 50) | [50,100) | [100,125] |
| $u$ | $p_{11}$ | $p_{12}$ | $p_{13}$ | $p_{14}$ |
| 1 | 0.213 | 0.434 | 0.281 | 0.072 |
| 2 | 0.354 | 0.475 | 0.148 | 0.023 |
|  | Cumulative distribution | | | |
| 1 | 0.213 | 0.647 | 0.928 | 1.000 |
| 2 | 0.354 | 0.829 | 0.967 | 1.000 |

Levina and Bickel (2001) proved that the earth movers' distance is the same as the Mallows' distance. Rüschendorf (2001) traces the history of this distance.

## Exercises

**4.1** Take the $Y$ = cholesterol histogram observations of Table 4.9. Verify the descriptive statistics for each region, and for each union and intersection pair of these regions, shown in Table 4.12. See Examples 4.9 and 4.10.

**4.2** Find the extended Gowda–Diday dissimilarities, the extended Ichino–Yaguchi distances, and the cumulative density function dissimilarities for the Iris species histogram data of Table 4.11. Compare the different measures.

**4.3** Refer to Example 4.12. Verify the details of Tables 4.12, 4.13, and 4.15.

**4.4** Refer to Example 4.14. Calculate the de Carvalho distances of Table 4.20.

**4.5** Refer to Example 4.16. Verify the Mallows' distance $d^2_M(1, 2) = 206.981$.

# 5

# General Clustering Techniques

## 5.1 Brief Overview of Clustering

As an exploratory tool, clustering techniques are frequently used as a means of identifying features in a data set, such as sets or subsets of observations with certain common (known or unknown) characteristics. The aim is to group observations into clusters (also called classes, or groups, or categories, etc.) that are internally as homogeneously as possible and as heterogeneous as possible across clusters. In this sense, the methods can be powerful tools in an initial data analysis, as a precursor to more detailed in-depth analyses.

However, the plethora of available methods can be both a strength and a weakness, since methods can vary in what specific aspects are important to that methodology, and since not all methods give the same answers on the same data set. Even then a given method can produce varying results depending on the underlying metrics used in its applications. For example, if a clustering algorithm is based on distance/dissimilarity measures, we have seen from Chapters 3 and 4 that different measures can have different values for the same data set. While not always so, often the distances used are Euclidean distances (see Eq. (3.1.8)) or other Minkowski distances such as the city block distance. Just as no one distance measure is universally preferred, so it is that no one clustering method is generally preferred. Hence, all give differing results; all have their strengths and weaknesses, depending on features of the data themselves and on the circumstances.

In this chapter, we give a very brief description of some fundamental clustering methods and principles, primarily as these relate to classical data sets. How these translate into detailed algorithms, descriptions, and applications methodology for symbolic data is deferred to relevant later chapters. In particular, we discuss clustering as partitioning methods in section 5.2 (with the treatment for symbolic data in Chapter 6) and hierarchical methods in section 5.3 (and in Chapter 7 for divisive clustering and Chapter 8 for agglomerative hierarchies for symbolic data). Illustrative examples are provided in section 5.4, where some of these differing methods are applied to the same data set, along with an example

*Clustering Methodology for Symbolic Data,* First Edition. Lynne Billard and Edwin Diday.
© 2020 John Wiley & Sons Ltd. Published 2020 by John Wiley & Sons Ltd.

using simulated data to highlight how classically-based methods ignore the internal variations inherent to symbolic observations. The chapter ends with a quick look at other issues in section 5.5. While their interest and focus is specifically targeted at the marketing world, Punj and Stewart (1983) provide a nice overview of the then available different clustering methods, including strengths and weaknesses, when dealing with classical realizations; Kuo et al. (2002) add updates to that review. Jain et al. (1999) and Gordon (1999) provide overall broad perspectives, mostly in connection with classical data but do include some aspects of symbolic data.

We have $m$ observations $\Omega$ with realizations described by $\mathbf{Y}_u = (Y_{u1}, \dots, Y_{up})$, $u = 1, \dots, m$, in $p$-dimensional space $\mathbb{R}^p$. In this chapter, realizations can be classical or symbolic-valued.

## 5.2    Partitioning

Partitioning methods divide the data set of observations $\Omega$ into non-hierarchical, i.e., not nested, clusters. There are many methods available, and they cover the gamut of unsupervised or supervised, agglomerative or divisive, and monothetic or polythetic methodologies. Lance and Williams (1967a) provide a good general introduction to the principles of partitions. Partitioning algorithms typically use the data as units (in $k$-means type methods) or distances (in $k$-medoids type methods) for clustering purposes.

**Definition 5.1**    **A partition**, $P$, of a set of observations $\Omega$ is a set of clusters $P = (C_1, \dots, C_K)$ which satisfies

(i)  $C_{v_1} \cap C_{v_2} = \phi$ where $\phi$ is the empty set, for all $v_1 \neq v_2 = 1, \dots, K$;
(ii) $\cup_v C_v = \Omega$. □

The basic partitioning method has essentially the following steps:

(1) Establish an initial partition of a pre-specified number of clusters $K$.
(2) Determine classification criteria.
(3) Determine a reallocation process by which each observation in $\Omega$ can be reassigned iteratively to a new cluster according to how the classification criteria are satisfied; this includes reassignment to its current cluster.
(4) Determine a stopping criterion.

The convergence of the basic partitioning method can be formalized as follows. Let $W(P)$ be a quality criterion defined on any partition $P = (C_1, \dots, C_K)$ of the initial population, and let $w(C_k)$ be defined as the quality criterion on the cluster $C_k$, $k = 1, \dots, K$, with

$$W(P) = \sum_{k=1}^{K} w(C_k). \qquad (5.2.1)$$

Both $W(P)$ and $w(C_k)$ have positive values, with the internal variation decreasing as the homogeneity increases so that the quality of the clusters increases. At the $i$th iteration, we start with the partition $P^{(i)} = (C_1^{(i)}, \dots, C_K^{(i)})$. One observation from cluster $C_k^{(i)}$, say, is reallocated to another cluster $C_k^{(i+1)}$, say, so that the new partition $P^{(i+1)} = (C_1^{(i+1)}, \dots, C_K^{(i+1)})$ decreases the overall criterion $W(P)$. Hence, in this manner, at each step $W(P^{(i)})$ decreases and converges since $W(P^{(i)}) > 0$.

There are many suggestions in the literature as to how the initial clusters might be formed. Some start with $K$ seeds, others with $K$ clusters. Of those that use seeds, representative methods, summarized by Anderberg (1973) and Cormack (1971), include:

(i) Select any $K$ observations from $\Omega$ subjectively.
(ii) Randomly select $K$ observations from $\Omega$.
(iii) Calculate the centroid (defined, e.g., as the mean of the observation values by variable) of $\Omega$ as the first seed and then select further seeds as those observations that are at least some pre-specified distance from all previously selected seeds.
(iv) Choose the first $K$ observations in $\Omega$.
(v) Select every $K$th observation in $\Omega$.

For those methods that start with initial partitions, Anderberg's (1973) summary includes:

(a) Randomly allocate the $m$ observations in $\Omega$ into $K$ clusters.
(b) Use the results of a hierarchy clustering.
(c) Seek the advice of experts in the field.

Other methods introduce "representative/supplementary" data as seeds. We observe that different initial clusters can lead to different final clusters, especially when, as is often the case, the algorithm used is based on minimizing an error sum of squares as these methods can converge to local minima instead of global minima. Indeed, it is not possible to guarantee that a global minima is attained unless all possible partitions are explored, which detail is essentially impossible given the very large number of possible clusters. It can be shown that the number of possible $K$ clusters from $m$ observations is the Stirling number of the second kind,

$$S^{(m)} = \frac{1}{K!} \sum_{k=0}^{K} (-1)^{K-k} \binom{K}{k} k^m.$$

Thus, for example, suppose $K = 3$. Then, when $m = 10$, $S^{(m)} = 9330$, and when $m = 20$, $S^{(m)} = 580, 606, 446$.

The most frequently used partitioning method is the so-called $k$-means algorithm and its variants introduced by Forgy (1965), MacQueen (1967), and Wishart (1969). The method calculates the mean vector, or centroid, value for each of the $K$ clusters. Then, the iterative reallocation step involves moving each observation, one at a time, to that cluster whose centroid value is closest; this includes the possibility that the current cluster is "best" for a given observation. When, as is usually the case, a Euclidean distance (see Eq. (3.1.8)) is used to measure the distance between an observation and the $K$ means, this implicitly minimizes the variance within each cluster. Forgy's method is non-adaptive in that it recalculates the mean vectors after a complete round (or pass) on all observations has been made in the reallocation step. In contrast, MacQueen's (1967) approach is adaptive since it recalculates the centroid values after each individual observation has been reassigned during the reallocation step. Another distinction is that Forgy continues the reiteration of these steps (of calculating the centroids and then passing through all observations at the reallocation step) until convergence, defined as when no further changes in the cluster compositions occur. On the other hand, MacQueen stops after the second iteration.

These $k$-means methods have the desirable feature that they converge rapidly. Both methods have piece-wise linear boundaries between the final cluster. This is not necessarily so for symbolic data, since, for example, interval data produce observations that are hypercubes in $\mathbb{R}^p$ rather than the points of classically valued observations; however, some equivalent "linearity" concept would prevail. Another variant uses a so-called forcing pass which tests out every observation in some other cluster in an effort to avoid being trapped in a local minima, a short-coming byproduct of these methods (see Friedman and Rubin (1967)). This approach requires an accommodation for changing the number of clusters $K$. This is particularly advantageous when outliers exist in the data set. Other variants involve rules as to when the iteration process stops, when, if at all, a reallocation process is implemented on the final set of clusters, and so on. Overall then, no completely satisfactory method exists. A summary of some of the many variants of the $k$-means approach is in Jain (2010).

While the criteria driving the reallocation step can vary from one approach to another, how this is implemented also changes from one algorithm to another. Typically though, this is achieved by a hill-climbing steepest descent method or a gradient method such as is associated with Newton–Raphson techniques. This $k$-means method is now a standard method widely used and available in many computer packages.

A more general partitioning method is the "dynamical clustering" method. This method is based on a "representation function" ($g$) of any partition with $g(P) = \mathbf{L} = (L_1, \dots, L_K)$. This associates a representation $L_k$ with each cluster $C_k$ of the partition $P = (C_1, \dots, C_K)$. Then, given $\mathbf{L} = (L_1, \dots, L_K)$, an allocation function $f$ with $f(\mathbf{L}) = P$ associates a cluster $C'_k$ of a new partition $P' = (C'_1, \dots, C'_K)$ with each representation of $L_k$ of $\mathbf{L}$. For example, for the case of a representation by density functions $(d_1, \ldots, d_K)$, each individual $u'$ is then allocated to the cluster $C'_k$ if $d_k(u') = \max\{d_1(u), \ldots, d_K(u)\}$ for this individual. In the case of representation by regression, each individual is allocated to the cluster $C'_k$ if this individual has the best fit to the regression $L_k$, etc. By alternating the use of these $g$ and $f$ functions, the method converges.

More formally, starting with the partition $P = (C_1, \ldots, C_K)$ of the population $\Omega$, the representation function $g$ produces a vector of representation $\mathbf{L} = (L_1, \ldots, L_K)$ such that $g(P) = \mathbf{L}$. Then, the allocation function $\mathbf{L}$ produces a new partition $P' = (C'_1, \ldots, C'_K)$ such that $f(\mathbf{L}) = P'$. Eq. (5.2.1) can be generalized to

$$W(P, \mathbf{L}) = \sum_{k=1}^{K} w(C_k, L_k) \tag{5.2.2}$$

where $w(C_k, L_k) \geq 0$ is a measure of the "fit" between each cluster $C_k$ and its representation $L_k$ and where $w(C_k, L_k)$ decreases as the fit increases. For example, if $L_k$ is a distribution function, then $w(C_k, L_k)$ could be the inverse of the likelihood of the cluster $C_k$ for distribution $L_k$.

That convergence of this more general dynamical clustering partitioning method pertains can be proved as follows. Suppose at the $i$th iteration we have the partition $P^{(i)} = (C_1^{(i)}, \ldots, C_K^{(i)})$ and suppose the representation vector is $\mathbf{L}^{(i)} = (L_1^{(i)}, \ldots, L_K^{(i)})$, and let $u_i = W(P^{(i)}, \mathbf{L}^{(i)})$ be the quality measure on $P$ at this $i$th iteration. First, by the reallocation of each individual belonging to a cluster $C_k^{(i)}$ in $P^{(i)}$ to a new cluster $C_k^{(i+1)}$ such that $f(P^{(i)}) = P^{(i+1)}$, the new partition $P^{(i+1)}$ obtained by this reallocation improves the fit with $L_k^{(i)}$ in order that we have $w(C_k^{(i+1)}, L_k^{(i)}) \leq w(C_k^{(i)}, L_k^{(i)}), k = 1, \ldots, K$. This implies

$$W(P^{(i+1)}, \mathbf{L}^{(i)}) = \sum_{k=1}^{K} w(C_k^{(i+1)}, L_k^{(i)}) \leq \sum_{k=1}^{K} w(C_k^{(i)}, L_k^{(i)}) = W(P^{(i)}, \mathbf{L}^{(i)}).$$
$$\tag{5.2.3}$$

After this $i$th iteration, we can now define at the next $(i + 1)$th iteration a new representation $\mathbf{L}^{(i+1)} = (L_1^{(i+1)}, \ldots, L_K^{(i+1)})$ with $g(P^{(i+1)}) = \mathbf{L}^{(i+1)}$ and where we now have a better fit than we had before for the $i$th iteration. This means that $w(C_k^{(i+1)}, L_k^{(i+1)}) \leq w(C_k^{(i+1)}, L_k^{(i)}), k = 1, \ldots, K$. Hence,

$$W(P^{(i+1)}, \mathbf{L}^{(i+1)}) \leq W(P^{(i)}, \mathbf{L}^{(i)}). \tag{5.2.4}$$

Therefore, since $W(P^{(i)}, \mathbf{L}^{(i)}) \geq 0$ decreases at each iteration, it must be converging. We note that a simple condition for the sequence $\{W(P^{(i)}, \mathbf{L}^{(i)})\}$ to be decreasing is that $W(P, g(P)) \leq W(P, L)$ for any partition $P$ and any representation $L$. Thus, in the reallocation process, a reallocated individual which decreases the criteria the most will by definition be that individual which gives the best fit.

As special cases, when $L_k$ is the mean of cluster $k$, we have the $k$-means algorithm, and when $L_k$ is the medoid, we have the $k$-medoid algorithm. Notice then that this implies that the $D$ and $D^k$ of Eq. (6.1.5) correspond, respectively, to the $W(P)$ and $w(C_k)$ of Eq. (5.2.1). Indeed, the representation of any cluster can be a density function (see Diday and Schroeder (1976)), a regression (see Charles (1977) and Liu (2016)), or more generally a canonical analysis (see Diday (1978, 1986a)), a factorial axis (see Diday (1972a,b)), a distance (see, e.g., Diday and Govaert (1977)), a functional curve (Diday and Simon (1976)), points of a population (Diday (1973)), and so on. For an overview, see Diday (1979), Diday and Simon (1976), and Diday (1989).

Many names have been used for this general technique. For example, Diday (1971a,b, 1972b) refers to this as a "nuées dynamique" method, while in Diday (1973), Diday et al. (1974), and Diday and Simon (1976), it is called a "dynamic clusters" technique, and Jain and Dubes (1988) call it a "square error distortion" method. Bock (2007) calls it an "iterated minimum partition" method and Anderberg (1973) refers to it as a "nearest centroid sorting" method; these latter two descriptors essentially reflect the actual procedure as described in Chapter 6. This dynamical partitioning method can also be found in, e.g., Diday et al. (1974), Schroeder (1976), Scott and Symons (1971), Diday (1979), Celeux and Diebolt (1985), Celeux et al. (1989), Symons (1981), and Ralambondrainy (1995).

Another variant is the "self-organizing maps" (SOM) method, or "Kohonen neural networks" method, described by Kohonen (2001). This method maps patterns onto a grid of units or neurons. In a comparison of the two methods, Bacão et al. (2005) concludes that the SOM method is less likely than the $k$-means method to settle on local optima, but that the SOM method converges to the same result as that obtained by the $k$-means method . Hubert and Arabie (1985) compare some of these methods.

While the $k$-means method uses input coordinate values as its data, the $k$-medoids method uses dissimilarities/distances as its input. This method is also known as the partitioning around mediods (PAM) method developed by Kaufman and Rousseeuw (1987) and Diday (1979). The algorithm works in the same way as does the $k$-means algorithm but moves observations into clusters based on their dissimilarities/distances from cluster medoids (instead of cluster means). The medoids are representative but actual observations inside the cluster. Some variants use the median, midrange, or mode value as the medoid. Since the basic input elements are dissimilarities/distances, the method easily extends to symbolic data, where now dissimilarities/distances can be calculated using the results of Chapters 3 and 4. The PAM method is computationally intensive so is generally restricted to small data sets. Kaufman and Rousseeuw (1986) developed a version for large data sets called clustering large applications (CLARA). This version takes samples from the full data set, applies PAM to each sample, and then amalgamates the results.

As a different approach, Goodall (1966) used a probabilistic similarity matrix to determine the reallocation step, while other researchers based this criterion on multivariate statistical analyses such as discriminant analysis. Another approach is to assume observations arise from a finite mixture of distributions, usually multivariate normal distributions. These model-based methods require the estimation, often by maximum likelihood methods or Bayesian methods, of model parameters (mixture probabilities, and the distributional parameters such as means and variances) at each stage, before the partitioning process continues, as in, for example, the expectation-maximization (EM) algorithm of Dempster et al. (1977). See, for example, McLachlan and Peel (2000), McLachlan and Basford (1988), Celeux and Diebolt (1985), Celeux and Govaert (1992, 1995), Celeux et al. (1989), and Banfield and Raftery (1993).

## 5.3 Hierarchies

In contrast to partitioning methods, hierarchical methods construct hierarchies which are nested clusters. Usually, these are depicted pictorially as trees or dendograms, thereby showing the relationships between the clusters, as in Figure 5.1. In this chapter, we make no distinction between monothetic methods (which focus on one variable at a time) or polythetic methods (which use all variables simultaneously) (see Chapter 7, where the distinctions are made).

**Definition 5.2** For a set of observations with descriptions $\mathbf{Y}$ in $\Omega$, a **hierarchy**, $\mathcal{H}$, consists of sub-clusters, also called nodes, $C_1, C_2, \ldots$, which satisfy

(i) $\Omega \in \mathcal{H}$

(ii) an observation $\mathbf{Y} \in \mathcal{H}$ for all $\mathbf{Y} \in \Omega$

(iii) for any two sub-clusters $C_{v_1}, C_{v_2} \in \mathcal{H}$, $C_{v_1} \cap C_{v_2} \subseteq \{\phi, C_{v_1}, C_{v_2}\}$ where $\phi$ is the empty set.

Since each node contains at least one observation, a corollary is the condition

(iv) $\phi \notin \mathcal{H}$. ☐

Note this definition of the hierarchy $\mathcal{H}$ includes all the nodes of the tree. In contrast, the final clusters only, with none of the intermediate sub-cluster/nodes, can be viewed as the bi-partition $P_r = (C_1, \ldots, C_r)$ (although the construction of a hierarchy $\mathcal{H}$ differs from that for a partition $P_r$). Thus for a partition, the intersection of two clusters is empty (e.g., clusters $C_1$ and $C_2$ of Figure 5.1) whereas the intersection of two nested clusters in a hierarchy is not necessarily empty (e.g., clusters $C_1$ and $C_3$ of Figure 5.1).

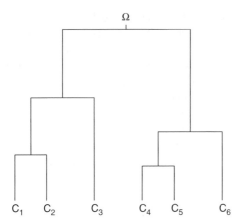

**Figure 5.1** Clusters $C_1, \ldots, C_6$ as a partition at the end of a hierarchy.

Hierarchical methodologies fall into two broad classes. Divisive clustering starts with the tree root $\Omega \equiv C_1$ containing all $m$ observations and divides this into two sub-clusters $C_2$ and $C_3$ and so on, down to a final set of $m$ clusters each with one observation. Thus, with reference to Figure 5.2, we start with one cluster $C_1$ (containing $m = 6$ observations) which is divided into $C_2$ and $C_3$, here each of three observations. Then, one of these (in this case $C_2$) is further divided into, here, $C_4$ and $C_8$. Then, cluster $C_3$ is divided into $C_5$ and $C_{11}$. The process continues until there are six clusters each containing just one observation.

Agglomerative methods, on the other hand, start with $m$ clusters, with each observation being a one-observation cluster at the bottom of the tree with successive merging of clusters upwards, until the final cluster consisting of all the observations ($\Omega$) is obtained. Thus, with reference to Figure 5.2, each of the six clusters $C_6, \ldots, C_{11}$ contains one observation only. The two clusters $C_9$ and $C_{10}$ are merged to form cluster $C_5$. Then, $C_6$ and $C_7$ merge to form cluster $C_4$. The

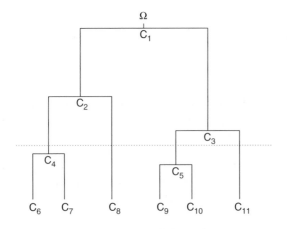

**Figure 5.2** Nodes-clusters $C_1, \ldots, C_{11}$ of a hierarchy.

next merging is when the clusters $C_5$ and $C_{11}$ merge to give $C_3$, or equivalently, at this stage (in two steps) the original single observation clusters $C_9$, $C_{10}$, and $C_{11}$ form the three-observation cluster of $C_3$. This process continues until all six observations in $\Omega$ form the root node $C_1 \equiv \Omega$.

Of course, for either of these methods, the construction can stop at some intermediate stage, such as between the formation of the clusters $C_4$ and $C_3$ in Figure 5.2 (indicated by the dotted line). In this case, the process stops with the four clusters $C_4$, $C_5$, $C_8$, and $C_{11}$. Indeed, such intermediate clusters may be more informative than might be a final complete tree with only one-observation clusters as its bottom layer. Further, these interim clusters can serve as valuable input to the selection of initial clusters in a partitioning algorithm.

At a given stage, how we decide which two clusters are to be merged in the agglomerative approach, or which cluster is to be bi-partitioned, i.e., divided into two clusters in a divisive approach, is determined by the particular algorithm adopted by the researcher. There are many possible criteria used for this.

The motivating principle behind any hierarchical construction is to divide the data set $\Omega$ into clusters which are internally as homogeneous as possible and externally as heterogeneous as possible. While some algorithms implicitly focus on maximizing the between-cluster heterogeneities, and others focus on minimizing the within-cluster homogeneities exclusively, most aim to balance both the between-cluster variations and within-cluster variations.

Thus, divisive methods search to find the best bi-partition at each stage, usually using a sum of squares approach which minimizes the total error sum of squares (in a classical setting) or a total within-cluster variation (in a symbolic setting). More specifically, the cluster $C$ with $m_c$ observations has a total within-cluster variation of $I(C)$ (say), described in Chapter 7 (Definition 7.1, Eq. (7.1.2)), and we want to divide the $m_c$ observations into two disjoint sub-clusters $C^{(1)}$ and $C^{(2)}$ so that the total within-cluster variations of both $C^{(1)}$ and $C^{(2)}$ satisfy $I(C^{(1)}) + I(C^{(2)}) < I(C)$. How this is achieved varies with the different algorithms, but all seek to find that division which minimizes $[I(C^{(1)}) + I(C^{(2)})]$. However, since these entities are functions of the distances (or dis/similarities) between observations in a cluster, different dissimilarities/distance functions will also produce differing hierarchies in general (though some naturally will be comparable for different dissimilarities/distance functions and/or different algorithms) (see Chapters 3 and 4).

In contrast, agglomerative methods seek to find the best merging of two clusters at each stage. How "best" is defined varies by the method, but typically involves a concept of merging the two "closest" observations. Most involve a dis/similarity (or distance) measure in some format. These include the following:

(i) Single-link (or "nearest neighbor") methods, where two clusters $C_{v_1}$ and $C_{v_2}$ are merged if the dissimilarity between them is the minimum of all such dissimilarities across all pairs $C_{v_1}$ and $C_{v_2}$, $v_1 \neq v_2 = 1, 2, \ldots$. Thus, with reference to Figure 5.2, at the first step, since each cluster corresponds to a single observation $u = 1, \ldots, m = 6$, this method finds those two observations for which the dissimilarity/distance measure, $d(u_1, u_2)$, is minimized; here $C_9$ and $C_{10}$ form $C_5$. At the next step, the process continues by finding the two clusters which have the minimum dissimilarity/distance between them. See Sneath and Sokal (1973).

(ii) Complete-link (or "farthest neighbor") methods are comparable to single-link methods except that it is the maximum dissimilarity/distance at each stage that determines the best merger, instead of the minimum dissimilarity/distance. See Sneath and Sokal (1973).

(iii) Ward's (or "minimum variance") methods, where two clusters, of the set of clusters, are merged if they have the smallest squared Euclidean distance between their cluster centroids. This is proportional to a minimum increase in the error sum of squares for the newly merged cluster (across all potential mergers). See Ward (1963).

(iv) There are several others, including average-link (weighted average and group average), median-link, flexible, centroid-link, and so on. See, e.g, Anderberg (1973), Jain and Dubes (1988), McQuitty (1967), Gower (1971), Sokal and Michener (1958), Sokal and Sneath (1963), MacNaughton-Smith et al. (1964), Lance and Williams (1967b) and Gordon (1987, 1999).

Clearly, as for divisive methods, these methods involve dissimilarity (or distance) functions in some form. A nice feature of the single-link and complete-link methods is that they are invariant to monotonic transformations of the dissimilarity matrices. The medium-link method also retains this invariance property; however, it has a problem in implementation in that ties can frequently occur. The average-link method loses this invariance property. The Ward criterion can retard the growth of large clusters, with outlier points tending to be merged at earlier stages of the clustering process. When the dissimilarity matrices are ultrametric (see Definition 3.4), single-link and complete-link methods produce the same tree. Furthermore, an ultrametric dissimilarity matrix will always allow a hierarchy tree to be built. Those methods based on Euclidean distances are invariant to orthogonal rotation of the variable axes.

In general, ties are a problem, and so usually algorithms assume no ties exist. However, ties can be broken by adding an arbitrarily small value $\delta$ to one distance; Jardine and Sibson (1971) show that, for the single-link method, the results merge to the same tree as this perturbation $\delta \to 0$. Also, reversals can occur for median and centroid constructions, though they can be prevented when the distance between two clusters exceeds the maximum tree height for each separate cluster (see, e.g., Gordon (1987)).

Gordon (1999), Punj and Stewart (1983), Jain and Dubes (1988), Anderberg (1973), Cormack (1971), Kaufman and Rousseeuw (1990), and Jain et al. (1999), among many other sources, provide an excellent coverage and comparative review of these methods, for both divisive and agglomerative approaches. While not all such comparisons are in agreement, by and large complete-link methods are preferable to single-link methods, though complete-link methods have more problems with ties. The real agreement between these studies is that there is no one best dissimilarity/distance measure, no one best method, and no one best algorithm, though Kaufman and Rousseeuw (1990) argue strenuously that the group average-link method is preferable. It all depends on the data at hand and the purpose behind the construction itself. For example, suppose in a map of roads it is required to find a road system, then a single-link would likely be wanted, whereas if it is a region where types of housing clusters are sought, then a complete-link method would be better. Jain et al. (1999) suggest that complete-link methods are more useful than single-link methods for many applications. In statistical applications, minimum variance methods tend to be used, while in pattern recognition, single-link methods are more often used. Also, a hierarchy is in one-to-one correspondence with an ultrametric dissimilarity matrix (Johnson, 1967).

A pyramid is a type of agglomerative construction where now clusters can overlap. In particular, any given object can appear in two overlapping clusters. However, that object cannot appear in three or more clusters. This implies there is a linear order of the set of objects in $\Omega$. To construct a pyramid, the dissimilarity matrix must be a Robinson matrix (see Definition 3.6). Whereas a partition consists of non-overlapping/disjoint clusters and a hierarchy consists of nested non-overlapping clusters, a pyramid consists of nested overlapping clusters. Diday (1984, 1986b) introduces the basic principles, with an extension to spatial pyramids in Diday (2008). They are constructed from dissimilarity matrices by Diday (1986b), Diday and Bertrand (1986), and Gaul and Schader (1994).

For most methods, it is not possible to be assured that the resulting tree is globally optimal, only that it is step-wise optimal. Recall that at each step/stage of the construction, an optimal criterion is being used; this does not necessarily guarantee optimally for the complete tree. Also, while step-wise optimality pertains, once that stage is completed (be this up or down the tree, i.e., divisive or agglomerative clustering), there is no allowance to go back to any earlier stage.

There are many different formats for displaying a hierarchy tree; Figure 5.1 is but one illustration. Gordon (1999) provides many representations. For each representation, there are $2^{m-1}$ different orders of the tree branches of the same hierarchy of $m$ objects. Thus, for example, a hierarchy on the three objects $\{1, 2, 3\}$ can take any of the $2^{m-1} = 4$ forms shown in Figure 5.3. However, a tree with the object 3 as a branch in-between the branches 1 and 2 is not possible.

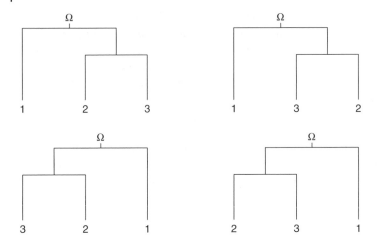

**Figure 5.3** Different orders for a hierarchy of objects $\{1, 2, 3\}$.

In contrast, there are only two possible orderings if a pyramid is built on these three objects, namely, $(\{1, 2\}, \{2, 3\})$ and $(\{3, 2\}, \{2, 1\})$.

A different format is a so-called dendogram, which is a hierarchy tree on which a measure of "height" has been added at each stage of the tree's construction.

**Definition 5.3** Let $u_1, u_2$ be two observations in the dataset $\Omega$. Then, the **height**, $h_{u_1 u_2}$, between them is defined as the smallest sub-cluster in the hierarchy $\mathcal{H}$ which contains both $u_1$ and $u_2$.  □

It is easy to show that (Gordon, 1999), for any two observations in $\Omega$ with realizations $\mathbf{Y}_u$, $u = 1, \ldots, m$, the height satisfies the ultrametric property, i.e.,

$$h_{u_1 u_2} \leq \max(h_{u_1 u_3}, h_{u_3 u_2}) \quad \text{for all } \mathbf{Y}_{u_1}, \mathbf{Y}_{u_2}, \mathbf{Y}_{u_3} \in \Omega. \tag{5.3.1}$$

One such measure is $1 - E_r$ where $E_r$ is the explained rate (i.e., the proportion of total variation in $\Omega$ explained by the tree) at stage $r$ of its construction (see section 7.4 for details).

There are many divisive and agglomerative algorithms in the literature for classical data, but few have been specifically designed for symbolic-valued observations. However, those (classical) algorithms based on dissimilarities/distances can sometimes be extended to symbolic data where now the symbolic dissimilarities/distances between observations are utilized. Some such extensions are covered in Chapters 7 and 8.

## 5.4    Illustration

We illustrate some of the clustering methods outlined in this chapter. The first three provide many different analyses on the same data set, showing how different methods can give quite different results. The fourth example shows how symbolic analyses utilizing internal variations are able to identify clusters that might be missed when using a classical analysis on the same data.

**Example 5.1**    The data in Table 5.1 were obtained by aggregating approximately 50000 classical observations, representing measurements observed for $m = 16$ airlines for flights arriving at and/or departing from a major airport hub. Of particular interest are the $p = 5$ variables, $Y_1$ = flight time between airports (airtime), $Y_2$ = time to taxi in to the arrival gate after landing (taxi-in time), $Y_3$ = time flight arrived after its scheduled arrival time (arrival-delay time), $Y_4$ = time taken to taxi out from the departure gate until liftoff (taxi-out time), and $Y_5$ = time flight was delayed after its scheduled departure time (departure-delay time). The histograms for each variable and each airline were found using standard methods and packages. Thus, there were $s_1 = s_2 = 12$ histogram sub-intervals for each of $Y_1$ and $Y_2$, $s_3 = s_5 = 11$ sub-intervals for $Y_3$ and $Y_5$, and $s_4 = 10$ for $Y_4$. The relative frequencies for each $k = 1, \ldots, s_j$, for each airline $u = 1, \ldots, 16$, are displayed in Table 5.1. Flights that were delayed in some manner by weather were omitted in this aggregation. The original data were extracted from Falduti and Taibaly (2004).

Let us calculate the extended Gowda–Diday dissimilarity matrix, from Eqs. (4.2.24)–Eq.(4.2.27), and shown in Table 5.2. By using this Gowda–Diday dissimilarity matrix, we can find the monothetic divisive clustering dendogram, as shown in Figure 5.4, which was built by using the methodology for divisive clustering for histogram-valued observations described in section 7.2.4. Figure 5.5 shows the tree constructed when using the polythetic algorithm of section 7.3, again based on the extended Gowda–Diday dissimilarity matrix. It is immediately clear that these trees are quite different from each other. For example, after the first cut, the airlines (2, 8) are part of the cluster that also includes airlines $\{1, 6, 9, 10, 11, 13, 14\}$ for the polythetic tree but it is not a part of this cluster in the monothetic tree.

The monothetic divisive algorithm, based on the normalized cumulative density dissimilarity matrix of Eq. (4.2.47) (shown in Table 5.3), produces the tree of Figure 5.6; the polythetic tree (not shown) is similar by the end though the divisions/splits occur at different stages of the tree construction. This tree is also quite different from the tree of Figure 5.4, though both are built using the monothetic algorithm. Not only are the final clusters different, the cut criteria at each stage of the construction differ from each other. Even at the first division, the Gowda–Diday-based tree has a cut criterion of "Is $Y_3 \leq 8.3$?", whereas for the tree of Figure 5.6 based on the cumulative density dissimilarity matrix,

**Table 5.1** Flights data (Example 5.1)

| $u\backslash Y_1$ | [0, 70) | [70, 110) | [110, 150) | [150, 190) | [190, 230) | [230, 270) | [270, 310) | [310, 350) | [350, 390) | [390, 430) | [430, 470) | [470, 540] |
|---|---|---|---|---|---|---|---|---|---|---|---|---|
| 1 | 0.00017 | 0.10568 | 0.33511 | 0.20430 | 0.12823 | 0.045267 | 0.07831 | 0.07556 | 0.02685 | 0.00034 | 0.00000 | 0.00000 |
| 2 | 0.24826 | 0.54412 | 0.20365 | 0.00397 | 0.00000 | 0.00000 | 0.00000 | 0.00000 | 0.00000 | 0.00000 | 0.00000 | 0.00000 |
| 3 | 0.77412 | 0.22451 | 0.00137 | 0.00000 | 0.00000 | 0.00000 | 0.00000 | 0.00000 | 0.00000 | 0.00000 | 0.00000 | 0.00000 |
| 4 | 0.16999 | 0.00805 | 0.42775 | 0.20123 | 0.01955 | 0.02606 | 0.00556 | 0.00556 | 0.00057 | 0.01169 | 0.09774 | 0.02626 |
| 5 | 0.13464 | 0.10799 | 0.01823 | 0.37728 | 0.35063 | 0.01122 | 0.00000 | 0.00000 | 0.00000 | 0.00000 | 0.00000 | 0.00000 |
| 6 | 0.70026 | 0.22415 | 0.07264 | 0.00229 | 0.00065 | 0.00000 | 0.00000 | 0.00000 | 0.00000 | 0.00000 | 0.00000 | 0.00000 |
| 7 | 0.26064 | 0.21519 | 0.34916 | 0.06427 | 0.02798 | 0.01848 | 0.03425 | 0.02272 | 0.00729 | 0.00000 | 0.00000 | 0.00000 |
| 8 | 0.17867 | 0.41499 | 0.40634 | 0.00000 | 0.00000 | 0.00000 | 0.00000 | 0.00000 | 0.00000 | 0.00000 | 0.00000 | 0.00000 |
| 9 | 0.28907 | 0.41882 | 0.28452 | 0.00683 | 0.00076 | 0.00000 | 0.00000 | 0.00000 | 0.00000 | 0.00000 | 0.00000 | 0.00000 |
| 10 | 0.00000 | 0.00000 | 0.00000 | 0.00000 | 0.03811 | 0.30793 | 0.34299 | 0.21494 | 0.08384 | 0.01220 | 0.00000 | 0.00000 |
| 11 | 0.51329 | 0.35570 | 0.11021 | 0.01651 | 0.00386 | 0.00000 | 0.00021 | 0.00000 | 0.00000 | 0.00000 | 0.00000 | 0.00000 |
| 12 | 0.39219 | 0.31956 | 0.19201 | 0.09442 | 0.00182 | 0.00000 | 0.00000 | 0.00000 | 0.00000 | 0.00000 | 0.00000 | 0.00000 |
| 13 | 0.00000 | 0.61672 | 0.36585 | 0.00348 | 0.00174 | 0.00000 | 0.00523 | 0.00174 | 0.00348 | 0.00000 | 0.00000 | 0.00174 |
| 14 | 0.07337 | 0.28615 | 0.19896 | 0.05956 | 0.04186 | 0.07898 | 0.10876 | 0.10315 | 0.04877 | 0.00043 | 0.00000 | 0.00000 |
| 15 | 0.76391 | 0.20936 | 0.01719 | 0.00645 | 0.00263 | 0.00048 | 0.00000 | 0.00000 | 0.00000 | 0.00000 | 0.00000 | 0.00000 |
| 16 | 0.48991 | 0.06569 | 0.26276 | 0.10803 | 0.02374 | 0.02810 | 0.01543 | 0.00594 | 0.00040 | 0.00000 | 0.00000 | 0.00000 |

| $u\backslash Y_2$ | [0, 5) | [5, 10) | [10, 15) | [15, 20) | [20, 25) | [25, 30) | [30, 35) | [35, 40) | [40, 50) | [50, 75) | [75, 100) | [100, 125] |
|---|---|---|---|---|---|---|---|---|---|---|---|---|
| 1 | 0.38451 | 0.37194 | 0.12289 | 0.04647 | 0.02513 | 0.01256 | 0.01015 | 0.00706 | 0.00929 | 0.00740 | 0.00207 | 0.00052 |
| 2 | 0.53916 | 0.32659 | 0.07813 | 0.02697 | 0.01150 | 0.00555 | 0.00278 | 0.00278 | 0.00377 | 0.00238 | 0.00040 | 0.00000 |
| 3 | 0.46365 | 0.45450 | 0.05853 | 0.01189 | 0.00640 | 0.00229 | 0.00137 | 0.00046 | 0.00091 | 0.00000 | 0.00000 | 0.00000 |
| 4 | 0.70353 | 0.23745 | 0.03526 | 0.01418 | 0.00364 | 0.00307 | 0.00134 | 0.00077 | 0.00057 | 0.00019 | 0.00000 | 0.00000 |

*(Continued)*

**Table 5.1** (Continued)

| u | [−40, −20) | [−20, 0) | [0, 20) | [20, 40) | [40, 60) | [60, 80) | [80, 100) | [100, 120) | [120, 140) | [140, 160) | [160, 200) | [200, 240] |
|---|---|---|---|---|---|---|---|---|---|---|---|---|
| 5 | 0.32539 | 0.49509 | 0.13604 | 0.03086 | 0.00982 | 0.00140 | 0.00140 | 0.00000 | 0.00000 | 0.00000 | 0.00000 | 0.00000 |
| 6 | 0.63514 | 0.24280 | 0.04647 | 0.02127 | 0.01734 | 0.01014 | 0.00687 | 0.00458 | 0.00622 | 0.00687 | 0.00098 | 0.00131 |
| 7 | 0.33017 | 0.47584 | 0.12142 | 0.03748 | 0.01865 | 0.00543 | 0.00339 | 0.00187 | 0.00288 | 0.00237 | 0.00034 | 0.00017 |
| 8 | 0.32277 | 0.46398 | 0.10663 | 0.03458 | 0.03458 | 0.00865 | 0.01441 | 0.00288 | 0.00288 | 0.00576 | 0.00288 | 0.00000 |
| 9 | 0.28907 | 0.51897 | 0.13050 | 0.02883 | 0.01214 | 0.01214 | 0.00531 | 0.00000 | 0.00228 | 0.00076 | 0.00000 | 0.00000 |
| 10 | 0.42378 | 0.32470 | 0.11433 | 0.04878 | 0.02591 | 0.02591 | 0.00610 | 0.00762 | 0.00915 | 0.00915 | 0.00000 | 0.00457 |
| 11 | 0.42152 | 0.37500 | 0.11321 | 0.04245 | 0.01672 | 0.01158 | 0.00793 | 0.00343 | 0.00386 | 0.00322 | 0.00107 | 0.00000 |
| 12 | 0.23559 | 0.60009 | 0.11621 | 0.02860 | 0.00908 | 0.00726 | 0.00227 | 0.00045 | 0.00000 | 0.00045 | 0.00000 | 0.00000 |
| 13 | 0.33972 | 0.46167 | 0.08711 | 0.03136 | 0.02787 | 0.02265 | 0.00174 | 0.00697 | 0.00697 | 0.01045 | 0.00348 | 0.00000 |
| 14 | 0.59732 | 0.29650 | 0.07078 | 0.01640 | 0.00691 | 0.00647 | 0.00302 | 0.00043 | 0.00129 | 0.00043 | 0.00000 | 0.00043 |
| 15 | 0.61733 | 0.32442 | 0.04201 | 0.00907 | 0.00430 | 0.00095 | 0.00095 | 0.00024 | 0.00048 | 0.00000 | 0.00024 | 0.00000 |
| 16 | 0.93154 | 0.05738 | 0.00594 | 0.00277 | 0.00119 | 0.00079 | 0.00040 | 0.00000 | 0.00000 | 0.00000 | 0.00000 | 0.00000 |

| u\$Y_3$ | [−40, −20) | [−20, 0) | [0, 20) | [20, 40) | [40, 60) | [60, 80) | [80, 100) | [100, 120) | [120, 140) | [140, 160) | [160, 200) | [200, 240] |
|---|---|---|---|---|---|---|---|---|---|---|---|---|
| 1 | 0.09260 | 0.38520 | 0.28589 | 0.09725 | 0.04854 | 0.03046 | 0.01773 | 0.01411 | 0.00637 | 0.00654 | 0.01532 | 0.00000 |
| 2 | 0.14158 | 0.45588 | 0.24886 | 0.07238 | 0.03272 | 0.01745 | 0.01170 | 0.00595 | 0.00377 | 0.00377 | 0.00595 | 0.00000 |
| 3 | 0.04298 | 0.44810 | 0.31367 | 0.09419 | 0.04161 | 0.02515 | 0.01280 | 0.00686 | 0.00640 | 0.00320 | 0.00503 | 0.00000 |
| 4 | 0.07359 | 0.41989 | 0.31909 | 0.11154 | 0.04312 | 0.01763 | 0.00652 | 0.00422 | 0.00211 | 0.00077 | 0.00153 | 0.00000 |
| 5 | 0.09537 | 0.45863 | 0.30014 | 0.07433 | 0.03226 | 0.01683 | 0.01403 | 0.00281 | 0.00281 | 0.00000 | 0.00281 | 0.00000 |
| 6 | 0.12958 | 0.41361 | 0.21008 | 0.09097 | 0.04450 | 0.02716 | 0.02094 | 0.01440 | 0.01276 | 0.00884 | 0.02716 | 0.00000 |
| 7 | 0.06054 | 0.44362 | 0.33475 | 0.08648 | 0.03510 | 0.01865 | 0.00797 | 0.00661 | 0.00356 | 0.00051 | 0.00220 | 0.00000 |
| 8 | 0.08934 | 0.44957 | 0.29683 | 0.07493 | 0.01729 | 0.03746 | 0.00865 | 0.00576 | 0.00576 | 0.00576 | 0.00865 | 0.00000 |
| 9 | 0.07967 | 0.36646 | 0.28376 | 0.10698 | 0.06070 | 0.03794 | 0.02883 | 0.00835 | 0.01366 | 0.00835 | 0.00531 | 0.00000 |
| 10 | 0.14024 | 0.30030 | 0.29573 | 0.18293 | 0.03659 | 0.01067 | 0.00762 | 0.00305 | 0.00152 | 0.00762 | 0.01372 | 0.00000 |

(Continued)

**Table 5.1** (Continued)

| | [0, 8] | [8, 16] | [16, 24] | [24, 32] | [32, 40] | [40, 48] | [48, 56] | [56, 64] | [64, 72] | [72, 80] | [80, 88] | [88, 96] |
|---|---|---|---|---|---|---|---|---|---|---|---|---|
| 11 | 0.07955 | 0.34627 | 0.27980 | 0.12393 | 0.06411 | 0.03967 | 0.02423 | 0.01844 | 0.00922 | 0.00493 | 0.00986 | 0.00000 |
| 12 | 0.03949 | 0.40899 | 0.33727 | 0.12483 | 0.04585 | 0.02224 | 0.00817 | 0.00635 | 0.00227 | 0.00136 | 0.00318 | 0.00000 |
| 13 | 0.07840 | 0.44599 | 0.21603 | 0.10627 | 0.04530 | 0.03310 | 0.01916 | 0.01394 | 0.00871 | 0.01220 | 0.02091 | 0.00000 |
| 14 | 0.07682 | 0.41951 | 0.27147 | 0.09840 | 0.03712 | 0.03151 | 0.01942 | 0.01122 | 0.00950 | 0.00604 | 0.01899 | 0.00000 |
| 15 | 0.10551 | 0.55693 | 0.22989 | 0.06493 | 0.02363 | 0.01074 | 0.00286 | 0.00143 | 0.00167 | 0.00095 | 0.00143 | 0.00000 |
| 16 | 0.05857 | 0.58726 | 0.23823 | 0.05540 | 0.02928 | 0.01187 | 0.01029 | 0.00317 | 0.00198 | 0.00079 | 0.00317 | 0.00000 |
| $u\backslash Y_4$ | [0, 8] | [8, 16] | [16, 24] | [24, 32] | [32, 40] | [40, 48] | [48, 56] | [56, 64] | [64, 72] | [72, 80] | [80, 88] | [88, 96] |
| 1 | 0.01687 | 0.43580 | 0.30878 | 0.12513 | 0.05370 | 0.02496 | 0.01136 | 0.00740 | 0.00499 | 0.01102 | 0.00000 | 0.00000 |
| 2 | 0.09974 | 0.40472 | 0.25223 | 0.13405 | 0.05870 | 0.02499 | 0.01190 | 0.00516 | 0.00357 | 0.00496 | 0.00000 | 0.00000 |
| 3 | 0.05761 | 0.41198 | 0.26932 | 0.14769 | 0.05624 | 0.03109 | 0.01280 | 0.00503 | 0.00137 | 0.00686 | 0.00000 | 0.00000 |
| 4 | 0.02779 | 0.57877 | 0.24722 | 0.08317 | 0.03910 | 0.01016 | 0.00479 | 0.00287 | 0.00211 | 0.00402 | 0.00000 | 0.00000 |
| 5 | 0.00561 | 0.33941 | 0.32959 | 0.17251 | 0.08555 | 0.03787 | 0.02104 | 0.00421 | 0.00421 | 0.00000 | 0.00000 | 0.00000 |
| 6 | 0.10995 | 0.46073 | 0.23920 | 0.10046 | 0.04221 | 0.01963 | 0.00916 | 0.00687 | 0.00393 | 0.00785 | 0.00000 | 0.00000 |
| 7 | 0.00967 | 0.37850 | 0.33593 | 0.14601 | 0.06427 | 0.02900 | 0.01509 | 0.00627 | 0.00526 | 0.01001 | 0.00000 | 0.00000 |
| 8 | 0.08934 | 0.46398 | 0.26225 | 0.10663 | 0.04900 | 0.01153 | 0.00865 | 0.00576 | 0.00288 | 0.00000 | 0.00000 | 0.00000 |
| 9 | 0.01897 | 0.34901 | 0.31259 | 0.16313 | 0.08877 | 0.03642 | 0.01669 | 0.00986 | 0.00152 | 0.00303 | 0.00000 | 0.00000 |
| 10 | 0.03963 | 0.39634 | 0.39024 | 0.10366 | 0.04116 | 0.01372 | 0.00610 | 0.00457 | 0.00152 | 0.00305 | 0.00000 | 0.00000 |
| 11 | 0.06003 | 0.42024 | 0.25772 | 0.13572 | 0.06111 | 0.02830 | 0.01715 | 0.00686 | 0.00536 | 0.00750 | 0.00000 | 0.00000 |
| 12 | 0.03404 | 0.45166 | 0.26146 | 0.11484 | 0.06128 | 0.03223 | 0.01816 | 0.01044 | 0.00681 | 0.00908 | 0.00000 | 0.00000 |
| 13 | 0.01742 | 0.38502 | 0.28920 | 0.15157 | 0.08537 | 0.04704 | 0.01568 | 0.00697 | 0.00000 | 0.00174 | 0.00000 | 0.00000 |
| 14 | 0.00691 | 0.33880 | 0.34916 | 0.15451 | 0.07682 | 0.02935 | 0.01942 | 0.00604 | 0.00604 | 0.01295 | 0.00000 | 0.00000 |
| 15 | 0.08761 | 0.53927 | 0.21246 | 0.09048 | 0.03438 | 0.01337 | 0.00668 | 0.00573 | 0.00263 | 0.00740 | 0.00000 | 0.00000 |
| 16 | 0.41472 | 0.48833 | 0.06055 | 0.02137 | 0.00831 | 0.00198 | 0.00119 | 0.00040 | 0.00119 | 0.00198 | 0.00000 | 0.00000 |

*(Continued)*

**Table 5.1** (Continued)

| $u\backslash Y_5$ | [−15, 5) | [5, 25) | [25, 45) | [45, 65) | [65, 85) | [85, 105) | [105, 125) | [125, 145) | [145, 165) | [165, 185) | [185, 225) | [225, 265) |
|---|---|---|---|---|---|---|---|---|---|---|---|---|
| 1 | 0.67762 | 0.16988 | 0.05714 | 0.03219 | 0.01893 | 0.01463 | 0.00878 | 0.00000 | 0.00361 | 0.00947 | 0.00775 | 0.00000 |
| 2 | 0.77414 | 0.12552 | 0.04144 | 0.01943 | 0.01368 | 0.00654 | 0.00694 | 0.00000 | 0.00416 | 0.00476 | 0.00337 | 0.00000 |
| 3 | 0.78235 | 0.10700 | 0.04161 | 0.02469 | 0.01783 | 0.01143 | 0.00549 | 0.00000 | 0.00366 | 0.00320 | 0.00274 | 0.00000 |
| 4 | 0.59448 | 0.27386 | 0.08317 | 0.02626 | 0.01016 | 0.00575 | 0.00287 | 0.00000 | 0.00134 | 0.00115 | 0.00096 | 0.00000 |
| 5 | 0.84993 | 0.07293 | 0.03086 | 0.01964 | 0.01683 | 0.00421 | 0.00281 | 0.00000 | 0.00000 | 0.00140 | 0.00140 | 0.00000 |
| 6 | 0.65249 | 0.14071 | 0.06872 | 0.04025 | 0.02749 | 0.01669 | 0.01407 | 0.00000 | 0.01014 | 0.01407 | 0.01538 | 0.00000 |
| 7 | 0.77650 | 0.14516 | 0.04036 | 0.01611 | 0.01051 | 0.00526 | 0.00305 | 0.00000 | 0.00085 | 0.00068 | 0.00153 | 0.00000 |
| 8 | 0.63112 | 0.24784 | 0.04323 | 0.02017 | 0.02882 | 0.00288 | 0.00865 | 0.00000 | 0.00865 | 0.00000 | 0.00865 | 0.00000 |
| 9 | 0.70030 | 0.12064 | 0.06297 | 0.04628 | 0.02049 | 0.01290 | 0.01897 | 0.00000 | 0.00986 | 0.00607 | 0.00152 | 0.00000 |
| 10 | 0.73323 | 0.16463 | 0.04726 | 0.01677 | 0.01220 | 0.00305 | 0.00457 | 0.00000 | 0.00152 | 0.00762 | 0.00915 | 0.00000 |
| 11 | 0.64537 | 0.15866 | 0.07654 | 0.04867 | 0.02744 | 0.01780 | 0.00858 | 0.00000 | 0.00686 | 0.00686 | 0.00322 | 0.00000 |
| 12 | 0.78711 | 0.12165 | 0.05311 | 0.01816 | 0.00772 | 0.00635 | 0.00227 | 0.00000 | 0.00136 | 0.00045 | 0.00182 | 0.00000 |
| 13 | 0.71080 | 0.12369 | 0.05749 | 0.03310 | 0.01916 | 0.00523 | 0.01045 | 0.00000 | 0.01742 | 0.01394 | 0.00871 | 0.00000 |
| 14 | 0.74234 | 0.09754 | 0.05697 | 0.03064 | 0.02201 | 0.01381 | 0.01252 | 0.00000 | 0.00216 | 0.00993 | 0.01208 | 0.00000 |
| 15 | 0.83600 | 0.10862 | 0.03032 | 0.01408 | 0.00573 | 0.00286 | 0.00095 | 0.00000 | 0.00072 | 0.00048 | 0.00024 | 0.00000 |
| 16 | 0.76573 | 0.13850 | 0.04432 | 0.02335 | 0.01464 | 0.00594 | 0.00356 | 0.00000 | 0.00079 | 0.00119 | 0.00198 | 0.00000 |

For each airline, $u = 1, \ldots, 16$:

$Y_1$ = flight time between airports (air time); $s_1 = 12$ histogram sub-intervals,

$Y_2$ = time to taxi in to the gate after landing (taxi-in time); $s_2 = 12$ histogram sub-intervals,

$Y_3$ = time arrived after scheduled arrival time (arrival-delay time); $s_3 = 11$ histogram sub-intervals,

$Y_4$ = time to taxi out from the gate until liftoff (taxi-out time); $s_4 = 10$ histogram sub-intervals,

$Y_5$ = delay time after scheduled departure time (departure-delay time); $s_5 = 11$ histogram sub-intervals.

**Table 5.2** Flights data: extended Gowda–Diday Matrix D (Example 5.1)

| u | 1 | 2 | 3 | 4 | 5 | 6 | 7 | 8 | 9 | 10 | 11 | 12 | 13 | 14 | 15 | 16 |
|---|---|---|---|---|---|---|---|---|---|----|----|----|----|----|----|----|
| 1 | 0 | 2.253 | 2.929 | 2.761 | 2.584 | 1.973 | 1.855 | 1.877 | 2.114 | 1.789 | 1.725 | 2.723 | 1.518 | 1.133 | 3.801 | 3.666 |
| 2 | 2.253 | 0 | 1.119 | 2.456 | 1.914 | 1.608 | 1.417 | 1.085 | 1.051 | 2.702 | 1.011 | 1.330 | 1.893 | 2.299 | 1.871 | 2.713 |
| 3 | 2.929 | 1.119 | 0 | 2.560 | 1.986 | 1.984 | 2.057 | 1.959 | 1.474 | 3.343 | 1.539 | 1.356 | 2.723 | 2.536 | 1.307 | 2.616 |
| 4 | 2.761 | 2.456 | 2.560 | 0 | 1.830 | 3.546 | 1.759 | 2.692 | 2.675 | 3.177 | 2.953 | 2.044 | 3.509 | 2.361 | 2.436 | 2.366 |
| 5 | 2.584 | 1.914 | 1.986 | 1.830 | 0 | 3.340 | 1.237 | 2.357 | 1.943 | 3.180 | 2.482 | 1.222 | 3.032 | 2.268 | 2.075 | 2.218 |
| 6 | 1.973 | 1.608 | 1.984 | 3.546 | 3.340 | 0 | 2.726 | 1.469 | 1.638 | 2.704 | 1.067 | 2.541 | 1.446 | 2.361 | 2.699 | 3.852 |
| 7 | 1.855 | 1.417 | 2.057 | 1.759 | 1.237 | 2.726 | 0 | 2.027 | 1.846 | 2.914 | 1.902 | 1.124 | 2.532 | 1.955 | 2.160 | 2.261 |
| 8 | 1.877 | 1.085 | 1.959 | 2.692 | 2.357 | 1.469 | 2.027 | 0 | 1.327 | 2.332 | 1.164 | 1.934 | 1.367 | 2.540 | 2.573 | 3.144 |
| 9 | 2.114 | 1.051 | 1.474 | 2.675 | 1.943 | 1.638 | 1.846 | 1.327 | 0 | 2.909 | 0.837 | 1.382 | 1.598 | 1.871 | 2.461 | 3.062 |
| 10 | 1.789 | 2.702 | 3.343 | 3.177 | 3.180 | 2.704 | 2.914 | 2.332 | 2.909 | 0 | 2.626 | 3.306 | 2.272 | 2.628 | 3.877 | 3.991 |
| 11 | 1.725 | 1.011 | 1.539 | 2.953 | 2.482 | 1.067 | 1.902 | 1.164 | 0.837 | 2.626 | 0 | 1.639 | 1.589 | 2.066 | 2.464 | 3.319 |
| 12 | 2.723 | 1.330 | 1.356 | 2.044 | 1.222 | 2.541 | 1.124 | 1.934 | 1.382 | 3.306 | 1.639 | 0 | 2.687 | 2.441 | 1.620 | 2.210 |
| 13 | 1.518 | 1.893 | 2.723 | 3.509 | 3.032 | 1.446 | 2.532 | 1.367 | 1.598 | 2.272 | 1.589 | 2.687 | 0 | 2.223 | 3.580 | 4.018 |
| 14 | 1.133 | 2.299 | 2.536 | 2.361 | 2.268 | 2.361 | 1.955 | 2.540 | 1.871 | 2.628 | 2.066 | 2.441 | 2.223 | 0 | 3.459 | 3.459 |
| 15 | 3.801 | 1.871 | 1.307 | 2.436 | 2.075 | 2.699 | 2.160 | 2.573 | 2.461 | 3.877 | 2.464 | 1.620 | 3.580 | 3.459 | 0 | 2.396 |
| 16 | 3.666 | 2.713 | 2.616 | 2.366 | 2.218 | 3.852 | 2.261 | 3.144 | 3.062 | 3.991 | 3.319 | 2.210 | 4.018 | 3.459 | 2.396 | 0 |

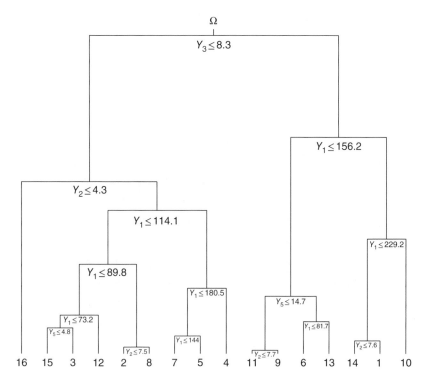

**Figure 5.4** Monothetic dendogram based on Gowda–Diday dissimilarities (with cuts) (Example 5.1).

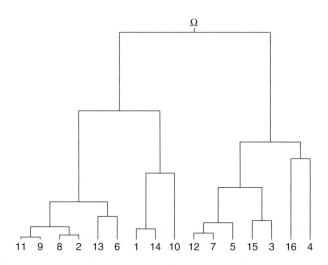

**Figure 5.5** Polythetic dendogram based on Gowda–Diday dissimilarities (Example 5.1).

**Table 5.3** Flights data: normalized cumulative density function Matrix **D** (Example 5.1)

| u | 1 | 2 | 3 | 4 | 5 | 6 | 7 | 8 | 9 | 10 | 11 | 12 | 13 | 14 | 15 | 16 |
|---|---|---|---|---|---|---|---|---|---|----|----|----|----|----|----|----|
| 1 | 0 | 0.261 | 0.318 | 0.210 | 0.166 | 0.317 | 0.204 | 0.221 | 0.230 | 0.256 | 0.251 | 0.262 | 0.179 | 0.111 | 0.392 | 0.385 |
| 2 | 0.261 | 0 | 0.115 | 0.241 | 0.201 | 0.158 | 0.115 | 0.082 | 0.113 | 0.446 | 0.125 | 0.094 | 0.155 | 0.258 | 0.141 | 0.206 |
| 3 | 0.318 | 0.115 | 0 | 0.304 | 0.270 | 0.126 | 0.174 | 0.164 | 0.153 | 0.520 | 0.111 | 0.110 | 0.212 | 0.309 | 0.093 | 0.237 |
| 4 | 0.210 | 0.241 | 0.304 | 0 | 0.207 | 0.330 | 0.199 | 0.216 | 0.287 | 0.367 | 0.307 | 0.251 | 0.299 | 0.206 | 0.292 | 0.270 |
| 5 | 0.166 | 0.201 | 0.270 | 0.207 | 0 | 0.350 | 0.132 | 0.210 | 0.208 | 0.346 | 0.280 | 0.192 | 0.220 | 0.179 | 0.295 | 0.306 |
| 6 | 0.317 | 0.158 | 0.126 | 0.330 | 0.350 | 0 | 0.255 | 0.160 | 0.167 | 0.542 | 0.104 | 0.195 | 0.188 | 0.318 | 0.144 | 0.269 |
| 7 | 0.204 | 0.115 | 0.174 | 0.199 | 0.132 | 0.255 | 0 | 0.136 | 0.125 | 0.406 | 0.174 | 0.092 | 0.158 | 0.196 | 0.220 | 0.216 |
| 8 | 0.221 | 0.082 | 0.164 | 0.216 | 0.210 | 0.160 | 0.136 | 0 | 0.114 | 0.417 | 0.128 | 0.118 | 0.116 | 0.260 | 0.190 | 0.231 |
| 9 | 0.230 | 0.113 | 0.153 | 0.287 | 0.208 | 0.167 | 0.125 | 0.114 | 0 | 0.464 | 0.079 | 0.099 | 0.086 | 0.214 | 0.235 | 0.287 |
| 10 | 0.256 | 0.446 | 0.520 | 0.367 | 0.346 | 0.542 | 0.406 | 0.417 | 0.464 | 0 | 0.487 | 0.461 | 0.414 | 0.312 | 0.562 | 0.552 |
| 11 | 0.251 | 0.125 | 0.111 | 0.307 | 0.280 | 0.104 | 0.174 | 0.128 | 0.079 | 0.487 | 0 | 0.112 | 0.139 | 0.270 | 0.189 | 0.262 |
| 12 | 0.262 | 0.094 | 0.110 | 0.251 | 0.192 | 0.195 | 0.092 | 0.118 | 0.099 | 0.461 | 0.112 | 0 | 0.158 | 0.261 | 0.166 | 0.211 |
| 13 | 0.179 | 0.155 | 0.212 | 0.299 | 0.220 | 0.188 | 0.158 | 0.116 | 0.086 | 0.414 | 0.139 | 0.158 | 0 | 0.194 | 0.288 | 0.323 |
| 14 | 0.111 | 0.258 | 0.309 | 0.206 | 0.179 | 0.318 | 0.196 | 0.260 | 0.214 | 0.312 | 0.270 | 0.261 | 0.194 | 0 | 0.378 | 0.369 |
| 15 | 0.392 | 0.141 | 0.093 | 0.292 | 0.295 | 0.144 | 0.220 | 0.190 | 0.235 | 0.562 | 0.189 | 0.166 | 0.288 | 0.378 | 0 | 0.185 |
| 16 | 0.385 | 0.206 | 0.237 | 0.270 | 0.306 | 0.269 | 0.216 | 0.231 | 0.287 | 0.552 | 0.262 | 0.211 | 0.323 | 0.369 | 0.185 | 0 |

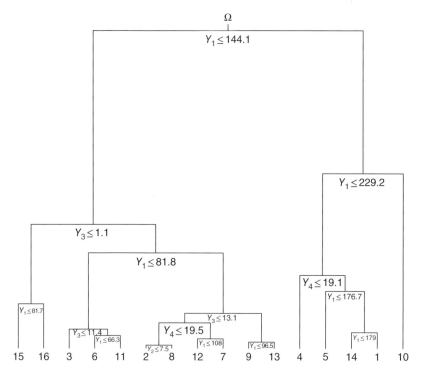

**Figure 5.6** Monothetic dendogram based on cumulative density function dissimilarities (with cuts) (Example 5.1).

the cut criterion is "Is $Y_1 \leq 144.06$?". A hint of why these trees differ can be found by comparing the different dissimilarity matrices in Tables 5.2 and 5.3.

When an average-link agglomerative algorithm (described in section 8.1) is applied to these same histogram data, the tree of Figure 5.7 emerges. As is apparent by comparison with the trees of Figures 5.4–5.6, this tree is different yet again. This tree is also different from that shown in Figure 5.8, obtained when using a complete-link agglomerative algorithm based on the extended Ichino–Yaguchi dissimilarities for histogram data.

In contrast, the number ($\sim 50000$) of individual observations is too large for any classical agglomerative algorithm to work on the classical data set directly. Were we to seek such a tree, then we could take the sample means (shown in Table 5.4) of the aggregated values for each airline and each variable to use in any classical agglomerative algorithm. The tree that pertains when using the complete-link method on the Euclidean distance between these means is the tree of Figure 5.9. While both Figures 5.8 and 5.9 were each built as a complete-link tree, the results are very different. These differences reflect the fact that when using just the means (as in Figure 5.9), a lot of the internal

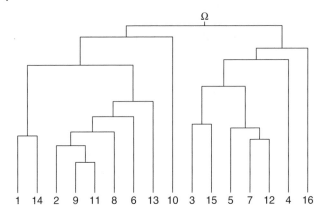

**Figure 5.7** Average-link agglomerative dendogram based on Gowda–Diday dissimilarities (histograms) (Example 5.1).

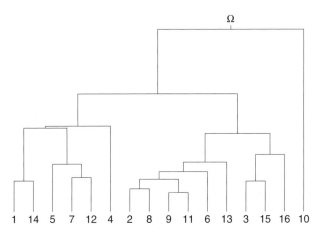

**Figure 5.8** Complete-link agglomerative dendogram based on Ichino–Yaguchi Euclidean dissimilarities (histograms) (Example 5.1).

variation information is lost. However, by aggregating the internal values for each airline into histograms, much of this lost information is retained, that is, the tree of Figure 5.8 uses all this information (both the central mean values plus internal measures of variations). This can be illustrated, e.g., by looking at the cluster on the far right-hand side of the tree of Figure 5.9. This tree is built on the means, and we see from Table 5.4 and Figure 5.10 (which shows the bi-plots of the sample means $\bar{Y}_1$ and $\bar{Y}_3$) that these three airlines (7, 13, and 16) have comparable $Y_1$ means. Furthermore, we also know from Table 5.4 that airline 13 has a much smaller sample variance for $Y_1$ while the corresponding sample variances for airlines 7 and 16 are much larger. There

**Table 5.4** Flights data: sample means and standard deviations (individuals) (Example 5.1)

| Airline | Means | | | | | Standard deviations | | | | |
|---|---|---|---|---|---|---|---|---|---|---|
| $u$ | $\bar{Y}_1$ | $\bar{Y}_2$ | $\bar{Y}_3$ | $\bar{Y}_4$ | $\bar{Y}_5$ | $S_1$ | $S_2$ | $S_3$ | $S_4$ | $S_5$ |
| 1 | 183.957 | 9.338 | 16.794 | 21.601 | 16.029 | 74.285 | 9.810 | 46.137 | 13.669 | 41.445 |
| 2 | 90.432 | 7.043 | 9.300 | 20.066 | 12.303 | 25.270 | 6.326 | 37.860 | 13.252 | 33.291 |
| 3 | 56.703 | 6.507 | 11.038 | 20.528 | 7.755 | 56.703 | 3.619 | 33.161 | 12.860 | 29.366 |
| 4 | 175.257 | 5.345 | 5.793 | 17.393 | 9.068 | 120.702 | 3.656 | 25.008 | 9.388 | 20.858 |
| 5 | 158.226 | 7.715 | 4.421 | 22.529 | 3.183 | 55.297 | 3.837 | 26.773 | 10.411 | 22.003 |
| 6 | 65.619 | 7.090 | 16.486 | 18.969 | 18.515 | 24.305 | 9.324 | 52.194 | 14.423 | 45.134 |
| 7 | 117.937 | 8.183 | 6.404 | 22.222 | 5.5431 | 70.147 | 6.032 | 26.700 | 13.618 | 20.885 |
| 8 | 98.906 | 9.111 | 8.713 | 17.966 | 12.662 | 29.935 | 8.244 | 37.140 | 9.569 | 34.142 |
| 9 | 93.355 | 8.355 | 14.342 | 22.028 | 11.693 | 28.190 | 5.063 | 38.516 | 11.175 | 34.701 |
| 10 | 293.591 | 9.780 | 10.522 | 19.562 | 9.075 | 41.300 | 12.197 | 37.942 | 10.888 | 32.951 |
| 11 | 73.689 | 8.100 | 17.840 | 20.941 | 14.171 | 30.956 | 7.060 | 40.788 | 13.040 | 34.866 |
| 12 | 89.934 | 8.114 | 10.997 | 21.323 | 5.180 | 36.885 | 4.324 | 29.889 | 13.757 | 23.727 |
| 13 | 107.329 | 9.464 | 15.856 | 21.533 | 12.907 | 32.546 | 9.779 | 50.134 | 10.806 | 41.337 |
| 14 | 178.243 | 6.223 | 14.059 | 23.269 | 12.754 | 99.283 | 4.968 | 44.831 | 14.289 | 40.294 |
| 15 | 57.498 | 5.540 | -0.581 | 17.582 | 1.593 | 24.836 | 3.671 | 22.611 | 12.703 | 15.712 |
| 16 | 102.514 | 3.761 | 3.762 | 10.877 | 8.203 | 61.890 | 2.239 | 26.478 | 8.042 | 22.390 |

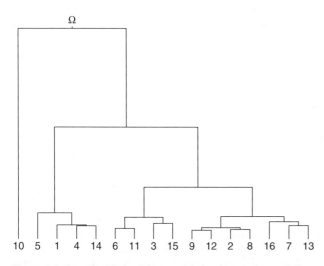

**Figure 5.9** Complete-link agglomerative dendogram (means) (Example 5.1).

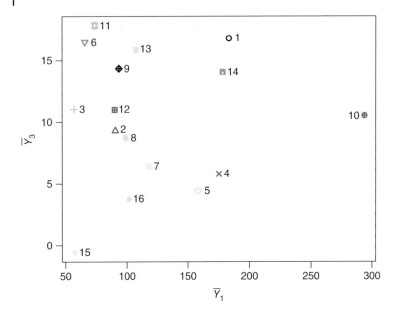

**Figure 5.10** Means $\bar{Y}_1 \times \bar{Y}_3$ (Example 5.2).

are similar comparisons between these airlines for $Y_2$ (see Table 5.4). However, these variances are not used in the tree construction here, only the means. On the other hand, both the means and the variances, through the internal variations, are taken into account when constructing the symbolic-based tree of Figure 5.8. In this tree, these particular airlines fall into three distinct parts (larger branches, left, middle, right) of the tree. □

**Example 5.2** Suppose now the airline data used in Example 5.1 were prepared as 5–95% quantile intervals, as shown in Table 5.5. Then, a symbolic partitioning on the 5–95% quantile intervals again using Euclidean distances produced the partition $P = (C_1, \ldots, C_5)$ with the $K = 5$ clusters, $C_1 = \{2, 3, 8, 9, 11, 12, 15\}$, $C_2 = \{4, 10\}$, $C_3 = \{1, 14\}$, $C_4 = \{6, 13\}$, and $C_5 = \{5, 7, 16\}$ (see section 6.3 for partitioning methods of interval data). This partition is superimposed on the bi-plot of the $\bar{Y}_1 \times \bar{Y}_3$ values in Figure 5.11(a) (with the uncircled observations forming the cluster $C_1$). A $k$-means partition on the means data of Table 5.4, shown in Figure 5.11(b) (with the un-circled observations forming the cluster $C_4$), for the $K = 5$ partition with $C_1 = \{15\}$, $C_2 = \{10\}$, $C_3 = \{3, 6, 11\}$, $C_4 = \{2, 7, 8, 9, 12, 13, 16\}$, and $C_5 = \{1, 4, 5, 14\}$, is different from the symbolic partitioning on the interval data. However, this partition on the means is consistent with the four major branches of the complete-link tree of Figure 5.9, which was also based on the means. The plot of Figure 5.10 suggests that $Y_1 =$ flight time is an important variable in

**Table 5.5** Flights data (5–95%) quantiles (Example 5.2)

| u | $Y_1$ | $Y_2$ | $Y_3$ | $Y_4$ | $Y_5$ | Number |
|---|---|---|---|---|---|---|
| 1 | [95, 339] | [3, 25] | [−24, 91] | [10, 43] | [−8, 79] | 6340 |
| 2 | [46, 128] | [2, 16] | [−28, 60] | [7, 41] | [−10, 55] | 5590 |
| 3 | [34, 87] | [3, 12] | [−19, 67] | [8, 42] | [−10, 57] | 2256 |
| 4 | [45, 462] | [2, 11] | [−22, 50] | [9, 34] | [−8, 45] | 5228 |
| 5 | [55, 219] | [4, 15] | [−24, 50] | [11, 44] | [−9, 41] | 726 |
| 6 | [39, 120] | [2, 21] | [−25, 120] | [7, 40] | [−10, 113] | 3175 |
| 7 | [36, 285] | [3, 18] | [−21, 54] | [10, 44] | [−8, 38] | 5969 |
| 8 | [38, 135] | [3, 22] | [−22, 74] | [7, 35] | [−6, 72] | 352 |
| 9 | [48, 135] | [4, 17] | [−23, 88] | [10, 44] | [−10, 85] | 1320 |
| 10 | [234, 365] | [3, 28] | [−30, 56] | [9, 35] | [−9, 48] | 665 |
| 11 | [36, 134] | [3, 20] | [−23, 92] | [8, 44] | [−11, 78] | 4877 |
| 12 | [43, 161] | [4, 15] | [−18, 55] | [9, 46] | [−10, 39] | 2294 |
| 13 | [85, 128] | [3, 26] | [−22, 115] | [10, 44] | [−10, 111] | 578 |
| 14 | [65, 350] | [3, 13] | [−23, 95] | [11, 46] | [−8, 86] | 2357 |
| 15 | [33, 100] | [2, 11] | [−23, 37] | [8, 36] | [−8, 28] | 4232 |
| 16 | [42, 230] | [2, 6] | [−20, 47] | [5, 20] | [0, 48] | 2550 |

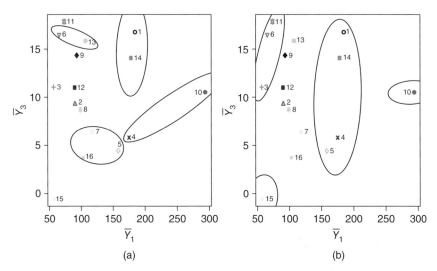

**Figure 5.11** Partitioning on airlines: (a) Euclidean distances and (b) *k*-means on means (Example 5.2).

the partitioning process. A $k$-means partition for $K = 5$ on the means data separates out airline $u = 15$ as a one-airline fifth partition. Figure 5.10 further suggests that $Y_3 =$ arrival-delay time is also a defining factor, at least for some airlines. Indeed, we see from the monothetic trees built on the histograms (and so using internal variations plus the means) that after flight time, it is this arrival-delay time that helps separate out the airlines. □

**Example 5.3**  In Figure 5.12, we have the pyramid built on the 5–95% quantile intervals of Table 5.5. This highlights four larger clusters that appear in the pyramid tree, specifically, $C_1 = \{5, 7, 14\}$, $C_2 = \{2, 3, 4, 5, 10, 12, 15, 16\}$, $C_3 = \{1, 10, 13\}$, and $C_4 = \{6, 8, 9, 11, 13\}$. These plots show the overlapping clusters characteristic of pyramids. For example, airline $u = 13$ is contained in both $C_3$ and $C_4$ clusters, whereas each airline is in its own distinct cluster from the other methodologies. Notice also that these three airlines in $C_3$ appear in different parts of the relevant trees and partition considered herein, e.g., in Figure 5.8 at the three-cluster stage, these airlines are in three different clusters. However, the tree in Figure 5.8 was based on the means only and ignored the additional information contained in the observation internal variations. □

**Example 5.4**  Figure 5.13 displays simulated individual classical observations drawn from bivariate normal distributions $N_2(\boldsymbol{\mu}, \boldsymbol{\Sigma})$. There are five samples each with $n = 100$ observations, and with $\boldsymbol{\mu}$ and $\boldsymbol{\Sigma}$ as shown in Table 5.6. Each of the samples can be aggregated to produce a histogram observation $\mathbf{Y}_u$, $u = 1, \ldots, 5$. When any of the divisive algorithms (see Chapter 7) are applied to these histogram data, three clusters emerge containing the observations $C_1 = \{1\}$, $C_2 = \{2, 3\}$, and $C_3 = \{4, 5\}$, respectively. In contrast, applying such algorithms to classical surrogates (such as the means), only two clusters emerge, namely, $C_1 = \{1\}$ and $C_2 = \{2, 3, 4, 5\}$, and the $k$-means algorithm on the individual

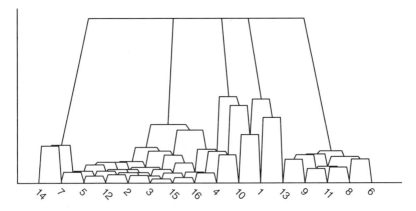

**Figure 5.12** Pyramid hierarchy on intervals (Example 5.3).

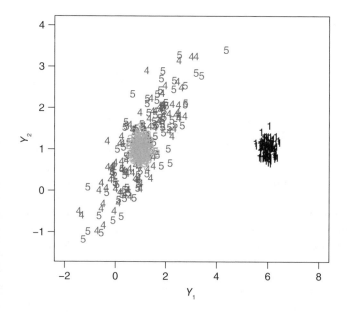

**Figure 5.13** Five simulated histograms, clusters identified (Example 5.4).

**Table 5.6** Simulated bivariate normal distribution parameters (Example 5.4)

| Sample | $\mu$ | $\Sigma$ |
|---|---|---|
| 1 | $(6, 1)$ | $\begin{pmatrix} 0.2 & 0 \\ 0 & 0.2 \end{pmatrix}$ |
| ....... | ..... | ............ |
| 2, 3 | $(1, 1)$ | $\begin{pmatrix} 0.2 & 0 \\ 0 & 0.2 \end{pmatrix}$ |
| ....... | ..... | ............ |
| 4, 5 | $(1, 1)$ | $\begin{pmatrix} 0.2 & 0.8 \\ 0.8 & 0.2 \end{pmatrix}$ |

observations gives the same two clusters. That is, the classical methods are unable to identify observations 2, ... , 5 as belonging to different clusters as the information on the internal variations is lost. □

## 5.5 Other Issues

The choice of an appropriate number of clusters is still unsolved for partitioning and hierarchical procedures. In the context of classical data, several researchers have tried to provide answers, e.g., Milligan and Cooper (1985), and Tibshirani and Walther (2005). One approach is to run the algorithm for several specific values of $K$ and then to compare the resulting approximate weight of evidence (AWE) suggested by Banfield and Raftery (1993) for hierarchies and Fraley and Raftery (1998) for mixture distribution model-based partitions. Kim and Billard (2011) extended the Dunn (1974) and Davis and Bouldin (1979) indices to develop quality indices in the hierarchical clustering context for histogram observations which can be used to identify a "best" value for $K$ (see section 7.4).

Questions around robustness and validation of obtained clusters are largely unanswered. Dubes and Jain (1979), Milligan and Cooper (1985), and Punj and Stewart (1983) provide some insights for classical data. Fräti et al. (2014) and Lisboa et al. (2013) consider these questions for the $k$-means approach to partitions. Hardy (2007) and Hardy and Baume (2007) consider the issue with respect to hierarchies for interval-valued observations.

Other topics not yet developed for symbolic data include, e.g., block modeling (see, e.g., Doreian et al. (2005) and Batagelj and Ferligoj (2000)), constrained or relational clustering where internal connectedness (such as geographical boundaries) is imposed (e.g., Ferligoj and Batagelj (1982, 1983, 1992)), rules-based algorithms along the lines developed by Reynolds et al. (2006) for classical data, and analyses for large temporal and spatial networks explored by Batagelj et al. (2014).

In a different direction, decision-based trees have been developed by a number of researchers, such as the classification and regression trees (CART) of Breiman et al. (1984) for classical data, in which a dependent variable is a function of predictor variables as in standard regression analysis. Limam (2005), Limam et al. (2004), Winsberg et al. (2006), Seck (2012), and Seck et al. (2010), have considered CART methods for symbolic data, but much more remains to be done.

In a nice review of model-based clustering for classical data, Fraley and Raftery (1998) introduce a partitioning approach by combining hierarchical clustering with the expectation-maximization algorithm of Dempster et al. (1977). A stochastic expectation-maximization algorithm was developed in Celeux and Diebolt (1985), a classification expectation-maximization algorithm was introduced by Celeux and Govaert (1992), and a Monte Carlo expectation-maximization algorithm was described by Wei and Tanner (1990). How these might be adapted to a symbolic-valued data setting are left as open problems.

Other exploratory methods include factor analysis and principal component analysis, both based on eigenvalues/eigenvectors of the correlation matrices,

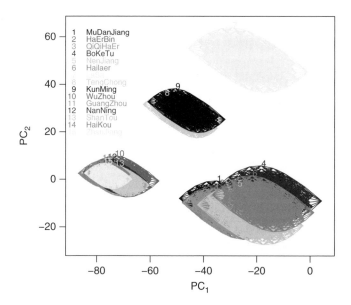

**Figure 5.14** Principal component analysis on China temperature data of Table 7.9.

or projection pursuit (e.g. Friedman, 1987). Just as different distance measures can produce different hierarchies, so can the different approaches (clustering *per se* compared to eigenvector-based methods) produce differing patterns. For example, in Chapter 7, Figure 7.6 shows the hierarchy that emerged when a monothethic divisive clustering approach (see section 7.2, Example 7.7) was applied to the temperature data set of Table 7.9. The resultant clusters are different from those obtained in Billard and Le-Rademacher (2012) when a principal component analysis was applied to those same data (see Figure 5.14).

# 6

# Partitioning Techniques

In Chapter 5, a brief introduction to the principles of techniques to partition a data set into distinct non-overlapping clusters was presented. In this chapter, how these partitions are obtained for symbolic data is studied.

Partitioning methodology is perhaps the most developed of all clustering techniques, at least for classical data, with many different approaches presented in the literature, starting from the initial and simplest approach based on the coordinate data by using variants of the $k$-means methods, or procedures based on dissimilarity/distance measures with the use of the $k$-medoids method, or their variations. This includes the more general dynamical partitioning of which the $k$-means and $k$-medoids methods are but special cases. Not all of these ideas have been extended to symbolic data, and certainly not to all types of symbolic data. Rather than trying to consider each technique where it may, or may not, have been applied to the different respective types of symbolic data, our approach in this chapter is to consider each different type of symbolic data in turn, and then to apply some of the partitioning methods to that data type where they exist. In some instances, we contrast different methods on the same type of symbolic data. For example, for interval data, we contrast the use of the simple $k$-means approach which uses only the interval means, with the more informative $k$-medoid (i.e., dynamical) approach using dissimilarities based on the entire interval between observations (see Examples 6.3–6.5)

Accordingly, multi-valued list observations are considered in section 6.2, interval-valued observations are considered in section 6.3, followed by histogram observations in section 6.4. In most methods developed thus far for symbolic data, it is assumed that all variables have the same type of observation. However, when the methodology is based on dissimilarity/distance measures, this is usually not a limitation. Hence, in section 6.5, an example of mixed-valued observations is considered. Then, in section 6.6, mixture distribution models are discussed. A short description of issues involved in calculating the final cluster representations is covered in section 6.7. Some other related issues and comments close out the chapter in section 6.8.

*Clustering Methodology for Symbolic Data*, First Edition. Lynne Billard and Edwin Diday.
© 2020 John Wiley & Sons Ltd. Published 2020 by John Wiley & Sons Ltd.

## 6.1 Basic Partitioning Concepts

Since most partitioning methods depend in some way on the $k$-means method or its variations, a brief description of the $k$-means method is first presented here, followed by a brief outline of the extension to the $k$-medoid and dynamical partitioning methods. Note that the $k$-means and $k$-medoids methods are in fact particular cases of the more general dynamical partitioning methods. That convergence occurs to some local optimal partition was shown in section 5.2.

The most commonly used partitioning procedure is the $k$-means method reviewed in section 5.2. Essentially, the method involves calculating dissimilarities/distances, usually Euclidean distances, between the mean of the observations for each cluster and the observations themselves. Most algorithms for this method use a version of that originally developed by MacQueen (1967), therefore we will describe this method in terms of the generic $k$-means algorithm. Then, adaptation to other forms, such as the $k$-medoids and dynamical partitions concepts, is briefly discussed.

We start with a set of observations $u = 1, \ldots, m$, in the sample space $\Omega$. The goal is to partition the $m$ observations in $\Omega$ into $K$ non-overlapping clusters, i.e., $\Omega \equiv (C_1, \ldots, C_K)$. For $k = 1, \ldots, K$, let the cluster $C_k$ contain observations $C_k = \{u_1, \ldots, u_{m_k}\}$, with $\sum_k m_k = m$. Let the mean of each cluster $C_k$ be represented by $\bar{\mathbf{Y}}^k = (\bar{Y}_1^k, \ldots, \bar{Y}_p^k), k = 1, \ldots, K$, where $\bar{Y}_j^k$ is the mean of the observations in $C_k$ for variable $Y_j, j = 1, \ldots, p$. The actual calculation of the mean varies depending on the type of data (see section 2.4.1). Thus, for example, if the observations are interval-valued with $Y_{uj} = [a_{uj}, b_{uj}], j = 1, \ldots, p, u = 1, \ldots, m$, then the mean of the observations in $C_k$ is, for $k = 1, \ldots, K$, from Eq. (2.4.2),

$$\bar{\mathbf{Y}}^k = (\bar{Y}_j^k), \quad \bar{Y}_j^k = \frac{1}{2m_k} \sum_{u=u_1}^{u_{m_k}} (a_{uj} + b_{uj}), \quad j = 1, \ldots, p. \tag{6.1.1}$$

The corresponding classical point observations are simply the means of each of the symbolic-valued observations. Thus, for example, for interval-valued observations, each $Y_{uj}$ value is replaced by its own mean $\bar{Y}_{uj}$ where, from Eq. (2.4.2),

$$\bar{Y}_{uj} = (a_{uj} + b_{uj})/2, \, j = 1, \ldots, p, \, u = 1, \ldots, m. \tag{6.1.2}$$

The Euclidean distance between an observation $\bar{\mathbf{Y}}_u = (\bar{Y}_{u1}, \ldots, \bar{Y}_{up})$ and the mean $\bar{\mathbf{Y}}^k$ of cluster $C_k$ is given by

$$E_u = \left[ \sum_{j=1}^p (\bar{Y}_{uj} - \bar{Y}_j^k)^2 \right]^{1/2}. \tag{6.1.3}$$

At the first iteration, the $K$ seeds constitute the $K$ means. Observations are successively moved into the cluster whose mean is closest, i.e., for each given observation $u$, the distance between the observation $u$ and the mean $\bar{\mathbf{Y}}^k$ is smallest. Then, the new cluster mean is calculated. The process is repeated, moving each observation from its current cluster into another cluster for which that observation is now closest to that mean. This is repeated until there is no more movement between clusters.

The within-cluster squared distance for cluster $C_k$ becomes

$$E^k = \sum_{u=u_1}^{u_{m_k}} \left[ \sum_{j=1}^{p} (\bar{Y}_{uj} - \bar{Y}_j^k)^2 \right]^{1/2}, \quad k = 1, \ldots, K, \tag{6.1.4}$$

where $\bar{Y}_j^k$ and $\bar{Y}_{uj}$ are as defined in Eqs. (6.1.1) and (6.1.2), respectively. The goal is that the total within-cluster distance $E = \sum_{k=1}^{K} E^k$ is minimized, for a given $K$. Formally, the essential steps of the basic $k$-means algorithm are as follows:

**Step 1:** Initialization.
Select $K$ initial clusters $(C_1^{(0)}, \ldots, C_K^{(0)}) \equiv P^{(0)}$.

**Step 2:** Representation.
Calculate the mean values $\bar{\mathbf{Y}}^k$(such as Eq. (6.1.1) for interval data) for the clusters $C_k^{(i)}, k = 1, \ldots, K$, at the $i$th iteration, $i = 0, 1, \ldots$.

**Step 3:** Allocation.
Calculate the distance $E_u$ from Eq. (6.1.3) for each $u = u_1, \ldots, u_{m_k}$ in each $C_k^{(i)}, k = 1, \ldots, K$. Move observation $u$ into that cluster $C_k^{(i+1)}$ for which this distance $E_u$ is smallest, i.e., reallocate observation $u$ into the cluster whose mean is closest to $u$ itself. If no observations are reallocated to a new cluster, i.e., if $C_k^{(i)} \equiv C_k^{(i+1)}$, for all $k = 1, \ldots, K$, go to step 4. Else return to step 2.

**Step 4:** Stopping rule.
Calculate the total within-cluster distance $E = \sum_{k=1}^{K} E^k$ from Eq. (6.1.4). If $E$ is less than some preassigned minimum value or if no observations are being reallocated to produce a different partition, then stop.

This algorithm is available in SAS ("proc fastclus" procedure) and the R-package ("kmeans" function) for classical data. Although most programs use the minimum within-cluster squared distances of Eq. (6.1.3) (as do we in the following), other criteria exist. These include maximizing $\mathbf{T}/\mathbf{W}$, maximizing the sum of the diagonal elements of $\mathbf{W}^{-1}\mathbf{B}$, and maximizing the largest eigenvalue of $\mathbf{W}^{-1}\mathbf{B}$, where $\mathbf{T}$ is the total scatter matrix of all observations, $\mathbf{B}$ is the between-clusters scatter matrix, and $\mathbf{W}$ is the total within-clusters scatter matrix. These criteria are compared in Friedman and Rubin (1967).

The partitioning around medoids (PAM) method, a term first coined by Kaufman and Rousseeuw (1987, 1990) but developed in Diday (1979), is also called the $k$-medoids method, and is similar to the $k$-means method except that instead of the actual observation values in $\mathbb{R}^p$, only distances, or dissimilarities, between observations in the data set are known. In this case, so-called "medoids" are the representative observations in a cluster (instead of the mean values in the $k$-means method), and unlike the $k$-means method where the means are not necessarily equal to any particular observation, a medoid always corresponds to a specific observation within a cluster. However, some variants use the median observation as the medoid of a cluster; in the following, we use the medoid as a particular observation in $C_k$, $k = 1, \ldots, K$. Since this method uses distances or dissimilarities between observations, though originally developed for classical point data it can also be used for symbolic data by using the distances/dissimilarities between the symbolic observations. Thus, inherent advantages or disadvantages of the PAM method apply equally to symbolic data analyses.

The goal is to find those $K$ clusters that minimize the total within distances

$$D = \sum_{k=1}^{K} D^k, \quad D^k = \sum_{u_2=1}^{m_k} \sum_{\substack{u_1=1 \\ u_1 > u_2}}^{m_k} d(u_1, u_2), \tag{6.1.5}$$

where $d(u_1, u_2)$ is the distance, or dissimilarity, between any two observations $u_1$ and $u_2$ in the cluster $C_k = \{u_1, \ldots, u_{m_k}\}$. Notice that $D^k$ in Eq. (6.1.5) and hence $D$ do not contain distances/dissimilarities between observations in different clusters, only the $d(u_1, u_2)$ within each $C_k$, $k = 1, \ldots, K$. The distances/dissimilarities are calculated according to the procedures of Chapters 3 and 4 for non-modal and modal valued observations, respectively.

The PAM algorithm begins with the selection of $K$ seeds corresponding to $K$ observations. Then, successively, each observation is assigned to a cluster according to which medoid that observation is closest. This algorithm can be found in SAS ("proc modeclus" procedure) and the R-package ("pam" function, or the variant "clara" function).

Usually, the medoid value is recalculated after all observations have been allocated or reallocated to a new cluster, at the end of a complete pass. These are called non-adaptive methods. If, as for some algorithms, the medoid values are recalculated after each observation is reallocated to its new cluster, the method is considered to be an adaptive process. For example, an adaptive dynamic clusters algorithm (an expanded version of the $k$-means partitioning method) is in Diday (1974) and Diday and Simon (1976).

The means of the $k$-means clusters are particular cases of the more general "prototype" or "centroid" values used to represent the clusters (step 2), i.e., for the pure $k$-means algorithm, the centroid/prototype is exactly the cluster means. However, when applying the more general dynamical partitioning method (see section 5.2), the "mean" (of step 2) may be some other "representation function," such as a probability density function, a feature such as the regression equation approach, or the like. Thus, for example, when using a regression-based method, the representation function (see section 5.2) $L_k$, $k = 1, \dots, K$, is a regression measure, instead of a mean, or, more generally a canonical analysis, a factorial axis, a distance, a functional curve, points of a population, and so on.

Likewise, rather than the particular total within-cluster distance $E$, the allocation/reallocation criteria of steps 3 and 4 may be a general function $W(\mathbf{L}^{(i)}, \mathbf{P}^{(i)})$ for a given partition $\mathbf{P}^{(i)} = (C_1^{(i)}, \dots, C_K^{(i)})$ with given centroids or prototypes representation $\mathbf{L}^{(i)}$ at the $i$th iteration, e.g., a likelihood function, as in section 6.6. That convergence occurs was proved in section 5.2. In the following, we will tend to use the term "centroid" to represent the $\mathbf{L}^{(i)}$ for the $i$th iteration. Thus, for example, in a pure $k$-means algorithm, the cluster centroids are the means (variously defined depending on the type of data), in the $k$-medoids method, the centroids would be the cluster medoids, in the Chavent and Lechevallier (2002) method for intervals in Example 6.6, medians are used, and so on.

## 6.2 Multi-valued List Observations

We first extend the partitioning methods for classical data to multi-valued list observations, primarily through illustrative examples. We recall that multi-valued list data were defined in Eqs. (2.2.1) and (2.2.2). Thus, for modal multi-valued list data, realizations are of the form

$$\mathbf{Y}_u = (\{Y_{ujk_j}, p_{ujk_j};\ k_j = 1, \dots, s_{uj}\},\ j = 1, \dots, p),\ \sum_{k_j=1}^{s_{uj}} p_{ujk_j} = 1,\ u = 1, \dots, m.$$

$$(6.2.1)$$

and non-modal multi-valued list observations are special cases of Eq. (6.2.1) with $p_{ujk_j} = 1/s_{uj}$ for observed categories and $p_{ujk_j} = 0$ for unobserved categories, from $\mathcal{Y}_j = \{Y_{j1}, \dots, Y_{js_j}\}, j = 1, \dots, p$ (see Definitions 2.1 and 2.2). These are lists of categories, i.e., they are qualitative values from a set of possible categorical values $\mathcal{Y} = \mathcal{Y}_1 \times \cdots \times \mathcal{Y}_p$. Except when a measure of (for example) a sample mean of list multi-valued observations can be found, the $k$-means method cannot be applied to these kinds of data. However, the $k$-medoid method can be applied to the corresponding distances/dissimilarities, as illustrated in Example 6.1.

**Example 6.1** Consider the data of Table 6.1, which have multi-valued list values for $m = 12$ counties and $p = 3$ variables with household characteristics as recorded in the US 2000 Census. Here, $Y_1$ = fuel type used in households takes possible values $\mathcal{Y}_1$ = {gas, electric, oil, coal/wood, none}, $Y_2$ = number of rooms in the house with possible values $\mathcal{Y}_2$ = {(1, 2), (3, 4, 5), ≥ 6}, and $Y_3$ = household income in 1000s of US\$ with income ranges of $\mathcal{Y}_3$ = {< 10, [10, 25), [25, 50), [50, 75), [75, 100), [100, 150), [150, 200), [≥ 200]}.

**Table 6.1** Household characteristics (Example 6.1)

| County | $Y_1$ = Fuel types | | | | | $Y_2$ = Number of rooms | | |
|---|---|---|---|---|---|---|---|---|
| $u$ | Gas | Electric | Oil | Wood | None | {1, 2} | {3, 4, 5} | {≥ 6} |
| 1 | 0.687 | 0.134 | 0.166 | 0.011 | 0.002 | 0.054 | 0.423 | 0.523 |
| 2 | 0.639 | 0.088 | 0.137 | 0.135 | 0.001 | 0.069 | 0.424 | 0.507 |
| 3 | 0.446 | 0.107 | 0.406 | 0.025 | 0.016 | 0.178 | 0.693 | 0.129 |
| 4 | 0.144 | 0.808 | 0.013 | 0.031 | 0.004 | 0.045 | 0.508 | 0.447 |
| 5 | 0.021 | 0.914 | 0.013 | 0.041 | 0.011 | 0.027 | 0.576 | 0.397 |
| 6 | 0.263 | 0.713 | 0.005 | 0.019 | 0.000 | 0.041 | 0.495 | 0.464 |
| 7 | 0.454 | 0.526 | 0.004 | 0.012 | 0.004 | 0.039 | 0.522 | 0.439 |
| 8 | 0.748 | 0.236 | 0.001 | 0.007 | 0.008 | 0.153 | 0.473 | 0.374 |
| 9 | 0.063 | 0.087 | 0.083 | 0.755 | 0.012 | 0.272 | 0.486 | 0.242 |
| 10 | 0.351 | 0.277 | 0.007 | 0.365 | 0.001 | 0.052 | 0.523 | 0.425 |
| 11 | 0.624 | 0.237 | 0.004 | 0.133 | 0.002 | 0.089 | 0.549 | 0.362 |
| 12 | 0.165 | 0.287 | 0.027 | 0.518 | 0.003 | 0.039 | 0.517 | 0.444 |

| County | $Y_3$ = Household income (US\$1000) | | | | | | | |
|---|---|---|---|---|---|---|---|---|
| $u$ | < 10 | [10, 25) | [25, 50) | [50, 75) | [75, 100) | [100, 150) | [150, 200) | ≥ 200 |
| 1 | 0.093 | 0.185 | 0.287 | 0.200 | 0.109 | 0.083 | 0.024 | 0.019 |
| 2 | 0.121 | 0.269 | 0.338 | 0.165 | 0.065 | 0.029 | 0.005 | 0.008 |
| 3 | 0.236 | 0.224 | 0.272 | 0.142 | 0.065 | 0.042 | 0.010 | 0.009 |
| 4 | 0.167 | 0.262 | 0.285 | 0.164 | 0.068 | 0.038 | 0.009 | 0.007 |
| 5 | 0.200 | 0.280 | 0.322 | 0.114 | 0.050 | 0.021 | 0.007 | 0.006 |
| 6 | 0.187 | 0.288 | 0.297 | 0.159 | 0.022 | 0.029 | 0.007 | 0.011 |
| 7 | 0.136 | 0.217 | 0.318 | 0.169 | 0.086 | 0.046 | 0.012 | 0.016 |
| 8 | 0.079 | 0.133 | 0.233 | 0.198 | 0.135 | 0.135 | 0.047 | 0.040 |
| 9 | 0.124 | 0.167 | 0.289 | 0.230 | 0.093 | 0.057 | 0.004 | 0.036 |
| 10 | 0.074 | 0.200 | 0.299 | 0.197 | 0.122 | 0.067 | 0.021 | 0.020 |
| 11 | 0.128 | 0.268 | 0.303 | 0.162 | 0.071 | 0.043 | 0.010 | 0.014 |
| 12 | 0.097 | 0.196 | 0.312 | 0.190 | 0.103 | 0.067 | 0.017 | 0.018 |

Thus, there are $s_1 = 5$, $s_2 = 3$, and $s_3 = 8$, categories, respectively, for each variable. Let us take $Y_1$ and $Y_2$ only, and let us use the extended multi-valued list Euclidean Ichino–Yaguchi distances calculated from Eq. (4.1.15). Those for $\gamma = 0.25$ give the distance matrix shown in Table 6.2.

Suppose the observations $u = 1$, $u = 4$, $u = 8$, and $u = 12$ are taken as $K = 4$ initial seeds or medoids. Then, from Table 6.2, we see that the respective distances for each observation from these seeds are as shown in Table 6.3 for the first iteration (column (i)). For example, the extended Euclidean Ichino–Yaguchi distance from the first seed to the observation $u = 5$ is $d(1, 5) = 1.250$, likewise for the other three seeds, $d(4, 5) = 0.211$,

**Table 6.2** Extended Euclidean Ichino–Yaguchi distances (Example 6.1)

| u | 1 | 2 | 3 | 4 | 5 | 6 | 7 | 8 | 9 | 10 | 11 | 12 |
|---|---|---|---|---|---|---|---|---|---|---|---|---|
| 1 | 0 | 0.188 | 0.715 | 1.052 | 1.250 | 0.887 | 0.611 | 0.338 | 1.207 | 0.760 | 0.415 | 0.313 |
| 2 | 0.188 | 0 | 0.727 | 1.092 | 1.275 | 0.944 | 0.678 | 0.443 | 1.027 | 0.645 | 0.313 | 0.887 |
| 3 | 0.715 | 0.727 | 0 | 1.163 | 1.298 | 1.039 | 0.791 | 0.744 | 1.138 | 0.884 | 0.715 | 1.115 |
| 4 | 1.052 | 1.092 | 1.163 | 0 | 0.211 | 0.180 | 0.465 | 0.926 | 1.250 | 0.811 | 0.882 | 0.783 |
| 5 | 1.250 | 1.275 | 1.298 | 0.211 | 0 | 0.383 | 0.655 | 1.107 | 1.294 | 0.983 | 1.047 | 0.957 |
| 6 | 0.887 | 0.944 | 1.039 | 0.180 | 0.383 | 0 | 0.295 | 0.758 | 1.287 | 0.657 | 0.732 | 0.787 |
| 7 | 0.611 | 0.678 | 0.791 | 0.465 | 0.655 | 0.295 | 0 | 0.479 | 1.293 | 0.534 | 0.452 | 0.794 |
| 8 | 0.338 | 0.443 | 0.744 | 0.926 | 1.107 | 0.758 | 0.479 | 0 | 1.267 | 0.625 | 0.226 | 0.898 |
| 9 | 1.207 | 1.027 | 1.138 | 1.250 | 1.294 | 1.287 | 1.293 | 1.267 | 0 | 0.789 | 1.101 | 0.572 |
| 10 | 0.760 | 0.645 | 0.884 | 0.811 | 0.983 | 0.657 | 0.534 | 0.625 | 0.789 | 0 | 0.422 | 0.280 |
| 11 | 0.415 | 0.313 | 0.715 | 0.883 | 1.047 | 0.732 | 0.452 | 0.226 | 1.101 | 0.422 | 0 | 0.699 |
| 12 | 0.313 | 0.887 | 1.115 | 0.783 | 0.957 | 0.787 | 0.794 | 0.898 | 0.572 | 0.280 | 0.699 | 0 |

**Table 6.3** Extended Euclidean Ichino–Yaguchi distances to medoids (Example 6.1)

| House | (i) Medoids: first iteration | | | | House | (ii) Medoids: second iteration | | | |
|---|---|---|---|---|---|---|---|---|---|
| u | u = 1 | u = 4 | u = 8 | u = 12 | u | u = 1 | u = 4 | u = 7 | u = 12 |
| 2 | **0.188** | 1.092 | 0.443 | 0.887 | 2 | **0.188** | 1.092 | 0.678 | 0.887 |
| 3 | **0.715** | 1.163 | 0.744 | 1.115 | 3 | **0.715** | 1.163 | 0.791 | 1.115 |
| 5 | 1.250 | **0.211** | 1.107 | 0.957 | 5 | 1.250 | **0.211** | 0.655 | 0.957 |
| 6 | 0.887 | **0.180** | 0.758 | 0.787 | 6 | 0.887 | **0.180** | 0.295 | 0.787 |
| 7 | 0.611 | **0.465** | 0.479 | 0.794 | 8 | **0.338** | 0.926 | 0.479 | 0.898 |
| 9 | 1.207 | 1.250 | 1.267 | **0.572** | 9 | 1.207 | 1.250 | 1.293 | **0.572** |
| 10 | 0.760 | 0.811 | 0.625 | **0.280** | 10 | 0.760 | 0.811 | 0.534 | **0.280** |
| 11 | 0.415 | 0.883 | **0.226** | 0.699 | 11 | **0.415** | 0.883 | 0.452 | 0.699 |

$d(8, 5) = 1.107$, and $d(12, 5) = 0.957$. Since observation $u = 5$ is closest to the seed $u = 4$, this observation is put into the same cluster as the seed $u = 4$. After placing all observations into the cluster which minimizes the distances from the initial seeds, we see that the initial partition $P^{(0)} = (C_1^{(0)}, \ldots, C_4^{(0)})$ has clusters $C_1^{(0)} = \{1, 2, 3\}$, $C_2^{(0)} = \{4, 5, 6, 7\}$, $C_3^{(0)} = \{9, 10, 12\}$, and $C_4^{(0)} = \{8, 11\}$. From Eq. (6.1.5), the within-cluster distances are $D^1 = 0.188 + 0.715 = 0.903$, $D^2 = 0.856$, $D^3 = 0.852$, and $D^4 = 0.226$, and hence the total within-cluster distance is $D = 2.837 \equiv D_{(1)}$, say.

Now suppose the medoids are taken to be the observations $u = 1$, $u = 4$, $u = 7$, and $u = 12$, i.e., $u = 8$ is replaced by $u = 7$. The corresponding distances are shown in Table 6.3 (column (ii)) with the closest medoid to each observation highlighted. Now, the new partition is $P^{(1)} = (C_1^{(1)}, \ldots, C_4^{(1)})$ with clusters $C_1^{(1)} = \{1, 2, 3, 8, 11\}$, $C_2^{(1)} = \{4, 5, 6\}$, $C_3^{(1)} = \{7\}$, and $C_4^{(1)} = \{9, 10, 12\}$. The within-cluster distances are $D^1 = 1.656$, $D^2 = 0.391$, $D^3 = 0$, and $D^4 = 0.852$, respectively, and hence the total within-cluster distance is $D = 2.899 \equiv D_{(2)}$. Since $D_{(1)} < D_{(2)}$, we conclude that of these two medoid choices, the first was better. The iteration continues successively swapping out specific observations until the optimal total within-cluster distance $D$ is obtained. In this case, the initial choice is the best.

Thus, using the $k$-medoids method to this distance matrix produces the four clusters containing, respectively, the counties $C_1 = \{1, 2, 3\}, C_2 = \{4, 5, 6, 7\}, C_3 = \{9, 10, 12\}$, and $C_4 = \{8, 11\}$. Indeed, the counties $u = 4, 5, 6$, and 7 are rural counties in the US South region, counties $u = 1, 2$, and 3 are in the North-East with $u = 3$ being a heavily urban area, and the remaining counties $u = 8, \ldots, 12$, are in the West.

Had we used the normalized extended multi-valued list Euclidean Ichino–Yaguchi distances calculated from Eqs. (4.1.15) and (3.1.9), displayed in Table 6.4, the same partition emerges (see Exercise 6.1). □

In a different direction, Korenjak-Černe et al. (2011) and Batagelj et al. (2015) developed an adapted leader's algorithm for partitioning for variables with a discrete distribution. For each cluster $C_k$, in the partition $P_K = (C_1, \ldots, C_K)$, there is a leader observation $T_k$ calculated by applying Eq. (2.4.1). Then, the cluster error, $E(C_k)$, with respect to $T_k$, is

$$E(C_k) = \sum_{u=1}^{m_k} d(u, T_k), \qquad (6.2.2)$$

where $d(u, T_k)$ is the distance between observations $u = 1, \ldots, m_k$ and $T_k$ in $C_k$, $k = 1, \ldots, K$. The algorithm is based on Euclidean distances and so is in effect also an adapted Ward method (see Chapter 8). The leaders $T_k$ are those that

**Table 6.4** Normalized extended Euclidean Ichino–Yaguchi distances (Example 6.1)

| $u$ | 1 | 2 | 3 | 4 | 5 | 6 | 7 | 8 | 9 | 10 | 11 | 12 |
|---|---|---|---|---|---|---|---|---|---|---|---|---|
| 1 | 0 | 0.133 | 0.505 | 0.744 | 0.884 | 0.627 | 0.432 | 0.239 | 0.853 | 0.537 | 0.293 | 0.221 |
| 2 | 0.133 | 0 | 0.514 | 0.772 | 0.901 | 0.667 | 0.479 | 0.314 | 0.726 | 0.456 | 0.221 | 0.627 |
| 3 | 0.505 | 0.514 | 0 | 0.822 | 0.918 | 0.734 | 0.560 | 0.526 | 0.805 | 0.625 | 0.506 | 0.788 |
| 4 | 0.744 | 0.772 | 0.822 | 0 | 0.149 | 0.128 | 0.329 | 0.655 | 0.884 | 0.574 | 0.624 | 0.554 |
| 5 | 0.884 | 0.901 | 0.918 | 0.149 | 0 | 0.271 | 0.463 | 0.783 | 0.915 | 0.695 | 0.740 | 0.676 |
| 6 | 0.627 | 0.667 | 0.734 | 0.128 | 0.271 | 0 | 0.209 | 0.536 | 0.910 | 0.465 | 0.517 | 0.556 |
| 7 | 0.432 | 0.479 | 0.560 | 0.329 | 0.463 | 0.209 | 0 | 0.338 | 0.914 | 0.377 | 0.319 | 0.561 |
| 8 | 0.239 | 0.314 | 0.526 | 0.655 | 0.783 | 0.536 | 0.338 | 0 | 0.896 | 0.442 | 0.160 | 0.635 |
| 9 | 0.853 | 0.726 | 0.805 | 0.884 | 0.915 | 0.910 | 0.914 | 0.896 | 0 | 0.558 | 0.779 | 0.405 |
| 10 | 0.537 | 0.456 | 0.625 | 0.574 | 0.695 | 0.465 | 0.377 | 0.442 | 0.558 | 0 | 0.299 | 0.198 |
| 11 | 0.293 | 0.221 | 0.506 | 0.624 | 0.740 | 0.517 | 0.319 | 0.160 | 0.779 | 0.299 | 0 | 0.495 |
| 12 | 0.221 | 0.627 | 0.788 | 0.554 | 0.676 | 0.556 | 0.561 | 0.635 | 0.405 | 0.198 | 0.495 | 0 |

satisfy

$$E_k(C_k) = \min_T \sum_{u=1}^{m_k} d(u, T). \qquad (6.2.3)$$

The dissimilarity for variable $Y_j$ is given by

$$d_j(u, T) = \sum_{k_j=1}^{s_j} w_{k_j} \delta(p_{ujk_j}, p_{Tjk_j}) \qquad (6.2.4)$$

where $w_{k_j}$ is the weight for the $k_j$th category $Y_{jk_j}$ of variable $Y_j$. While Korenjak-Černe et al. (2011) consider many dissimilarity measures, we limit attention to the squared Euclidean distance

$$d_j(u, T) = \frac{1}{2} \sum_{k_j=1}^{s_j} w_{k_j} (p_{ujk_j} - p_{Tjk_j})^2. \qquad (6.2.5)$$

For all variables, the dissimilarity is

$$d(u, T) = \sum_{j=1}^{p} \alpha_j d_j(u, T) \qquad (6.2.6)$$

where $\alpha_j \geq 0$ is the weight associated with variable $Y_j, j = 1, \ldots, p$, and $\sum_j \alpha_j = 1$. When variables are assumed to be of equal importance, then $\alpha_j = 1/p$. This is illustrated in the following example.

**Example 6.2** Consider the same household characteristics data of Table 6.1, for the two multi-valued list variables $Y_1$ = fuel types and $Y_2$ = number of bedrooms in the house for $m = 12$ counties. Suppose $K = 3$.

Let the initial partition be $P^{(0)} = (C_1^{(0)}, C_2^{(0)}, C_3^{(0)})$ with $C_1^{(0)} = \{1, 2, 3, 4\}$, $C_2^{(0)} = \{5, 6, 7, 8\}$, and $C_3^{(0)} = \{9, 10, 11, 12\}$. First, we calculate the leader $T_k$ for these clusters $C_k^{(0)}$, $k = 1, 2, 3$. From Eq. (2.4.1), the leader probabilities $p_{Tjk_j}$ are as given in Table 6.5. For example, in cluster $C_1$, the leader probability for the first category (gas) of $Y_1$ is $p_{T11} = (0.687 + 0.639 + 0.446 + 0.144)/4 = 0.479$.

From Eq. (6.2.5), we obtain the distance between each observation $u = 1, \ldots, m$, and each leader $T_k$, $k = 1, 2, 3$. We assume the weights $w_{kj} \equiv 1$. These distances are summarized in Table 6.6. For example, when $u = 1$ and

Table 6.5 Initial cluster leaders: households (Example 6.2)

| Leader | $Y_1$ = Fuel types | | | | | $Y_2$ = Number of rooms | | |
|--------|------|----------|-----|------|------|--------|----------|-----|
| $k$ | Gas | Electric | Oil | Wood | None | $\{1, 2\}$ | $\{3, 4, 5\}$ | $\geq 6$ |
| $T_1$ | 0.479 | 0.284 | 0.181 | 0.050 | 0.006 | 0.087 | 0.512 | 0.401 |
| $T_2$ | 0.371 | 0.597 | 0.006 | 0.020 | 0.006 | 0.065 | 0.517 | 0.418 |
| $T_3$ | 0.301 | 0.222 | 0.030 | 0.443 | 0.004 | 0.113 | 0.519 | 0.368 |

Table 6.6 Leader distances (Example 6.2)

| County | Distances from leader $T_k$ in $C_k$ | | | |
|--------|-----------|-----------|-----------|------------------|
| $u$ | $D_{T_1}$ | $D_{T_2}$ | $D_{T_3}$ | $\min_k(D_{T_k})$ |
| 1 | 0.023 | 0.090 | 0.100 | 0.023 |
| 2 | 0.023 | 0.094 | 0.067 | 0.023 |
| 3 | 0.050 | 0.133 | 0.111 | 0.050 |
| 4 | 0.105 | 0.024 | 0.137 | 0.024 |
| 5 | 0.161 | 0.057 | 0.183 | 0.057 |
| 6 | 0.067 | 0.007 | 0.109 | 0.007 |
| 7 | 0.024 | 0.003 | 0.078 | 0.003 |
| 8 | 0.029 | 0.071 | 0.099 | 0.029 |
| 9 | 0.195 | 0.244 | 0.054 | 0.054 |
| 10 | 0.037 | 0.056 | 0.005 | 0.005 |
| 11 | 0.016 | 0.053 | 0.051 | 0.016 |
| 12 | 0.086 | 0.097 | 0.010 | 0.010 |

**Table 6.7** Within-cluster variations (Example 6.2)

| | Cluster $C_k$ | | |
|---|---|---|---|
| Variable | $C_1$ | $C_2$ | $C_3$ |
| | | First iteration | |
| $Y_1$ | 0.320 | 0.267 | 0.208 |
| $Y_2$ | 0.081 | 0.011 | 0.031 |
| $(Y_1, Y_2)$ | 0.201 | 0.139 | 0.120 |
| | | Last iteration | |
| $Y_1$ | 0.099 | 0.092 | 0.074 |
| $Y_2$ | 0.081 | 0.003 | 0.030 |
| $(Y_1, Y_2)$ | 0.090 | 0.048 | 0.052 |

$k = 1$, from Eq. (6.2.5), we have

for $Y_1$, $\quad d_1(1, T_1) = [(0.687 - 0.479)^2 + \cdots + (0.002 - 0.006)^2]/2 = 0.0338$,

for $Y_2$, $\quad d_2(1, T_1) = [(0.054 - 0.087)^2 + \cdots + (0.523 - 0.401)^2]/2 = 0.0119$.

Hence, from Eq. (6.2.6), $d(1, T_1) = [0.0338 + 0.0119]/2 = 0.02228$.

Now, each observation is reallocated to its new cluster based on these distances. Table 6.6 also shows the minimum distance between each observation and each leader. Thus, for example, observation $u = 4$ is closest to the leader $T_2$, so this observation is moved into cluster $C_2$. After reallocation for all observations, the partitioning becomes $P^{(1)} = (C_1^{(1)}, C_2^{(1)}, C_3^{(1)})$ with $C_1^{(1)} = \{1, 2, 3, 8, 11\}$, $C_2^{(1)} = \{4, 5, 6, 7\}$, and $C_3^{(1)} = \{9, 10, 12\}$.

This process is repeated until no observations are reallocated. In this case, the partition obtained after the first iteration is the final partition.

The within-cluster variations can be calculated from Eq. (6.2.2) for each partition. These are displayed in Table 6.7 for the initial partition and for the final partition. For the initial partition, the total within-cluster variation is, from Table 6.7, $0.201 + 0.139 + 0.120 = 0.460$, in contrast to the final total within-cluster variation of $0.090 + 0.048 + 0.052 = 0.190$. $\qquad \square$

## 6.3 Interval-valued Data

Let us now consider partitioning of interval-valued observations, primarily through illustrative examples. Again, we seek a partition $P = (C_1, \ldots, C_K)$. We recall that from Definition 2.3 and Eq. (2.2.3) we have that an interval-valued observation $u$ has realization

$$\mathbf{Y}_u = ([a_{u1}, b_{u1}], \dots, [a_{up}, b_{up}]), \quad u = 1, \dots, m. \tag{6.3.1}$$

The observation mean $\bar{Y}_u$ is given by Eq. (6.1.2), $u = 1, \dots, m$, and the cluster $C_k = \{u_1, \dots, u_{m_k}\}$ has cluster mean given by Eq. (6.1.1), $k = 1, \dots, K$.

For interval data, the $k$-means method has to be adapted somewhat as it is based on classical coordinate values. We show in Example 6.3 the results obtained when the intervals are replaced by their sample means, usually the midpoints. Clearly, this partitioning fails to account for internal variations within the data intervals. One approach to include the internal information is to base the partitioning procedure on dissimilarities/distances between symbolic valued observations. The PAM algorithm is one method to achieve this. Therefore, in Example 6.5, dissimilarities are used to generate the appropriate partitions. Note that unlike the $k$-means method, which is limited to means, the PAM method allows for the choice of any useful dissimilarity between observations.

**Example 6.3** Take the China temperature and elevation data means of Table 6.8, which gives the midpoint values for each variable and each observation for the interval data of Table 3.11, at each of $m = 10$ stations, obtained from Eq. (6.1.2). Here, $Y_1 =$ January temperature, $Y_2 =$ April temperature, $Y_3 =$ July temperature, $Y_4 =$ October temperature, and $Y_5 =$ elevation. Both non-standardized means (a) and standardized means (b) are shown.

**Table 6.8** Means for China data (Examples 6.3 and 6.4)

| | | (a) Non-standardized means | | | | | (b) Standardized means | | | | |
|---|---|---|---|---|---|---|---|---|---|---|---|
| $u$ | Station | $\bar{Y}_{u1}$ | $\bar{Y}_{u2}$ | $\bar{Y}_{u3}$ | $\bar{Y}_{u4}$ | $\bar{Y}_{u5}$ | $\bar{Y}_{u1}$ | $\bar{Y}_{u2}$ | $\bar{Y}_{u3}$ | $\bar{Y}_{u4}$ | $\bar{Y}_{u5}$ |
| 1 | MuDanJia | −12.95 | 6.55 | 21.75 | 6.85 | 4.82 | −0.816 | −0.817 | −0.228 | −0.781 | −0.537 |
| 2 | HaErBin | −14.80 | 6.05 | 22.50 | 6.15 | 3.44 | −0.938 | −0.888 | −0.069 | −0.862 | −0.595 |
| 3 | BoKeTu | −19.45 | 2.50 | 17.95 | 2.45 | 14.78 | −1.244 | −1.385 | −1.033 | −1.293 | −0.117 |
| 4 | NenJiang | −21.95 | 5.25 | 20.55 | 4.15 | 4.84 | −1.409 | −1.000 | −0.482 | −1.095 | −0.536 |
| 5 | LaSa | 0.30 | 9.05 | 17.00 | 10.05 | 73.16 | 0.057 | −0.467 | −1.235 | −0.408 | 2.343 |
| 6 | TengChon | 9.60 | 17.10 | 20.10 | 19.00 | 32.96 | 0.669 | 0.662 | −0.578 | 0.634 | 0.649 |
| 7 | KunMing | 9.70 | 16.90 | 20.65 | 16.05 | 37.82 | 0.676 | 0.634 | −0.461 | 0.291 | 0.854 |
| 8 | WuZhou | 13.85 | 19.85 | 29.00 | 23.40 | 2.38 | 0.949 | 1.047 | 1.309 | 1.147 | −0.640 |
| 9 | NanNing | 14.60 | 21.00 | 29.65 | 23.60 | 1.44 | 0.998 | 1.209 | 1.446 | 1.170 | −0.680 |
| 10 | ShanTou | 15.50 | 19.55 | 29.10 | 23.85 | 0.02 | 1.058 | 1.005 | 1.330 | 1.199 | −0.739 |

(i) Take the non-standardized means. Suppose we seek a partition with $K = 3$ clusters. Let the initial seeds be the observations $u = 3, 5, 10$, for clusters $C_1, C_2, C_3$, respectively, i.e., $\text{Seed}_1 \equiv \text{Centroid}_1 \equiv u = 3$, $\text{Seed}_2 \equiv \text{Centroid}_2 \equiv u = 5$, and $\text{Seed}_3 \equiv \text{Centroid}_3 \equiv u = 10$. The Euclidean distance between each seed and each observation, obtained from Eq. (6.1.3), is shown in Table 6.9(i). Then, each observation is successively allocated to the cluster to which it is closest as adjudged from the Euclidean distance. The shortest distance between a given station $u$ and the three seeds is highlighted. For example, observation $u = 1$ is closest to the seed corresponding to $u = 3$ and so is added to that cluster. This gives the initial partition as $P^{(0)} = (C_1^{(0)}, C_2^{(0)}, C_3^{(0)})$ with $C_1^{(0)} = \{1, 2, 3, 4\}$, $C_2^{(0)} = \{5, 7\}$, and $C_3^{(0)} = \{6, 8, 9, 10\}$. Then, the centroids are calculated for each of these clusters; they are, respectively, $\text{Centroid}_1 = \bar{Y}^1 = (\bar{Y}_1^1, \ldots, \bar{Y}_5^1) = (-17.2875, 5.0875, 20.6875, 4.9, 6.97)$, $\text{Centroid}_2 = \bar{Y}^2 = (5, 12.975, 18.825, 13.05, 55.49)$, and $\text{Centroid}_3 = \bar{Y}^3 = (13.3875, 19.375, 26.9625, 22.4625, 9.2)$. The Euclidean distances from these centroids to each observation $u$ are calculated and are shown in Table 6.9(ii). The new partition is now $P^{(1)} = (C_1^{(1)}, C_2^{(1)}, C_3^{(1)})$ with $C_1^{(1)} = \{1, 2, 3, 4\}$, $C_2^{(1)} = \{5, 6, 7\}$, and $C_3^{(1)} = \{8, 9, 10\}$, with $\text{Centroid}_1 = (-17.2875, 5.0875, 20.6875, 4.9, 6.97)$, $\text{Centroid}_2 = (6.533, 14.35, 19.25, 15.033, 47.98)$, and $\text{Centroid}_3 = (14.65, 20.133, 29.25, 23.6167, 1.28)$, respectively. The Euclidean distances between each observation and this third set of centroids are given in Table 6.9(iii).

This process is repeated until there is no change in the allocations. In this case, this occurs on the third iteration, as seen in Table 6.9. The final partition is therefore $P^{(3)} \equiv P = (C_1, C_2, C_3)$ with $C_1 = \{1, 2, 3, 4\}, C_2 = \{5, 6, 7\}$, and

**Table 6.9** Euclidean distances to centroids (Example 6.3)

| Station | (i) First iteration | | | (ii) Second iteration | | | (iii) Third iteration | | |
|---|---|---|---|---|---|---|---|---|---|
| $u$ | Centroid$_1$ | Centroid$_2$ | Centroid$_3$ | Centroid$_1$ | Centroid$_2$ | Centroid$_3$ | Centroid$_1$ | Centroid$_2$ | Centroid$_3$ |
| 1 | **13.844** | 69.893 | 36.667 | **5.523** | 54.570 | 33.886 | **5.523** | 48.749 | 36.002 |
| 2 | **14.043** | 71.717 | 38.326 | **4.942** | 56.660 | 35.934 | **4.942** | 50.964 | 37.696 |
| 3 | **0.000** | 62.449 | 48.087 | **9.266** | 49.779 | 43.312 | **9.266** | 45.583 | 47.241 |
| 4 | **11.057** | 72.281 | 45.732 | **5.185** | 58.597 | 42.939 | **5.185** | 53.622 | 45.038 |
| 5 | 62.449 | **0.000** | 77.638 | 68.893 | **19.028** | 67.986 | 68.893 | **27.035** | 76.352 |
| 6 | 40.818 | 43.093 | **35.077** | 41.736 | **24.141** | 25.360 | 41.736 | **16.094** | 33.81 |
| 7 | 42.192 | **38.056** | 40.022 | 44.090 | **19.028** | 30.328 | 44.090 | **11.079** | 38.747 |
| 8 | 46.094 | 75.049 | **2.932** | 40.248 | 56.186 | **7.210** | 40.248 | 48.252 | **1.428** |
| 9 | 47.580 | 76.385 | **2.301** | 41.601 | 57.502 | **8.535** | 41.601 | 49.566 | **0.969** |
| 10 | 48.087 | 77.638 | **0.000** | 41.980 | 58.759 | **9.760** | 41.980 | 50.817 | **1.651** |

**Table 6.10** Cluster Euclidean distances for China data (Examples 6.3 and 6.4)

| Station | (a) Non-standardized | | | | (b) Standardized | | | |
|---|---|---|---|---|---|---|---|---|
| | $K = 3$ | | $K = 4$ | | $K = 3$ | | $K = 4$ | |
| $u$ | $C_k$ | Distance | $C_k$ | Distance | $C_k$ | Distance | $C_k$ | Distance |
| 1 | 1 | 5.523 | 1 | 5.523 | 1 | 0.484 | 1 | 0.484 |
| 2 | 1 | 4.492 | 1 | 4.941 | 1 | 0.486 | 1 | 0.486 |
| 3 | 1 | 9.266 | 1 | 9.266 | 1 | 0.824 | 1 | 0.824 |
| 4 | 1 | 5.185 | 1 | 5.185 | 1 | 0.334 | 1 | 0.334 |
| 5 | 2 | 27.035 | 2 | 0.000 | 2 | 0.000 | 2 | 0.000 |
| 6 | 2 | 16.094 | 3 | 2.858 | 3 | 1.467 | 3 | 0.209 |
| 7 | 2 | 11.079 | 3 | 2.858 | 3 | 1.596 | 3 | 0.209 |
| 8 | 3 | 1.428 | 4 | 1.428 | 3 | 0.927 | 4 | 0.100 |
| 9 | 3 | 0.969 | 4 | 0.969 | 3 | 1.099 | 4 | 0.148 |
| 10 | 3 | 1.651 | 4 | 1.651 | 3 | 1.027 | 4 | 0.120 |

**Table 6.11** Final cluster centroids for China data (Examples 6.3 and 6.4)

| $k$ | (a) Non-standardized | | | | | (b) Standardized | | | | |
|---|---|---|---|---|---|---|---|---|---|---|
| | $\bar{Y}_1$ | $\bar{Y}_2$ | $\bar{Y}_3$ | $\bar{Y}_4$ | $\bar{Y}_5$ | $\bar{Y}_1$ | $\bar{Y}_2$ | $\bar{Y}_3$ | $\bar{Y}_4$ | $\bar{Y}_5$ |
| | | | | | $K = 3$ clusters | | | | | |
| 1 | −17.288 | 5.088 | 20.688 | 4.900 | 6.970 | −1.102 | −1.022 | −0.453 | −1.008 | −0.447 |
| 2 | 6.533 | 14.350 | 19.250 | 15.033 | 47.980 | 0.057 | −0.467 | −1.235 | −0.408 | 2.343 |
| 3 | 14.650 | 20.133 | 29.250 | 23.617 | 1.280 | 0.870 | 0.911 | 0.609 | 0.888 | −0.111 |
| | | | | | $K = 4$ clusters | | | | | |
| 1 | −17.288 | 5.088 | 20.688 | 4.900 | 6.970 | −1.102 | −1.022 | −0.453 | −1.008 | −0.447 |
| 2 | 0.300 | 9.050 | 17.000 | 10.050 | 73.160 | 0.057 | −0.467 | −1.235 | −0.408 | 2.343 |
| 3 | 9.650 | 17.000 | 20.375 | 17.525 | 35.390 | 0.672 | 0.648 | −0.519 | 0.462 | 0.751 |
| 4 | 14.650 | 20.133 | 29.250 | 23.617 | 1.280 | 1.002 | 1.087 | 1.362 | 1.172 | −0.686 |

$C_3 = \{8, 9, 10\}$. The Euclidean distances between the observations and their respective cluster centroids are summarized in Table 6.10(a). The final cluster centroids are given in Table 6.11(a).

(ii) Suppose now we seek a partition of $K = 4$ clusters with initial seeds being the observations $u = 3, 5, 8, 9$, respectively. The same steps as were followed to find the $K = 3$ partition apply. In this case, the final partition gives the clusters $C_1 = \{1, 2, 3, 4\}$, $C_2 = \{5\}$, $C_3 = \{6, 7\}$, and $C_4 = \{8, 9, 10\}$. The resulting distances from each observation to its corresponding cluster centroid are shown

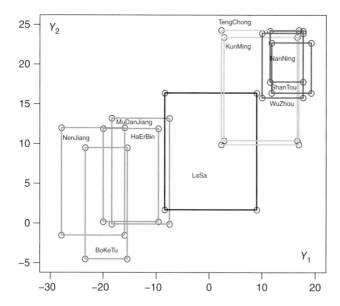

**Figure 6.1** Partition for China temperature data: $k$-means and non-standardized means (plots on $(Y_1, Y_2)$ axes) (Example 6.3).

in Table 6.10(a). Since the sum of these distances is $E = 31.821$ compared with the sum of distances $E = 122.262$ for $K = 3$ clusters, it is clear that the partition with $K = 4$ clusters is the better partition for these data. The final cluster centroids are in Table 6.11(a). Figure 6.1 displays the hypercube/rectangle of the non-standardized observations for the two variables $(Y_1, Y_2)$; the clusters from the $k$-means method on the non-standardized means are evident. (A plot of the means themselves is similar to that for the standardized means of Figure 6.2, except for a change in scales on the axes.)                                    □

**Example 6.4**   Let us now apply the $k$-means method to the standardized data of Table 6.8(b). When $K = 3$, we find the partition $P = (C_1, C_2, C_3)$ with $C_1 = \{1, 2, 3, 4\}$, $C_2 = \{5\}$, and $C_3 = \{6, 7, 8, 9, 10\}$, and the sum of the distances $E = 8.869$, from Table 6.10(b). When $K = 4$, the partition becomes $P = (C_1, \ldots, C_4)$ with $C_1 = \{1, 2, 3, 4\}$, $C_2 = \{5\}$, $C_3 = \{6, 7\}$, and $C_4 = \{8, 9, 10\}$, with total distances $E = 2.914$. The cluster centroids are displayed in Table 6.11(b). Again, it is clear that the partition with $K = 4$ clusters has the better number of clusters for this partitioning. The details are left to the reader. These partitions are illustrated graphically in Figure 6.2. Here, the unbroken lines identify the $K = 3$ partition, and the further partitioning of the stations $\{6, 7, 8, 9, 10\}$ into $C_3 = \{6, 7\}$ and $C_4 = \{8, 9, 10\}$ when $K = 4$ is indicated by the dotted line. The final partition is the same as for the

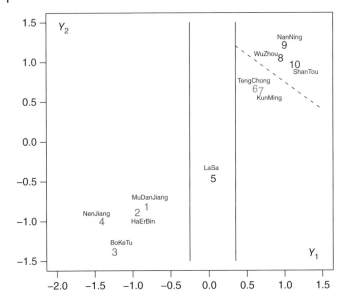

**Figure 6.2** Partition for China temperature data: $k$-means and standardized means (plots on $(Y_1, Y_2)$ axes) (Example 6.4).

non-standardized values when $K = 4$, but not when $K = 3$, as seen in Figure 6.2 and Table 6.10. □

When the observation values are unknown but the dissimilarities/distances between them are known, the $k$-medoids method can be used. We recall that any dissimilarity measure can be used, and not necessarily the usual Euclidean distances or Manhattan distances. The method applied to interval data is illustrated in the following example.

**Example 6.5** Take the China temperature data of Table 3.11, and the corresponding Gowda–Diday dissimilarities, Euclidean Gowda–Diday dissimilarities, and Mahalanobis Gowda–Diday dissimilarities of Table 6.12. Consider first obtaining a partition from the Mahalanobis Gowda–Diday dissimilarities using the $k$-medoid method.

Let $K = 4$ and suppose the initial seeds are the observations $u = 2, 4, 6, 8$, respectively. Then, from Table 6.12, we obtain the respective Mahalanobis Gowda–Diday dissimilarities from these seeds for each observation, as shown in Table 6.13(i). Thus, for example, for the observation $u = 5$, the dissimilarities are 1.359, 1.281, 1.827, and 3.661, respectively, from these four seeds. This observation is closest to the seed $u = 4$ and so is allocated to a cluster containing both $u = 4$ and $u = 5$. This process is repeated for all non-seed observations. Thus, at the end of the first iteration, we have the partition

**Table 6.12** Gowda–Diday dissimilarities for China stations (Example 6.5)

| Station pair | Dissimilarity by variable $Y_j$ | | | | | $\sum d_j$ $d$ | Euclidean distance | Mahalanobis distance |
|---|---|---|---|---|---|---|---|---|
| | $d_1$ | $d_2$ | $d_3$ | $d_4$ | $d_5$ | | | |
| 1-2 | 0.370 | 0.251 | 0.192 | 0.146 | 0.019 | 0.978 | 0.955 | 0.055 |
| 1-3 | 1.112 | 0.650 | 0.781 | 0.684 | 0.136 | 3.363 | 11.312 | 0.648 |
| 1-4 | 1.133 | 0.239 | 0.328 | 0.509 | 0.000 | 2.209 | 4.879 | 0.279 |
| 1-5 | 1.417 | 0.450 | 1.059 | 0.644 | 0.934 | 4.504 | 20.294 | 1.162 |
| 1-6 | 1.267 | 1.257 | 0.881 | 1.532 | 0.385 | 5.322 | 28.309 | 1.621 |
| 1-7 | 1.239 | 1.258 | 0.442 | 1.609 | 0.451 | 4.999 | 24.989 | 1.431 |
| 1-8 | 1.207 | 1.660 | 1.182 | 1.515 | 0.033 | 5.597 | 31.329 | 1.794 |
| 1-9 | 1.239 | 1.716 | 1.449 | 1.574 | 0.046 | 6.024 | 36.293 | 2.078 |
| 1-10 | 1.221 | 1.740 | 1.476 | 1.563 | 0.066 | 6.066 | 36.791 | 2.107 |
| 2-3 | 0.927 | 0.736 | 0.898 | 0.581 | 0.155 | 3.297 | 10.876 | 0.623 |
| 2-4 | 1.031 | 0.326 | 0.470 | 0.394 | 0.019 | 2.240 | 5.018 | 0.287 |
| 2-5 | 1.446 | 0.608 | 1.158 | 0.707 | 0.953 | 4.872 | 23.735 | 1.359 |
| 2-6 | 1.265 | 1.366 | 0.915 | 1.537 | 0.404 | 5.487 | 30.102 | 1.724 |
| 2-7 | 1.238 | 1.346 | 0.583 | 1.665 | 0.470 | 5.302 | 28.108 | 1.609 |
| 2-8 | 1.189 | 1.529 | 1.103 | 1.528 | 0.014 | 5.363 | 28.762 | 1.647 |
| 2-9 | 1.221 | 1.586 | 1.367 | 1.586 | 0.027 | 5.787 | 33.488 | 1.917 |
| 2-10 | 1.206 | 1.603 | 1.393 | 1.576 | 0.047 | 5.825 | 33.916 | 1.942 |
| 3-4 | 0.821 | 0.468 | 0.668 | 0.313 | 0.136 | 2.406 | 5.788 | 0.331 |
| 3-5 | 1.393 | 0.876 | 0.462 | 1.034 | 0.798 | 4.563 | 20.820 | 1.192 |
| 3-6 | 1.270 | 1.500 | 1.126 | 1.524 | 0.249 | 5.669 | 32.135 | 1.840 |
| 3-7 | 1.246 | 1.521 | 0.870 | 1.608 | 0.315 | 5.560 | 30.907 | 1.770 |
| 3-8 | 1.094 | 1.691 | 1.458 | 1.551 | 0.170 | 5.964 | 35.549 | 2.035 |
| 3-9 | 1.125 | 1.750 | 1.541 | 1.604 | 0.182 | 6.202 | 38.479 | 2.203 |
| 3-10 | 1.118 | 1.755 | 1.578 | 1.596 | 0.202 | 6.249 | 39.052 | 2.236 |
| 4-5 | 1.357 | 0.603 | 0.977 | 0.859 | 0.934 | 4.730 | 22.371 | 1.281 |
| 4-6 | 1.293 | 1.349 | 0.843 | 1.576 | 0.384 | 5.445 | 29.653 | 1.698 |
| 4-7 | 1.272 | 1.369 | 0.349 | 1.685 | 0.451 | 5.126 | 26.284 | 1.505 |
| 4-8 | 1.327 | 1.664 | 1.347 | 1.584 | 0.034 | 5.956 | 35.454 | 2.030 |
| 4-9 | 1.358 | 1.721 | 1.445 | 1.640 | 0.046 | 6.210 | 38.568 | 2.208 |
| 4-10 | 1.348 | 1.737 | 1.492 | 1.631 | 0.066 | 6.274 | 39.363 | 2.254 |
| 5-6 | 1.073 | 1.010 | 1.416 | 1.600 | 0.550 | 5.649 | 31.912 | 1.827 |
| 5-7 | 1.134 | 1.104 | 1.162 | 1.550 | 0.483 | 5.433 | 29.523 | 1.690 |
| 5-8 | 1.724 | 1.760 | 1.661 | 1.884 | 0.968 | 7.997 | 63.942 | 3.661 |
| 5-9 | 1.756 | 1.866 | 1.745 | 1.955 | 0.981 | 8.303 | 68.919 | 3.946 |

*(Continued)*

**Table 6.12** (Continued)

| Station pair | Dissimilarity by variable $Y_j$ | | | | | $\sum d_j$ $d$ | Euclidean distance | Mahalanobis distance |
|---|---|---|---|---|---|---|---|---|
| | $d_1$ | $d_2$ | $d_3$ | $d_4$ | $d_5$ | | | |
| 5-10 | 1.690 | 1.910 | 1.781 | 1.937 | 1.000 | 8.318 | 69.196 | 3.962 |
| 6-7 | 0.120 | 0.212 | 0.582 | 0.751 | 0.066 | 1.731 | 2.998 | 0.172 |
| 6-8 | 1.163 | 1.080 | 1.466 | 0.866 | 0.418 | 4.993 | 24.941 | 1.428 |
| 6-9 | 1.390 | 1.385 | 1.322 | 1.119 | 0.431 | 5.647 | 31.892 | 1.826 |
| 6-10 | 1.326 | 1.351 | 1.278 | 1.109 | 0.450 | 5.514 | 30.395 | 1.740 |
| 7-8 | 1.119 | 0.988 | 1.430 | 1.255 | 0.485 | 5.277 | 27.835 | 1.594 |
| 7-9 | 1.353 | 1.329 | 1.315 | 1.257 | 0.497 | 5.751 | 33.073 | 1.894 |
| 7-10 | 1.289 | 1.239 | 1.334 | 1.233 | 0.517 | 5.612 | 31.486 | 1.803 |
| 8-9 | 0.421 | 0.546 | 0.465 | 0.463 | 0.013 | 1.908 | 3.643 | 0.209 |
| 8-10 | 0.430 | 0.465 | 0.603 | 0.395 | 0.032 | 1.925 | 3.705 | 0.212 |
| 9-10 | 0.396 | 0.433 | 0.237 | 0.117 | 0.019 | 1.202 | 1.446 | 0.083 |

**Table 6.13** Mahalanobis Gowda–Diday distances (Example 6.5)

| Station $u$ | (i) Seeds: first iteration | | | | Station $u$ | (ii) Seeds: second iteration | | | |
|---|---|---|---|---|---|---|---|---|---|
| | $u=2$ | $u=4$ | $u=6$ | $u=8$ | | $u=3$ | $u=4$ | $u=6$ | $u=8$ |
| 1 | **0.055** | 0.279 | 1.621 | 1.794 | 1 | 0.648 | **0.279** | 1.621 | 1.794 |
| 3 | 0.623 | **0.331** | 1.840 | 2.035 | 2 | 0.623 | **0.287** | 1.724 | 1.647 |
| 5 | 1.359 | **1.281** | 1.827 | 3.661 | 5 | 1.192 | **1.281** | 1.827 | 3.661 |
| 7 | 1.609 | 1.505 | **0.172** | 1.594 | 7 | 1.770 | 1.505 | **0.172** | 1.594 |
| 9 | 1.917 | 2.208 | 1.826 | **0.209** | 9 | 2.203 | 2.208 | 1.826 | **0.209** |
| 10 | 1.942 | 2.254 | 1.740 | **0.212** | 10 | 2.236 | 2.254 | 1.740 | **0.212** |

$P^{(1)} = (C_1^{(1)}, \dots, C_4^{(1)})$ with the four clusters $C_1^{(1)} = \{1,2\}$, $C_2^{(1)} = \{3,4,5\}$, $C_3^{(1)} = \{6,7\}$, and $C_4^{(1)} = \{8,9,10\}$. The within-cluster dissimilarities are, from Eq. (6.1.5), $D^1 = 0.055$, $D^2 = 1.612$, $D^3 = 0.172$, and $D^4 = 0.421$, with total within-cluster distance $D = 2.260 \equiv D_{(1)}$, say. Had we used the seeds $u = 3,4,6,8$, instead, we obtain the dissimilarities as shown in Table 6.13(ii). In this case, the clusters are $C_1^{(1)} = \{3\}$, $C_2^{(1)} = \{1,2,4,5\}$, $C_3^{(1)} = \{6,7\}$, and $C_4^{(1)} = \{8,9,10\}$. The within cluster dissimilarities are, respectively, $D^1 = 0$, $D^2 = 1.891$, $D^3 = 0.172$, and $D^4 = 0.421$; hence, the total within cluster distance is $D = 2.484 \equiv D_{(2)} > D_{(1)} = 2.260$.

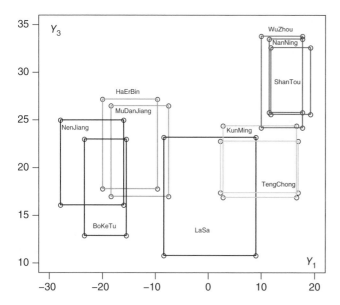

**Figure 6.3** Partition for China temperature (non-standardized) data: $k$-medoids (plots on $(Y_1, Y_3)$ axes) (Example 6.5).

After swapping out all possible seeds, we find that, in this case, the first iteration gave the best results. That is, the obtained partition is $P = (C_1, \ldots, C_4)$ with $C_1 = \{1, 2\}$, $C_2 = \{3, 4, 5\}$, $C_3 = \{6, 7\}$, and $C_4 = \{8, 9, 10\}$. This is illustrated in Figure 6.3, which plots the $Y_1, Y_3$ axes values; the clusters are evident. Notice that station $u = 5$ (LaSa) is now clustered with stations $u = 3$ (BoKeTu) and $u = 4$ (NenJiang) in $C_2$, in contrast to the $k$-means procedure which put this station into its own one-observation cluster in Figure 6.2. This difference is partly reflective of the fact that the $k$-means method, by only using the interval means, is not utilizing the interval internal variations, whereas the dissimilarity measures of the $k$-medoid method do use this internal variation information.

The same partition pertains when using the Gowda–Diday dissimilarities. However, when based on the Euclidean Gowda–Diday dissimilarities, the best partitioning gives $P = (C_1, C_2)$ with $C_1 = \{1, \ldots, 7\}$ and $C_2 = \{8, 9, 10\}$ (see Exercise 6.5). Notice that the partition based on the dissimilarities differs from those found in Examples 6.3 and 6.4 when using only the interval means. □

The next example is based on the tenets of the dynamical partitioning approach of Chavent and Lechevallier (2002) in which centroids expressed in terms of medians (referred to as prototypes in their article) are established for each cluster at each stage, and where the distances between intervals are calculated by the Hausdorff method of Eq. (3.3.4). An advantage of this approach is that only $m \times K$ distances need to be calculated, in contrast to the $m(m-1)$ needed when using the medoid approach, as in Example 6.5.

**Example 6.6** Consider the same China temperature data of Table 3.11 used in Example 6.5. We form a partition based on an adaptive method for intervals using the Hausdorff distance. As before, we use the five variables $Y_1$ = January temperature, $Y_2$ = April temperature, $Y_3$ = July temperature, $Y_4$ = October temperature, and $Y_5$ = elevation. Take $K = 3$. Suppose an initial partition is formed as $P^{(0)} = (C_1^{(0)}, C_2^{(0)}, C_2^{(0)})$ with $C_1^{(0)} = \{1, 2, 3\}$, $C_2^{(0)} = \{4, 5, 6\}$, and $C_3^{(0)} = \{7, 8, 9, 10\}$, respectively. The centroids for each cluster are defined as $[\alpha_k - \beta_k, \alpha_k + \beta_k]$, where $\alpha_k$ = median observation mean and $\beta_k$ = median half-range of the observations in $C_k$, $k = 1, 2, 3$. The respective centroids are as shown in Table 6.14(i). The Hausdorff distances between the $m = 10$ stations and each cluster centroid are found from Eq. (3.3.4). These are shown in Table 6.15(i).

Take the first observation, $u = 1$. It is clear from Table 6.15(i) that this observation is closest to the centroid of $C_1^{(0)}$ than to any of the other centroids, i.e.,

$$d(u = 1, l_1) = \min_k \{d(u = 1, l_k) = \min\{3.400, 51.940, 75.618\} = 3.400,$$

where $l_k$ is the centroid of $C_k^{(0)}$, $k = 1, 2, 3$. Thus, this observation stays in $C_1$. However, observation $u = 4$ is closest to the centroid of $C_1^{(0)}$ rather than the centroid of its current cluster $C_2^{(0)}$, since

$$d(u = 4, l_1) = \min_k \{d(u = 4, l_1), d(u = 4, l_2), d(u = 4, l_3)\} = 12.720.$$

**Table 6.14** Centroids (Example 6.6)

| Iteration | Centroid | January $Y_1$ | April $Y_2$ | July $Y_3$ | October $Y_4$ | Elevation $Y_5$ |
|---|---|---|---|---|---|---|
| (i) First | 1 | [−20.0, −9.6] | [−0.6, 12.7] | [17.0, 26.5] | [−0.2, 12.5] | 4.82 |
| | 2 | [−7.0, 7.6] | [1.85, 16.25] | [15.65, 24.55] | [3.3, 16.8] | 32.96 |
| | 3 | [10.45, 18.0] | [16.075, 23.325] | [25.25, 32.85] | [19.95, 27.05] | 1.8275 |
| (ii) Second | 1 | [−22.45, −11.8] | [−1.05, 12.35] | [16.425, 25.875] | [−1.25, 11.55] | 4.83 |
| | 2 | [2.3, 16.9] | [9.7, 24.1] | [16.35, 23.85] | [11.55, 20.55] | 37.82 |
| | 3 | [10.9, 18.3] | [16.65, 23.05] | [25.25, 32.95] | [20.15, 27.05] | 1.275 |

**Table 6.15** Hausdorff centroid distances (Example 6.6)

| u | (i) Centroids for $P^{(0)}$ | | | (ii) Centroids for $P^{(1)}$ | | |
|---|---|---|---|---|---|---|
| | $C_1^{(0)}$ | $C_2^{(0)}$ | $C_3^{(0)}$ | $C_1^{(1)}$ | $C_2^{(1)}$ | $C_3^{(1)}$ |
| 1 | **3.400** | 51.940 | 75.618 | **7.735** | 81.900 | 77.395 |
| 2 | **2.980** | 58.020 | 75.538 | **7.515** | 88.180 | 77.315 |
| 3 | **27.660** | 58.680 | 103.678 | **23.375** | 89.040 | 105.455 |
| 4 | **12.720** | 62.320 | 90.638 | **8.135** | 93.280 | 92.415 |
| 5 | 103.040 | **48.500** | 137.558 | 105.955 | **69.740** | 139.335 |
| 6 | 84.640 | **30.300** | 60.958 | 87.605 | **9.060** | 62.835 |
| 7 | 84.900 | **33.560** | 65.318 | 87.965 | **2.600** | 67.195 |
| 8 | 75.540 | 86.680 | **3.378** | 80.125 | 66.840 | **4.855** |
| 9 | 82.910 | 93.450 | **4.493** | 87.445 | 72.410 | **2.615** |
| 10 | 82.800 | 93.340 | **4.583** | 87.335 | 72.100 | **3.105** |

Proceeding in this manner, each observation is allocated to its new cluster (where, as for observation $u = 1$, the allocation can be to its current cluster). Hence, we obtain the partitions $P^{(1)} = (C_1^{(1)}, C_2^{(1)}, C_3^{(1)})$ with $C_1^{(1)} = \{1, 2, 3, 4\}$, $C_2^{(1)} = \{5, 6, 7\}$, and $C_3^{(1)} = \{8, 9, 10\}$.

The centroids for the clusters in $P^{(1)}$ are shown in Table 6.14(ii). Again, using Eq. (3.3.4), we calculate the Hausdorff distances between each observation and the respective centroids; these are given in Table 6.15(ii). From these values, it is seen that no observations are reallocated to a different cluster, i.e., all clusters remain the same. Hence, the process stops. The final partition is therefore $P = (C_1, C_2, C_3)$ with $C_1 = \{1, 2, 3, 4\}$, $C_2 = \{5, 6, 7\}$, and $C_3 = \{8, 9, 10\}$. If we look at the $Y_1 \times Y_2$ bi-plots for these data (in Figure 6.1), these obtained partitions are not unreasonable. □

## 6.4 Histogram Observations

Consider now the application of partitioning methods to histogram-valued observations, primarily through illustrative examples. We recall that histogram-valued observations, as described in section 2.2.4, have realizations

$$\mathbf{Y}_u = (\{[a_{ujk_j}, b_{ujk_j}), p_{ujk_j}; \ k_j = 1, \dots, s_{uj}\},$$

$$\sum_{k_j=1}^{s_{uj}} p_{ujk_j} = 1, \ j = 1, \dots, p), \ u = 1, \dots, m. \tag{6.4.1}$$

The partition $P$ has clusters $P = (C_1, \ldots, C_K)$. To apply the $k$-means method, it is necessary to compute the observation means, from Eq. (2.4.3), as

$$\bar{\mathbf{Y}}_u = (\bar{Y}_{uj}), \qquad \bar{Y}_{uj} = \frac{1}{2} \sum_{k=1}^{s_{uj}} (a_{ujk} + b_{ujk}) p_{ujk}, \qquad u = 1, \ldots, m, \qquad (6.4.2)$$

and the mean (i.e., centroid) of a cluster $C_k = \{u_1, \ldots, u_{m_k}\}$ is, from Eq. (2.4.3),

$$\bar{\mathbf{Y}}^k = (\bar{Y}_j^k), \qquad \bar{Y}_j^k = \frac{1}{2m_k} \sum_{u=u_1}^{u_{m_k}} \sum_{k=1}^{s_{uj}} (a_{ujk} + b_{ujk}) p_{ujk}, \qquad k = 1, \ldots, K. \quad (6.4.3)$$

For the $k$-medoids method, dissimilarities are calculated by using the various dissimilarities/distances for histogram data described in Chapter 4.

Both the $k$-means method and the $k$-medoid method are applied to the Fisher (1936) iris data in Examples 6.7 and 6.8, respectively. In each case, we omit the details of the methodology itself.

**Example 6.7**  Consider the histogram-valued observations for the iris data of Table 4.10. To apply the $k$-means method, we need the observation means, given in Table 4.19, for the respective variables $Y_1$ = sepal width, $Y_2$ = sepal length, $Y_3$ = petal width, and $Y_4$ = petal length. It is easy to show that these means are, for $u = 1 \equiv$ Verscolor, $\bar{\mathbf{Y}}_1 = (2.820, 5.976, 1.360, 4.430)$, for $u = 2 \equiv$ Virginica, $\bar{\mathbf{Y}}_2 = (3.072, 6.600, 2.056, 5.725)$, and for $u = 3 \equiv$ Setosa, $\bar{\mathbf{Y}}_3 = (3.468, 5.080, 0.272, 1.510)$, respectively. Then, by applying the $k$-means method, two clusters emerged, namely, $C_1 = \{$ Versicolor, Virginica $\}$ and $C_2 = \{$ Setosa $\}$; these are consistent with Figure 4.2(b).  $\square$

**Example 6.8**  Let us now apply the $k$-medoids method to the Fisher (1936) iris data of Table 4.10. Suppose we take the normalized Euclidean extended de Carvalho distances with the comparison function $cf_5$, calculated from Eqs. (4.2.37)–(4.2.43) and Eq. (3.1.9) (see Example 4.14 and Table 4.20(c)). Thus, the distance matrix is

$$\mathbf{D} = \begin{bmatrix} 0 & 0.659 & 0.592 \\ 0.659 & 0 & 0.489 \\ 0.592 & 0.489 & 0 \end{bmatrix}.$$

Applying the $k$-medoids method, we obtain the partition $P = (C_1, C_2)$ with clusters $C_1 = \{$ Versicolor $\}$ and $C_2 = \{$ Virginica, Setosa $\}$. However, when the normalized Euclidean extended de Carvalho distances with the comparison function $cf_1$ (see Table 4.20, right-most column) are used, the partition gives clusters $C_1 = \{$ Versicolor, Virginica $\}$ and $C_2 = \{$ Setosa $\}$.  $\square$

Another approach is to calculate distances based on the sample cumulative histogram functions, described by Košmelj and Billard (2011, 2012). That is, instead of using the histogram values directly, the inverse of the corresponding cumulative distribution function (or, simply, distribution function) $F_u$ for each observation $u = 1, \ldots, m$, is used. This approach adapts the $k$-means method and uses Mallows' distance (see Eq. (4.2.48) and Definition 4.18) between two distribution functions. At each iteration, a centroid histogram is calculated for each cluster. The Mallows' distance between an observation $u$ and the centroid value $F_c$, for a given variable $Y_j, j = 1, \ldots, p$, is

$$d^2_{Mj}(u, c) = d^2_{Mj}(F_{uj}, F_{cj}) = \int_0^1 \left[ F_{uj}^{-1}(t) - F_{cj}^{-1}(t) \right]^2 dt \qquad (6.4.4)$$

where, in Eq. (6.4.4), $F_{vj}^{-1}(\cdot)$, $v = u, c$, is the inverse distribution function for the observation $u$ and the centroid $c$ histograms, respectively.

It can be shown that the Mallows' distance between the histogram observations $u$ and $v$ satisfies, for variable $Y_j$,

$$d^2_{Mj}(u, v) = \sum_{k=1}^{s_j} \pi_k \left[ (c_{ujk} - c_{vjk})^2 + (1/3)(r_{ujk} - r_{vjk})^2 \right] \qquad (6.4.5)$$

where, in Eq. (6.4.5), the histogram sub-intervals $[a_{vjk}, b_{vjk}]$ have been transformed into the sub-interval centers $c_{vjk} = (a_{vjk} + b_{vjk})/2$ and radii $r_{vjk} = (b_{vjk} - a_{vjk})/2$, for $v = u, v$ and $k \equiv k_j = 1, \ldots, s_j$, and $s_j$ is the number of obtained quantile sub-intervals. The weights $\pi_k = w_k - w_{k-1}$ where the quantile sub-intervals are $[w_{k-1}, w_k]$.

Under the assumption that variables are independent, the Mallows' distance is

$$d^2_M(u, v) = \sum_{j=1}^p d^2_{Mj}(u, v). \qquad (6.4.6)$$

When variables are not independent, Eq. (6.4.6) is replaced by its Mahalanobis counterpart.

**Example 6.9** This approach is applied to the data of Table 6.16, which records the histogram values for $Y_1 = $ male income and $Y_2 = $ female income for each of $m = 12$ counties. Thus, the table shows the $p_{ujk_j}, j = 1, 2, k = 1, 2, 3, 4$, $u = 1, \ldots, m$, of Eq. (6.4.1), for the histogram sub-intervals $[a_{ujk}, b_{ujk}) = [0, 25), [25, 50), [50, 100), [100, 125]$ in US$1000s, for $k = 1, 2, 3, 4$, respectively. For example, 21.3% of males in county $u = 1$ have incomes under US$25000.

Notice that by applying Eqs. (6.4.4) and (6.4.5) where $u \equiv (u = 1)$ and $c \equiv (u = 2)$, we can calculate the distances between counties for each variable. Thus, for example, the distance between the first two counties for males ($Y_1$) is $d^2_{M1}(1, 2) = 206.981$ and for females ($Y_2$) is $d^2_{M2}(1, 2) = 178.381$. Hence, over both variables, under the independence assumption, the distance is

**Table 6.16** County income by gender (Example 6.9)

| County | $Y_1$ males | | | | $Y_2$ females | | | | Number of males | Number of females |
|---|---|---|---|---|---|---|---|---|---|---|
| $u$ | $P_{11}$ | $P_{12}$ | $P_{13}$ | $P_{14}$ | $P_{21}$ | $P_{22}$ | $P_{23}$ | $P_{24}$ | $N_1$ | $N_2$ |
| 1 | 0.213 | 0.434 | 0.281 | 0.072 | 0.329 | 0.506 | 0.149 | 0.016 | 56190 | 43139 |
| 2 | 0.354 | 0.475 | 0.148 | 0.023 | 0.615 | 0.328 | 0.052 | 0.005 | 7662 | 4519 |
| 3 | 0.333 | 0.460 | 0.184 | 0.023 | 0.359 | 0.490 | 0.138 | 0.013 | 156370 | 133458 |
| 4 | 0.407 | 0.407 | 0.156 | 0.030 | 0.723 | 0.230 | 0.034 | 0.013 | 3357 | 2110 |
| 5 | 0.510 | 0.385 | 0.085 | 0.020 | 0.732 | 0.262 | 0.006 | 0.000 | 1324 | 790 |
| 6 | 0.430 | 0.422 | 0.115 | 0.033 | 0.716 | 0.258 | 0.015 | 0.011 | 1970 | 1081 |
| 7 | 0.327 | 0.461 | 0.163 | 0.049 | 0.574 | 0.353 | 0.064 | 0.009 | 6832 | 5309 |
| 8 | 0.168 | 0.355 | 0.347 | 0.130 | 0.233 | 0.477 | 0.246 | 0.044 | 263225 | 187369 |
| 9 | 0.255 | 0.429 | 0.251 | 0.065 | 0.437 | 0.476 | 0.087 | 0.000 | 231 | 126 |
| 10 | 0.277 | 0.335 | 0.339 | 0.049 | 0.406 | 0.440 | 0.136 | 0.018 | 6069 | 3291 |
| 11 | 0.303 | 0.430 | 0.217 | 0.050 | 0.486 | 0.388 | 0.115 | 0.011 | 26477 | 17551 |
| 12 | 0.219 | 0.388 | 0.327 | 0.066 | 0.423 | 0.407 | 0.150 | 0.020 | 5827 | 3676 |

$d_M^2(1,2) = d_{M1}^2(1,2) + d_{M2}^2(1,2) = 385.362$. These Mallows' distances between all counties by variable and their totals are shown in Table 6.17.

Suppose, after an iteration of the $k$-means method, observations $u = 2$ and $u = 8$ are in one cluster. Therefore, it is necessary to calculate the centroid histogram of this cluster. In Figure 6.4, plots of these two distributions are shown for males (a) and females (b) separately. Also shown are the respective centroids distributions. Notice the nodes (sub-interval end points) of the centroid histogram correspond to distribution nodes for both $u = 2$ and $u = 8$, i.e., the cumulative probabilities are, respectively, for males (0, 0.168, 0.354, 0.523, 0.829, 0.870, 0.977, 1.00) and for females (0, 0.233, 0.615, 0.710, 0.943, 0.956, 0.995, 1.0). These give us the quantile sub-intervals $[w_{k-1}, w_k]$ and hence $\pi_k$, e.g., for males, the $k = 2$ sub-interval is $[w_1, w_2] = [0.168, 0.354]$ and hence $\pi_2 = 0.186$ used in Eq. (6.4.5). Now we can calculate the distances between observations $u = 1, \ldots, m$, and the centroids (using Eq. (6.4.5) directly) for each $Y_j, j = 1, 2$, and thence the distances across all variables.

The final $K = 3$ partition gives $P = (C_1, C_2, C_3)$ with $C_1 = \{1, 3, 9, 10, 11, 12\}$, $C_2 = \{8\}$, and $C_3 = \{2, 4, 5, 6, 7\}$. The cluster centroid histograms are shown in Figure 6.5; now the nodes correspond to all nodes for all distributions in a given cluster. □

**Table 6.17** Income histograms: Mallows' distances (Example 6.9)

| County u | 1 | 2 | 3 | 4 | 5 | 6 | 7 | 8 | 9 | 10 | 11 | 12 |
|---|---|---|---|---|---|---|---|---|---|---|---|---|
| | | | | | | $d^2_{M1}(u_1, u_2)$ Males | | | | | | |
| 1 | 0 | 206.981 | 144.142 | 201.541 | 448.794 | 277.531 | 111.937 | 86.641 | 9.068 | 9.207 | 49.880 | 4.558 |
| 2 | 206.981 | 0 | 7.448 | 4.984 | 57.859 | 9.810 | 23.126 | 532.711 | 139.658 | 195.594 | 64.264 | 246.275 |
| 3 | 144.142 | 7.448 | 0 | 8.598 | 98.425 | 29.077 | 7.511 | 431.767 | 88.057 | 135.960 | 30.693 | 178.495 |
| 4 | 201.541 | 4.984 | 8.598 | 0 | 62.174 | 9.221 | 17.584 | 528.567 | 131.437 | 188.949 | 56.101 | 242.513 |
| 5 | 448.794 | 57.859 | 98.425 | 62.174 | 0 | 29.407 | 138.209 | 891.917 | 346.196 | 422.351 | 219.586 | 502.806 |
| 6 | 277.531 | 9.810 | 29.077 | 9.221 | 29.407 | 0 | 45.914 | 644.638 | 196.058 | 262.593 | 102.509 | 323.848 |
| 7 | 111.937 | 23.126 | 7.511 | 17.584 | 138.209 | 45.914 | 0 | 377.966 | 61.311 | 111.448 | 15.126 | 146.425 |
| 8 | 86.641 | 532.711 | 431.767 | 528.567 | 891.917 | 644.638 | 377.966 | 0 | 146.724 | 94.151 | 257.507 | 22.621 |
| 9 | 9.068 | 139.658 | 88.057 | 131.437 | 346.196 | 196.058 | 61.311 | 146.724 | 0 | 14.874 | 17.438 | 22.621 |
| 10 | 9.207 | 195.594 | 135.960 | 188.949 | 422.351 | 262.593 | 111.448 | 94.151 | 14.874 | 0 | 50.184 | 6.868 |
| 11 | 49.880 | 64.264 | 30.693 | 56.101 | 219.586 | 102.509 | 15.126 | 257.507 | 17.438 | 50.184 | 0 | 74.456 |
| 12 | 4.558 | 246.275 | 178.495 | 242.513 | 502.806 | 323.848 | 146.425 | 58.850 | 22.621 | 6.868 | 74.456 | 0 |

(*Continued*)

Table 6.17 (Continued)

| County u | 1 | 2 | 3 | 4 | 5 | 6 | 7 | 8 | 9 | 10 | 11 | 12 |
|---|---|---|---|---|---|---|---|---|---|---|---|---|
| | | | | | $d^2_{M2}(u_1, u_2)$ Females | | | | | | | |
| 1 | 0 | 178.381 | 2.693 | 265.425 | 394.259 | 306.079 | 127.035 | 100.004 | 66.635 | 6.976 | 37.590 | 9.128 |
| 2 | 178.381 | 0 | 141.787 | 15.345 | 71.772 | 27.648 | 7.026 | 497.795 | 40.288 | 132.860 | 65.443 | 149.799 |
| 3 | 2.693 | 141.787 | 0 | 221.569 | 341.142 | 259.510 | 95.999 | 130.724 | 45.386 | 2.731 | 21.450 | 7.263 |
| 4 | 265.425 | 15.345 | 221.569 | 0 | 68.390 | 12.736 | 33.283 | 640.012 | 95.931 | 206.611 | 120.426 | 225.853 |
| 5 | 394.259 | 71.772 | 341.142 | 68.390 | 0 | 41.590 | 113.838 | 823.390 | 163.208 | 336.216 | 227.820 | 362.032 |
| 6 | 306.079 | 27.648 | 259.510 | 12.736 | 41.590 | 0 | 55.424 | 701.012 | 116.144 | 247.904 | 155.066 | 269.778 |
| 7 | 127.035 | 7.026 | 95.999 | 33.283 | 113.838 | 55.424 | 0 | 411.702 | 21.656 | 87.265 | 34.152 | 101.640 |
| 8 | 100.004 | 497.795 | 130.724 | 640.012 | 823.390 | 701.012 | 411.702 | 0 | 296.212 | 140.451 | 234.298 | 125.987 |
| 9 | 66.635 | 40.288 | 45.386 | 95.931 | 163.208 | 116.144 | 21.656 | 296.212 | 0 | 47.463 | 20.234 | 62.328 |
| 10 | 6.976 | 132.860 | 2.731 | 206.611 | 336.216 | 247.904 | 87.265 | 140.451 | 47.463 | 0 | 14.774 | 1.968 |
| 11 | 37.590 | 65.443 | 21.450 | 120.426 | 227.820 | 155.066 | 34.152 | 234.298 | 20.234 | 14.774 | 0 | 21.250 |
| 12 | 9.128 | 149.799 | 7.263 | 225.853 | 362.032 | 269.778 | 101.640 | 125.989 | 62.328 | 1.968 | 21.250 | 0 |

(Continued)

**Table 6.17** (Continued)

| County u | 1 | 2 | 3 | 4 | 5 | 6 | 7 | 8 | 9 | 10 | 11 | 12 |
|---|---|---|---|---|---|---|---|---|---|---|---|---|
| | | | | | | $d_M^2(u_1, u_2)$ | | | | | | |
| 1 | 0 | 385.362 | 146.835 | 466.966 | 843.053 | 583.611 | 238.972 | 186.644 | 75.703 | 16.182 | 87.470 | 13.686 |
| 2 | 385.362 | 0 | 149.235 | 20.329 | 129.631 | 37.458 | 30.152 | 1030.506 | 179.946 | 328.454 | 129.707 | 396.074 |
| 3 | 146.835 | 149.235 | 0 | 230.167 | 439.568 | 288.587 | 103.510 | 562.491 | 133.443 | 138.690 | 52.142 | 185.758 |
| 4 | 466.966 | 20.329 | 230.167 | 0 | 130.565 | 21.956 | 50.867 | 1168.579 | 227.368 | 395.560 | 176.527 | 468.367 |
| 5 | 843.053 | 129.631 | 439.568 | 130.565 | 0 | 70.997 | 252.047 | 1715.307 | 509.404 | 758.567 | 447.407 | 864.838 |
| 6 | 583.611 | 37.458 | 288.587 | 21.956 | 70.997 | 0 | 101.338 | 1345.650 | 312.202 | 510.496 | 257.575 | 593.626 |
| 7 | 238.972 | 30.152 | 103.510 | 50.867 | 252.047 | 101.338 | 0 | 789.668 | 82.967 | 198.713 | 49.278 | 248.065 |
| 8 | 186.644 | 1030.506 | 562.491 | 1168.579 | 1715.307 | 1345.650 | 789.668 | 0 | 442.936 | 234.602 | 491.805 | 184.838 |
| 9 | 75.703 | 179.946 | 133.443 | 227.368 | 509.404 | 312.202 | 82.967 | 442.936 | 0 | 62.337 | 37.672 | 84.949 |
| 10 | 16.182 | 328.454 | 138.690 | 395.560 | 758.567 | 510.496 | 198.713 | 234.602 | 62.337 | 0 | 64.958 | 8.835 |
| 11 | 87.470 | 129.707 | 52.142 | 176.527 | 447.407 | 257.575 | 49.278 | 491.805 | 37.672 | 64.958 | 0 | 95.706 |
| 12 | 13.686 | 396.074 | 185.758 | 468.367 | 864.838 | 593.626 | 248.065 | 184.838 | 84.949 | 8.835 | 95.706 | 0 |

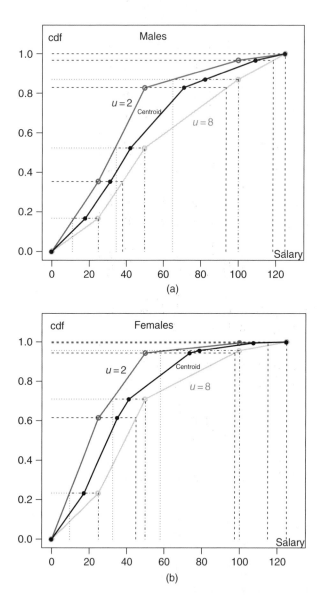

**Figure 6.4** Centroid distributions showing cumulative density function (cdf) against salary for $u = 2$ and $u = 8$: (a) males and (b) females (Example 6.9).

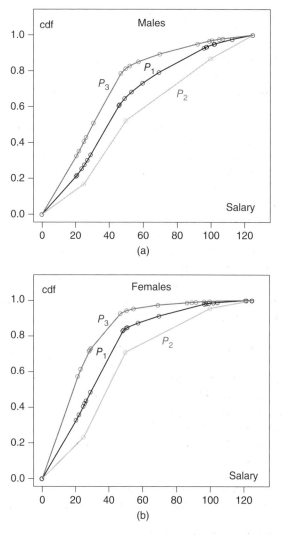

**Figure 6.5** Centroid distributions showing cumulative density function (cdf) against salary for $K = 3$ partitions: (a) males and (b) females (Example 6.9).

## 6.5 Mixed-valued Observations

Most methods so far in symbolic analysis assume that all variables are of the same type, i.e., all are multi-valued lists (modal, or non-modal), all intervals, all histograms, and so forth. However, when the analytical procedures are based on dissimilarities/distances between observations, it is possible to partition observations of mixed variable types. This is illustrated briefly in the following examples.

**Example 6.10** Consider the household characteristics data of Table 6.1. In Example 6.1, the first two variables, $Y_1$ = fuel type and $Y_2$ = number of rooms in the house, were taken. These are both modal multi-valued list variables. Let us now include the third variable, a histogram-valued variable, $Y_3$ = household income. See Example 6.1 for details of these variables. For computational purposes, let us suppose the lower limit of the first histogram sub-interval is 0, and the upper limit of the last sub-interval is 250.

Suppose we use the extended Ichino–Yaguchi dissimilarities. These dissimilarities $d_i$ corresponding to the two multi-valued variables $Y_j, j = 1, 2$, respectively, were calculated for Example 6.1 using Eq. (4.1.15). For the histogram variable $Y_3$, we use Eq. (4.2.29) which in turn uses Eqs. (4.2.15), (4.2.21) and (4.2.22). Since, for these data, the histogram sub-intervals are all the same across all observations, it is not necessary to transform these histograms.

The distance between two observations, from Eq. (4.2.29), is based on the standard deviations of each observation and on the union and the intersection of those two observations. Without giving the details, the means, variances, and standard deviations of each observation are provided in Table 6.18, and those for all union and intersection pairs are provided in Table 6.25. Then, the extended Ichino–Yaguchi dissimilarities $d_j, j = 1, 2, 3$, and their sum over all three variables $d = d_1 + d_2 + d_3$, are shown in Table 6.26. The Euclidean extended Ichino–Yaguchi distances (ED) from Eq. (3.1.8), and the normalized (by $1/p$) Euclidean extended Ichino–Yaguchi distances (NED) from Eq. (3.1.9), each with weights $w_j = 1, j = 1, 2, 3$, are also shown in Table 6.26.

When we perform a $k$-medoid partitioning on the Euclidean extended Ichino–Yaguchi distances, we obtain the $K = 2$ partition $P = (C_1, C_2)$ with clusters, $C_1 = \{1, 7, \ldots, 12\}$ and $C_2 = \{2, \ldots, 6\}$. Furthermore, the same partitions were obtained when the extended Ichino–Yaguchi distances and when the normalized Euclidean extended Ichino–Yaguchi distances were used. We omit the details (see Exercise 6.2). □

When sub-intervals of histogram-valued observations are treated as discrete values, the Korenjak-Černe et al. (2011) and Batagelj et al. (2015) adapted

**Table 6.18** Statistics for $Y_3$ households (Example 6.10)

| House | Mean | Variance | Standard deviation | House | Mean | Variance | Standard deviation |
|---|---|---|---|---|---|---|---|
| $u$ | $\bar{Y}$ | $S^2$ | $S$ | $u$ | $\bar{Y}$ | $S^2$ | $S$ |
| 1 | 55.353 | 2069.742 | 45.494 | 7 | 45.940 | 1671.866 | 40.888 |
| 2 | 40.288 | 1135.459 | 33.697 | 8 | 69.748 | 2965.311 | 54.455 |
| 3 | 38.888 | 1459.054 | 38.198 | 9 | 52.818 | 2173.303 | 46.619 |
| 4 | 40.208 | 1298.032 | 36.028 | 10 | 54.620 | 1965.372 | 44.333 |
| 5 | 34.675 | 1054.894 | 32.479 | 11 | 43.305 | 1545.418 | 39.312 |
| 6 | 36.300 | 1248.393 | 35.333 | 12 | 51.903 | 1874.106 | 43.291 |

leader's algorithm for partitioning for variables with a discrete distribution also applies to mixed variable types, as shown in the following example.

**Example 6.11** Consider again the household characteristics data of Table 6.1 and all three variables (as in Example 6.10). In Example 6.2, the first two variables were used. Here, we use all three variables. Suppose we seek a partition with $K = 3$ clusters, and suppose we set the initial partition as $P^{(0)} = (C_1^{(0)}, C_2^{(0)}, C_3^{(0)})$ with $C_1^{(0)} = \{1, 2, 3, 4\}$, $C_2^{(0)} = \{5, 6, 7, 8\}$, and $C_3^{(0)} = \{9, 10, 11, 12\}$. The distances between the leaders for this initial partition and each observation are shown in Table 6.19.

Following the same procedure as used in Example 6.2, we obtain the final partition to be $P = (C_1, C_2, C_3)$ with $C_1 = \{1, 2, 3, 8, 11\}$, $C_2 = \{4, 5, 6, 7\}$, and $C_3 = \{9, 10, 12\}$.  □

## 6.6 Mixture Distribution Methods

We now consider a case where each data value is a distribution function $F_{uj}(y_{uj})$ where $F_{uj}(y_{uj}) = P(Y_{uj} \le y_{uj})$, $u = 1, \ldots, m, j = 1, \ldots, p$. The goal is to partition these $m$ distributions into $K$ clusters.

As for the partitioning methods discussed thus far, after some initial set of clusters has been defined, an allocation/reallocation stage will move observations in and out of the given clusters until some stability, or convergence, occurs. Mixture distribution methods typically define the allocation/reallocation process in terms of maximizing a likelihood function that is based on a mixture model of underlying probability density functions. Each

**Table 6.19** Leader distances $(Y_1, Y_2, Y_3)$ (Example 6.11)

| County | Distances from leader $T_k$ in $C_k$ | | | |
|---|---|---|---|---|
| $u$ | $D_{T_1}$ | $D_{T_2}$ | $D_{T_3}$ | $\min_k(D_{T_k})$ |
| 1 | 0.017 | 0.061 | 0.067 | 0.017 |
| 2 | 0.016 | 0.064 | 0.046 | 0.016 |
| 3 | 0.035 | 0.090 | 0.077 | 0.035 |
| 4 | 0.070 | 0.017 | 0.093 | 0.017 |
| 5 | 0.109 | 0.040 | 0.126 | 0.040 |
| 6 | 0.046 | 0.006 | 0.076 | 0.006 |
| 7 | 0.016 | 0.002 | 0.052 | 0.002 |
| 8 | 0.025 | 0.052 | 0.069 | 0.025 |
| 9 | 0.131 | 0.165 | 0.037 | 0.037 |
| 10 | 0.026 | 0.039 | 0.003 | 0.003 |
| 11 | 0.011 | 0.036 | 0.035 | 0.011 |
| 12 | 0.059 | 0.066 | 0.007 | 0.007 |

observation $F_u(\cdot)$ is constructed from the random variable $\mathbf{Y}_u$, $u = 1, \ldots, m$, with values in $\mathbb{R}^p$.

First, we recall the classical mixture decomposition process where the data values are points. Then, the $m$ observations in $\Omega$ fall into $K$ distinct patterns/clusters, $P = (C_1, \ldots, C_K)$. The probability density function of observations described by $\mathbf{Y}_u$, $u = 1, \ldots, m_k$, in cluster $C_k$ is assumed to be $\phi_k(\mathbf{y}_u; \boldsymbol{\alpha}_k)$, $k = 1, \ldots, K$, where $\boldsymbol{\alpha}_k$ is the set of parameters associated with the probability density function of the $k$th partition. Thus, for example, if the probability density function $\phi_k(\cdot)$ is a multivariate normal distribution, then $\boldsymbol{\alpha}_k = (\boldsymbol{\mu}_k, \boldsymbol{\Sigma}_k)$ where $\boldsymbol{\mu}_k$ is a $p$-vector of means and $\boldsymbol{\Sigma}_k$ is a $p \times p$ variance–covariance matrix. Then, an allocation/reallocation process in terms of maximizing a likelihood takes place. See Diday and Schroeder (1976) where the representation of any class in the searched partition in the dynamic clustering method is a density function.

For the Diday and Schroeder (1976) approach, an observation is itself either in or not in a given cluster, and so is a non-fuzzy approach and leads to an exact partition. In contrast, in the standard mixture decomposition method (of Friedman and Rubin, 1967), the $m$ observations fall into $K$ fuzzy patterns/clusters $P = (C_1, \ldots, C_K)$ with each point observation having a probability of falling into a given sub-cluster rather that fully being in some particular sub-cluster. For this model, the probability density function of an observation in $\Omega$ is assumed to be the mixture

$$\phi(\mathbf{y}_u; \boldsymbol{\alpha}) = \sum_{k=1}^{K} p_k \phi_k(\mathbf{y}_u; \boldsymbol{\alpha}_k), \quad \sum_{k=1}^{K} p_k = 1, \tag{6.6.1}$$

where the mixing parameter, $p_k \geq 0$, is the weight associated with the distribution $\phi_k(\mathbf{y}_u; \boldsymbol{\alpha}_k)$, in the cluster $C_k$, and where $\boldsymbol{\alpha}_k$ is the set of parameters for the probability density function $\phi_k(\cdot)$, $k = 1, \ldots, K$. The complete set of parameters $\boldsymbol{\alpha} = (p_k, \boldsymbol{\alpha}_k, \ k = 1, \ldots, K)$ is estimated from the observations, often by the maximum likelihood estimation method. The partitioning problem then is to find those clusters in $\Omega$ for which Eq. (6.6.1) holds. When the distributions $\phi_k(\mathbf{y}_u; \boldsymbol{\alpha}_k)$ are multivariate normally distributed as $N(\boldsymbol{\mu}_k, \boldsymbol{\Sigma}_k)$ with $\boldsymbol{\Sigma}_k = \mathbf{I}\sigma_k$, then this mixture distribution partitioning is an extension of the $k$-means method.

In the case where the data values are distributions, some work has been done. We briefly consider one approach applied to distributional data (see Vrac (2002) and Vrac et al. (2012)), based on the exact approach of Diday and Schroeder (1976).

Suppose the data consist of observations described by the cumulative distribution functions, or simply distributions, $\mathbf{F}_u = (F_{u1}, \ldots, F_{up})$, $u = 1, \ldots, m$, where $F_j$ is the distribution associated with the random variable $Y_j$, $j = 1, \ldots, p$, with parameter $\theta_{uj}$. The $F_j$ can be (i) a known distribution, such as an exponential distribution with parameter $\beta = 1$, say, (ii) a known distribution whose parameters are unknown, such as a normal distribution with unknown parameters $(\mu_j, \sigma_j^2)$ estimated by the sample values $\hat{\mu}_j = \bar{Y}_j$ and $\hat{\sigma}_j^2 = S^2$, say, or (iii) a completely unknown distribution with unknown parameters to be estimated from a subset of $r$ observations by, for example, kernel density estimation techniques, such as would pertain if each observation $u$ is the result of the aggregation of $r$ other values. In practice, the data will usually emerge from either (ii) or (iii). For notational simplicity, we write $F_{uj}$, $u = 1, \dots, m$, $j = 1, \dots, p$, regardless of whether it is a known or an estimated value $\hat{F}_{uj}$. The parameters $\theta_j$ of these data distributions are different from the parameters $\alpha_k$, $k = 1, \dots, K$, of the mixing model of Eq. (6.6.1).

We need the concepts of a distribution function of distributions and the joint distribution function of distributions. For mathematical descriptive simplicity, we take the number of variables as $p = 1$; see Vrac et al. (2012) for the relevant formulae when $p > 1$.

**Definition 6.1**  Let $(F_1, \dots, F_m) = \mathcal{F}$ be a random sample of $m$ distributions associated with the single random variable $Y$. Then, a **distribution function of distribution values** at the point $Z$ is

$$G_Z(z) = \mathbb{P}(F(Z) \le z), \quad \text{for all } z \in \mathbb{R}, \tag{6.6.2}$$

where the domain of $Z$ corresponds to the domain of $\mathcal{F}$.  □

The function $G_Z(z)$ can be empirically modeled by

$$G_Z^e(z) = c(F_u \in \mathcal{F}; F_u(Z) \le z, u = 1, \dots, m)/c(\mathcal{F}) \tag{6.6.3}$$

where $c(A)$ is the cardinality equal to the number of elements in $A$.

**Example 6.12**  Suppose we have the $m = 4$ distributions $\{F_u, u = 1, \dots, 4\}$ shown in Figure 6.6. Then, the empirical distribution $G_Z^e(z) \equiv G_Z(z)$ is the percentage of distributions taking a value smaller than or equal to $z_i$ at the point $Z_i$. In this example, from Eq. (6.6.3), $G_{Z_1}(0.2) = 3/4$ and $G_{Z_1}(0.7) = 1$ for $Z_1 = 6$, and $G_{Z_2}(0.2) = 0$ and $G_{Z_2}(0.7) = 2/4 = 1/2$ for $Z_2 = 13.5$.  □

Instead of an empirical approach, this distribution can be estimated by a parametric approach. For example, $G(z)$ can follow a beta distribution

$$f(z; \boldsymbol{\alpha}) = \frac{z^{a_1-1}(1-z)^{a_2-1}}{\int_0^1 z^{a_1-1}(1-z)^{a_2-1}dz}, \quad 0 < z < 1, \tag{6.6.4}$$

with parameters $\boldsymbol{\alpha} = (a_1, a_2)$ with $a_i > 0$, $i = 1, 2$, so that

$$G(z; \boldsymbol{\alpha}) = \int_0^z f(t; \boldsymbol{\alpha})dt \tag{6.6.5}$$

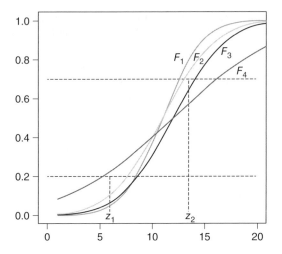

**Figure 6.6** Distribution of distributions (Example 6.12).

and where the parameters $\alpha$ are estimated by traditional techniques. Other parametric distributions can also be used instead of Eq. (6.6.4).

**Definition 6.2** A **joint distribution function of distribution values**, $H_Z(\cdot)$, is defined at the point $\mathbf{Z} = (Z_1, \dots, Z_n)$, as

$$H_{\mathbf{Z}}(z_1, \dots, z_n) = \mathbb{P}(F_u \in \mathcal{F}; F_u(Z_1) \leq z_1, \dots, F_u(Z_n) \leq z_n,\ u = 1, \dots, m). \tag{6.6.6}$$

$\square$

Since many different joint distributions can give the same set of marginal distributions, knowing the marginal distributions alone is insufficient to describe dependencies between the variables. However, the use of copula theory enables us to gain some measure of the underlying dependencies. In particular, by Sklar's (1959) theorem, there exists a copula $C(\cdot)$ linking the marginal distributions $G_{Z_i}(z_i)$, $i = 1, \dots, n$, and the joint distribution $H_{\mathbf{Z}}(\mathbf{z})$ such that, for all $(z_1, \dots, z_n) \in \mathbb{R}^n$,

$$H_{Z_1, \dots, Z_n}(z_1, \dots, z_n) = C(G_{Z_1}(z_1), \dots, G_{Z_n}(z_n)). \tag{6.6.7}$$

If the distributions $G_{Z_i}(z_i)$, $i = 1, \dots, n$, are continuous, then $C$ is unique. An in-depth treatment of copulas can be found in Nelsen (2007). Also, each of the three functions, $H(\cdot)$, $C(\cdot)$, $G(\cdot)$, in Eq. (6.6.7) can be a parametric function (and so include a parameter vector) or a non-parametric function.

The mixture decomposition of Eq. (6.6.1) can be applied to the joint distribution $H_{\mathbf{Z}}(z_1, \dots, z_n) \equiv H(z_1, \dots, z_n)$ to give

$$H(z_1, \dots, z_n; \boldsymbol{\gamma}) = \sum_{k=1}^{K} p_k H_k(z_1, \dots, z_n; \boldsymbol{\gamma}_k). \tag{6.6.8}$$

Then, applying Eq. (6.6.7) to Eq. (6.6.8), we obtain

$$H(z, \dots, z_n; \gamma) = \sum_{k=1}^{K} p_k C_k (G_{Z_1}^k(z_1; \alpha_1^k), \dots, G_{Z_n}^k(z_n; \alpha_n^k); \beta_k) \qquad (6.6.9)$$

where in Eq. (6.6.9) we write that the marginal distribution $G_{Z_i}^k(\cdot)$ has parameter $\alpha_i^k$, $i = 1, \dots, n$, associated with variable $Y$, and the copula $C_k(\cdot)$ has parameter $\beta_k$, for the $k$th cluster, $k = 1, \dots, K$; the $p_k$ are the mixing parameters. If we write $h(\cdot) \equiv h(z_1, \dots, z_n; \gamma_k)$, then we can show that the probability density function associated with the distribution function $H(\cdot) \equiv H(z_1, \dots, z_n; \gamma)$ becomes

$$h(z_1, \dots, z_n; \gamma) = \sum_{k=1}^{K} p_k \frac{\partial^n H_k}{\partial z_1 \dots \partial z_n} = \sum_{k=1}^{K} p_k \left\{ \prod_{i=1}^{n} \frac{dG_{Z_i}^k(z_i; \alpha_i^k)}{dz_i} \right\}$$

$$\times \frac{\partial^n}{\partial z_1 \dots \partial z_n} C_k(G_{Z_1}^k(z_1; \alpha_1^k), \dots, G_{Z_n}^k(z_n; \alpha_n^k); \beta_k)$$

$$(6.6.10)$$

where $\gamma = (\alpha_i^k, \beta_k, p_k, i = 1, \dots, n, k = 1, \dots, K)$ is the full set of parameters.

There are many possible criteria, $W(P, \gamma)$, that can be used to determine the best partition of $\Omega$ into $P = (C_1, \dots, C_K)$. The likelihood function of the mixture of Eq. (6.6.10) is one such criterion. Another criterion is its log-likelihood function,

$$W(P, \gamma) = \sum_{u=1}^{m} \ln \left[ \sum_{k=1}^{K} p_k h_k(F_u(Z_1), \dots, F_u(Z_n); \gamma_k) \right]; \qquad (6.6.11)$$

this $W(P, \gamma)$ is also called a mixture criterion (see, e.g., Symons (1981)).

The partition is formed as follows:

**Step 1:** Initialize the partition as $P^{(0)} = (C_1^{(0)}, \dots, C_K^{(0)})$.

**Step 2:** Let the partition after the $r$th iteration be $P^{(r)} = (C_1^{(r)}, \dots, C_K^{(r)})$. Then, the allocation step consists of two successive and iterative steps.

**Step 2(i):** Estimate the parameters of the mixture distribution in Eq. (6.6.10) by maximizing the criterion $W(P, \gamma)$ of Eq. (6.6.11), based on $P^{(r)}$, to give $C_k^{(r+1)}$ and $\gamma_k^{(r+1)}$; and

**Step 2(ii):** Obtain the new partition $P^{(r+1)} = \{C_k^{(r+1)}, k = 1, \dots, K\}$ where the new cluster $C_k^{(r+1)}$

$$C_k^{(r+1)} = \{F_u; p_k^{(r+1)} h_k(F_u; \gamma_k^{(r+1)}) \ge p_v^{(r+1)} h_v(F_u; \gamma_v^{(r+1)})$$

$$\text{for all } v \neq k, v = 1, \dots, K\}. \qquad (6.6.12)$$

**Step 3:** When $|W(P^{(r+1)}, \gamma^{(r+1)}) - W(P^{(r)}, \gamma^{(r)})| < \epsilon$, for some preassigned small value of $\epsilon$, the process stops.

The algorithm starts with the criterion $W(P, \gamma)$ of Eq. (6.6.11), which in turn is based on Eq. (6.6.10). This means that in addition to estimating the parameters $\gamma = (\alpha_i^k, \beta_k, p_k, i = 1, \ldots, n, k = 1, \ldots, K)$, it is also necessary to identify a suitable distribution of distributions $G(\cdot)$ function and a copula $C(\cdot)$ function.

There are many approaches to estimating $G(\cdot)$. For example, non-parametric estimation includes empirical estimation such as that of Eq. (6.6.3), or one of the many possible kernel density estimations, described in, for example, Silverman (1986). Or, $G_{Z_i}^k(z_i; \alpha_i^k), i = 1, \ldots, n, k = 1, \ldots, K$, can be estimated by a parametric approach such as the use of the Dirichlet distribution which for $p = 1$ becomes the beta distribution of Eq. (6.6.4), with parameters estimated in many ways such as maximum likelihood estimation or method of moments estimation.

The copula functions $C_k(z_1, z_2, \ldots; \beta_k), k = 1, \ldots, K$, can also be estimated by non-parametric methods or they can be specified as taking some specific class of copulas, such as the Frank (1979) family of Archimedean copulas, given by, for $n = 2$, and $(z_1, z_2) \in [0, 1]^n$,

$$C(z_1, z_2; \beta) = (ln\beta)^{-1} ln\{1 + [(\beta^{z_1} - 1)(\beta^{z_2} - 1)]/(\beta - 1)\} \qquad (6.6.13)$$

for $\beta > 0$ and $\beta \neq 1$ with

$$\frac{\partial^2 C}{\partial z_1 \partial z_2} = \frac{(\beta - 1)\beta^{z_1 + z_2} ln\beta}{[(\beta - 1) + (\beta_1^z - 1)(\beta_2^z - 1)]^2}. \qquad (6.6.14)$$

The parameter $\beta$ is estimated in the usual way such as by the maximum likelihood method or the method of moments. Other classes include the Clayton (1978), Genest-Ghoudi (1994), and Ali-Mikhail-Haq (1978) classes, among other Archimedean classes described in Nelsen (2007).

Finally, the mixing parameters are typically estimated from

$$\hat{p}_k = \frac{c(P_k)}{c(F)}. \qquad (6.6.15)$$

This algorithm is a variant of the $k$-means method, and as such is an example of a dynamical partitioning, applied to mixture distributions. It converges to a local optimum solution in a finite number of iterations (see Vrac et al. (2012)).

**Example 6.13** To illustrate this methodology, humidity distributions were obtained from the 24 altitude values measured at each of the $m = 16200$ latitude–longitude grid points, set $2°$ apart, at a specific point in time (midnight on 15 December 1999). (The data are extracted from a larger atmospheric dataset covering the globe from the European Center for Medium-range Weather Forecasts (ECMWF) located in Reading, UK.)

The humidity profiles, $F_u$, $u = 1, \ldots, m$, were estimated by the Parzen method for the kernal density estimation approach (see, e.g., Silverman

(1986)). From these, for each $k = 1, \ldots, K$, the distributions $G^k_{Z_i}(z_i)$ were assumed to follow a beta distribution (see Eqs. (6.6.4) and (6.6.5)), and were calculated at each of $i = 1, 2$ (i.e., $n = 2$) values. A Frank copula for $C(\cdot)$ (see Eq. (6.6.13)) was assumed. The partition with $K = 7$ clusters that emerged is shown in Figure 6.7 (from Vrac (2002)). The probability density functions $f_k(\cdot)$ of the humidity for each region, $k = 1, \ldots, 7$, are shown in Figure 6.8 (from Vrac (2002)). The corresponding estimated values for the copula parameters,

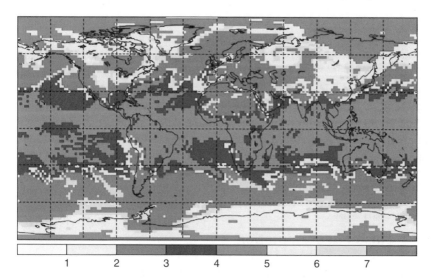

**Figure 6.7** Model-based partition for humidity: Frank copula (Example 6.13).

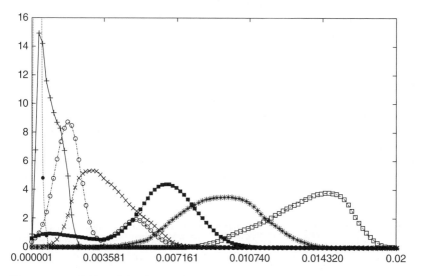

**Figure 6.8** Model-based distributions for humidity: Frank copula (Example 6.13).

**Table 6.20** Parameter estimates for $K = 7$ clusters: humidity (Example 6.13)

| $k$ Region | $\beta_k$ | $a_1^k - x_1$ | $a_2^k - x_1$ | $a_1^k - x_2$ | $a_2^k - x_2$ |
|---|---|---|---|---|---|
| 1 South Polar | 0.2000 | 1.7727 | 36.192 | 70.000 | 24.485 |
| 2 Temperate | 0.0169 | 6.2930 | 742.66 | 30.005 | 13.470 |
| 3 Sub-tropical | 0.6191 | $1 \times 10^{-6}$ | 12.617 | 16.619 | 20.368 |
| 4 Tropical | 0.4450 | $1 \times 10^{-6}$ | 12.617 | 29.425 | 48.242 |
| 5 Sub-temperate | 0.1000 | $1 \times 10^{-6}$ | 12.617 | 6.8909 | 5.8870 |
| 6 Sub-polar | 0.0206 | $1 \times 10^{-6}$ | 12.617 | 38.932 | 14.840 |
| 7 North Polar | 0.0178 | 2.2154 | 23.266 | 70.000 | 18.847 |

$\beta_k$, $k = 1, \ldots, 7$, are shown in Table 6.20 (from Vrac (2002)); the differing values for these estimates reflect the fact that different regions themselves have different humidity patterns. Also shown are estimates for the beta distributions parameters $\alpha^k = (a_1^k, a_2^k)$ for each of the $n = 2$ values $Z_1 = z_1$ and $Z_2 = z_2$ used in the analysis, for each cluster $k = 1, \ldots, K$. More complete details can be found in Vrac (2002) and Vrac et al. (2012). □

## 6.7 Cluster Representation

As indicated in Chapter 1, summary representations of the final clusters can be determined for all cluster nodes of all partitions and hierarchies described in Chapters 5–8. There are many ways to characterize the features of these cluster nodes and many possible approaches to calculating their representations. Some such considerations are illustrated in this chapter only.

The easiest formulation of a node is simply to provide a listing of the observations contained in any one cluster. For classical observations, the cluster is frequently described by the average of the observations contained in that cluster. Certainly, that is also possible for symbolic valued observations. In this case, the cluster means can be calculated from Eq. (2.4.1) for multi-valued list data, from Eq. (2.4.2) for interval-valued data, and from Eq. (2.4.3) for histogram-valued data. Another possibility, also derived from classical methods, is to take a representative observation such as a median value of those in a given cluster.

The following examples provide a mechanism for calculating a representation that itself is symbolic valued.

**Example 6.14** Consider the partitioning of Example 6.1, based on the two multi-valued list variables $Y_1$ and $Y_2$. The $m = 12$ observations were partitioned into clusters $C_1 = \{1, 2, 3\}$, $C_2 = \{4, 5, 6, 7\}$, $C_3 = \{9, 10, 12\}$, and $C_4 = \{8, 11\}$.

**Table 6.21** Cluster representations (Example 6.14)

| Cluster | $Y_1$ = Fuel types | | | | | $Y_2$ = Number of rooms | | |
|---|---|---|---|---|---|---|---|---|
| | Gas | Electric | Oil | Wood | None | {1,2} | {3,4,5} | ≥ 6 |
| $C_1$ | 0.591 | 0.110 | 0.236 | 0.057 | 0.006 | 0.100 | 0.513 | 0.387 |
| $C_2$ | 0.220 | 0.740 | 0.009 | 0.026 | 0.005 | 0.038 | 0.525 | 0.437 |
| $C_3$ | 0.193 | 0.217 | 0.039 | 0.546 | 0.005 | 0.121 | 0.509 | 0.370 |
| $C_4$ | 0.686 | 0.236 | 0.003 | 0.070 | 0.005 | 0.121 | 0.511 | 0.368 |

Then, from Eq. (2.4.1) and Table 6.1, it is easy to show, for $C_1$ say, that a representation observation for variable $Y_1$ is

$$Y_1^{(C_1)} = \{\text{gas}, 0.591; \text{electric}, 0.110; \text{oil}, 0.236; \text{wood}, 0.057; \text{none}, 0.006\}.$$

Likewise, the representation for each variable and each cluster in the partition can be found, as shown in Table 6.21. □

**Example 6.15** Take now the interval-valued observations of Table 3.11 and Example 6.5 for temperatures at $m = 10$ China stations. The partition gave clusters $C_1 = \{1, 2\}, C_2 = \{3, 4, 5\}, C_3 = \{6, 7\}$, and $C_4 = \{8, 9, 10\}$. Consider the first cluster $C_1$ and variable $Y_1$. For the observations in $C_1$, from Table 3.11, we see that $Y_1$ takes the values $[-18.4, -7.5]$ and $[-20.0, -9.6]$, respectively. If we set our representation as the interval $[-20.0, -7.5]$, the assumption that values across a symbolic valued interval are uniformly distributed cannot be sustained, since for the overlapping portion, $[-18.5, -9.6]$, there is a contribution from each of the two individual intervals. While this uniformity assumption may hold for each of the individual intervals, it certainly cannot hold for their union.

A similar problem pertains if a representation interval were obtained by averaging the respective endpoints, to give in this case the interval $[-19.2, -8.55]$; again an assumption of uniformity across this interval cannot be sustained. Neither can it be sustained for the midpoint-range approach proposed in Chavent and Lechevallier (2002) and de Carvalho et al. (2006). In this latter case, in a given cluster, for each variable $j$, the median $m_j$ of the observation midpoints is taken, along with the median $l_j$ of the observation half-lengths, with the representation interval being $[m_j - l_j, m_j + l_j]$. Likewise, taking the representation endpoints as the median of the cluster observations endpoints, as suggested in de Souza and de Carvalho (2004), has the same drawbacks.

Instead, the representation value for $Y_1$ is more accurately expressed by the histogram formed by the individual intervals. In this case, one representation value for $Y_1$ in $C_1$ is the histogram $\{[-20.0, -18.4), 0.077; [-18.4, -9.6), 0.827;$

**Table 6.22** Cluster representations for $Y_1, \ldots, Y_4$ (Example 6.15)

| Cluster | $Y_1$ | $Y_2$ |
|---|---|---|
| $C_1$ | $\{[-20, -15), 396; [-15, -10),$ $0.470; [-10, -5], 0.134\}$ | $\{[-0.2, 4.8), 0.381; [4.8, 9.8),$ $0.402; [9.8, 14], 0.217\}$ |
| $C_2$ | $\{[-30, -15), 0.667; [-15, 0),$ $0.161; [0, 10], 0.172\}$ | $\{[-4.5, 2.5), 0.284; [2.5, 9.5),$ $0.498; [9.5, 16.5], 0.218\}$ |
| $C_3$ | $\{[2.0, 7.5), 0.349; [7.5, 12.5),$ $0.352; [12.5, 17.5], 0.299\}$ | $\{[9, 14), 0.281; [14, 20), 0.439; [20,$ $25], 0.280\}$ |
| $C_4$ | $\{[10, 13), 0.265; [13, 16), 0.426;$ $[16, 20], 0.309\}$ | $\{[15, 18), 0.186; [18, 22), 0.584;$ $[22, 25], 0.230\}$ |

| | $Y_3$ | $Y_4$ |
|---|---|---|
| $C_1$ | $\{[15, 20), 0.275; [20, 25), 0.529;$ $[25, 30], 0.196\}$ | $\{[-0.2, 5), 0.381; [5, 10), 0.397;$ $[10, 15], 0.222\}$ |
| $C_2$ | $\{[10, 15), 0.182; [15, 20), 0.446;$ $[20, 25], 0.372\}$ | $\{[-4, 4), 0.420; [4, 13), 0.470; [13,$ $19], 0.110\}$ |
| $C_3$ | $\{[15, 20), 0.447; [20, 25), 0.553\}$ | $\{[12, 16), 0.330; [16, 20), 0.476;$ $[20, 24], 0.194\}$ |
| $C_4$ | $\{[24, 28), 0.341; [28, 32), 0.503;$ $[32, 36], 0.156\}$ | $\{[19, 22), 0.274; [22, 25), 0.416;$ $[25, 28], 0.310\}$ |

$[-9.6, -7.5], 0.096\}$. Note, however, there are many possible histogram representations that can be calculated here, depending on the actual histogram sub-intervals chosen. One such histogram representation for all variables and all partitions is displayed in Table 6.22. These histograms are constructed using the relative frequency histogram methods of Chapter 2 for interval data.

Since variable $Y_5$ = elevation is classically valued, the representation for this case is simply the sample mean of the corresponding values in each cluster. These are, respectively, $\bar{Y}_5 = 4.13, 30.93, 23.59$, and $1.28$ for clusters $C_1, C_2, C_3$, and $C_4$. □

**Example 6.16** As an example involving histogram observations, let us consider the obtained partition from Example 6.7. The representations for each cluster are the histograms constructed for histogram observations following the relative histogram methods of Chapter 2. For cluster $C_1$ = {*Versicolor, Virginica*}, the resulting representations for cluster $C_1$ = {*Versicolor, Virginica*} for $Y_1$ = sepal width, $Y_2$ = sepal length, $Y_3$ = petal width, and $Y_4$ = petal length are, respectively,

$$Y_1 = \{[1.5, 2.0), 0.020; [2.0, 2.5), 0.133; [2.5, 3.0), 0.432; [3.0, 3.5), 0.335;$$
$$[3.5, 4.0], 0.080\},$$

**Table 6.23** Cluster representations: households (Example 6.17)

| Cluster | $Y_1$ = Fuel types | | | | | $Y_2$ = Number of rooms | | |
|---|---|---|---|---|---|---|---|---|
| k | Gas | Electric | Oil | Wood | None | {1, 2} | {3, 4, 5} | ≥ 6 |
| $C_1$ | 0.629 | 0.160 | 0.143 | 0.062 | 0.006 | 0.109 | 0.512 | 0.379 |
| $C_2$ | 0.220 | 0.740 | 0.009 | 0.026 | 0.005 | 0.038 | 0.525 | 0.437 |
| $C_3$ | 0.193 | 0.217 | 0.039 | 0.546 | 0.005 | 0.121 | 0.509 | 0.370 |

| | $Y_3$ = Household income ($1000) | | | | | | | |
|---|---|---|---|---|---|---|---|---|
| | < 10 | [10, 25) | [25, 50) | [50, 75) | [75, 100) | [100, 150) | [150, 200) | ≥ 200 |
| $P_1$ | 0.131 | 0.216 | 0.287 | 0.174 | 0.089 | 0.066 | 0.019 | 0.018 |
| $P_2$ | 0.172 | 0.262 | 0.305 | 0.152 | 0.056 | 0.034 | 0.009 | 0.010 |
| $P_3$ | 0.098 | 0.188 | 0.300 | 0.206 | 0.106 | 0.064 | 0.014 | 0.024 |

$Y_2 = \{[4.5, 5.5), 0.137; [5.5, 6.5), 0.470; [6.5, 7.5), 0.310; [7.5, 8.5], 0.083\},$

$Y_3 = \{[1.0, 1.5), 0.480; [1.5, 2.0), 0.290; [2.0, 2.5), 0.2075; [2.5, 3.0],$
$\quad 0.0225\},$

$Y_4 = \{[3.0, 4.0), 0.160; [4.0, 5.0), 0.455; [5.0, 6.0), 0.240; [6.0, 7.0), 0.125;$
$\quad [7.0, 8.0], 0.020\}.$

The second cluster $C_2 = \{Setosa\}$ consists of itself; hence, the representation is itself, as shown in Table 4.10. □

**Example 6.17** These principles can be combined to calculate the corresponding representations for the mixed variable household data of Example 6.11. For example, the representation for $Y_2$ = number of bedrooms in cluster $C_3 = \{9, 10, 12\}$ for $Y_2 = \{3, 4, 5\}$ bedrooms has relative frequency $p_{t22} = (0.486 + 0.523 + 0.517)/3 = 0.121$. The results are summarized in Table 6.23. □

Note that while these representations refer to the final clusters, these representations can be akin to, but are not necessarily the same as, the centroid values used for sub-clusters in the course of the construction of the final hierarchy.

## 6.8 Other Issues

Other work specifically defined for symbolic data can be found in the literature. These include de Carvalho et al. (2006), who consider an $L_2$ distance for interval observations, Chavent et al. (2006), who use Hausdorff distances for intervals, de Sousa and de Carvalho (2004), , de Carvalho et al. (2008) and de Carvalho and Lechevallier (2009), who apply city-block distances to intervals, and de

Carvalho et al. (2004), who use Chebychev distances on intervals; the distances are usually based on the interval vertices (or, equivalently, the center-range values) so that for the $k$-means method, the representation of the obtained clusters (in step 2) is based on the means of the vertices. Verde and Irpino (2007), Irpino et al. (2006), Košmelj and Billard (2012), and Irpino et al. (2014) consider Mallows' distance for histogram observations, while Korenjak-Černe et al. (2011) and Batagelj et al. (2015) extend the $k$-means leaders approach to discrete distributions.

A promising new application relates to the work by Lechevallier et al. (2010) on clustering dissimilarity tables, motivated by categorizing documents. In a different direction not covered in this chapter, some algorithms are based on interval end points, e.g., de Souza et al. (2004), de Souza and de Carvalho (2004), Irpino and Verde (2008), and de Carvalho and Lechevallier (2009), among others. Bock (2003) and Hajjar and Hamdan (2013) considered a partitioning of interval data using self-organizing maps (e.g., Kohonen, 2001).

There are many unanswered questions surrounding the use of partitioning procedures. A major issue relates to what might be the optimal number of clusters $K$. Since usually it is not possible to know how many clusters is optimal, typically an algorithm is run several times for several possible $K$ clusters. One way to select a final $K$ value is to use the approximate weight of evidence (AWE) criterion discussed in Banfield and Raftery (1993). An example of this approach is its use in the determination that $K = 7$ in the partitioning of Example 6.13. Another approach is to use the Bayes information criteria (BIC) of Schwarz (1978). Caliński and Harabasz (1974) introduced a statistic based on the between and within sums of squares. Fraley and Raftery (1998) also study this question for mixture models.

As illustrated in some of the examples in this chapter, the use of different initial seeds, the use of different dissimilarity or distance measures, the use of different algorithms, and the use of standardized or non-standardized observations, among other approaches, can produce different clusters in the overall partition. There are also questions related to the treatment of outliers and influential observations (see, e.g., Cheng and Milligan (1996)), variable selection (see, e.g., Foulkes et al. (1988)), and the use of weighted observations, validation and cross-validation, sensitivity to choices of seeds, algorithms, and distance measures, among others. A comprehensive summary of such issues can be found in Steinley (2006) and Jain (2010). How these questions should be resolved, if at all, remains as an outstanding issue for future research. Except where otherwise indicated, references in this section refer to studies based on classical observations. How these answers might differ for symbolic-valued observations compared to classical observations is also unknown at this time.

# Exercises

**6.1** Verify that when the normalized Euclidean Ichino–Yaguchi distances (provided in Table 6.4) are used in the $k$-medoids method applied to the household data of Table 6.1, using $Y_1$ and $Y_2$ only, the best $K = 4$ partitioning is $P = (C_1, \dots, C_4)$ with $C_1 = \{1, 2, 3\}$, $C_2 = \{4, 5, 6\}$, $C_3 = \{7, 11\}$, and $C_4 = \{9, 10, 11\}$. Show that when $K = 3$, the partition is $P = (C_1, C_2, C_3)$ with clusters $C_1 = \{1, 2, 4, 5, 6, 7, 8, 10, 11, 12\}$, $C_2 = \{3\}$, and $C_3 = \{9\}$.

**6.2** Refer to Example 6.10. Verify the distances given in Table 6.26 are as shown when all three variables are taken in the household characteristics data of Table 6.1. How does the partitioning change when all three variables are used compared with the partition obtained when using just $Y_1$ and $Y_2$?

**6.3** Take the China weather station non-standardized data of Table 6.8. Show that when we seek $K = 4$ clusters, but with initial seeds being observations $u = 1, 2, 5, 9$, we obtain the partition with clusters $C_1 = \{1, 2, 4\}$, $C_2 = \{3\}$, $C_3 = \{5, 6, 7\}$, and $C_4 = \{8, 9, 10\}$, with the sum of these distances as $E = 79.774$. How can you explain the difference between this result and that obtained in Example 6.3? What does this tell you about the $k$-means method in general?

**6.4** Take the China weather station data of Table 6.8. Find the $k$-means clusters based on the four temperature variables $(Y_1, \dots, Y_4)$ for each of $K = 3$ and $K = 4$ clusters for the non-standardized and the standardized data, respectively. How do you explain the differences you obtained between these partitions, if any?

**6.5** With reference to Example 6.5, show that application of the $k$-medoids method to the China weather data for all $p = 5$ variables produces the partitioning $P = (C_1, \dots, C_4)$ with $C_1 = \{1, 2\}$, $C_2 = \{3, 4, 5\}$, $C_3 = \{6, 7\}$, and $C_4 = \{8, 9, 10\}$ based on the Gowda–Diday dissimilarities, but the partition is $P = (C_1, C_2)$ with clusters $C_1 = \{1, \dots, 7\}$ and $C_2 = \{8, 9, 10\}$ when using the Euclidean Gowda–Diday distances.

**6.6** Refer to the China data of Example 6.5. Now find the partition that results from using the Hausdorff distances instead of the Gowda–Diday dissimilarities.

**Table 6.24** Number of flights by airline (Exercise 6.10).

| Airline $u$ | $n_u$ | Airline $u$ | $n_u$ |
|---|---|---|---|
| 1 | 6341 | 6 | 4877 |
| 2 | 5228 | 7 | 2294 |
| 3 | 726 | 8 | 5590 |
| 4 | 3175 | 9 | 2256 |
| 5 | 665 | 10 | 4232 |

**6.7** Re-do the $k$-means method applied to the China data in Example 6.4 using an adaptive $k$-means procedure.

**6.8** Consider the four temperature variables ($Y_1$ = January, $Y_2$ = April, $Y_3$ = July, $Y_4$ = October) given in Table 3.11 and used in Example 6.4 for weather stations in China.

(a) Compare the outcomes when using the $k$-means method (given in Table 6.8) as in Example 6.4 with those of the $k$-medoid method as in Example 6.5 and the Chavent–Lechevalier method as in Example 6.6. Take $K = 3$.

(b) Now run a $k$-means analysis on the vertices (given in Table 3.11). Add the results of this analysis to your comparisons of part (a).

**6.9** For the flights data set of Table 5.1, apply the $k$-means method to determine a partition with (a) $K = 4$ and (b) $K = 5$ clusters. Compare your results with those of Figure 5.9. (You can use the observation means of Table 5.4.)

**6.10** Table 6.24 gives the number $n_u$ of flights that were aggregated for each airline $u = 1, \dots, 16$, of Table 5.1. Re-do Exercise 6.9 where each observation $u$ is weighted proportionally to $n_u$.

**6.11** Repeat Exercise 6.9 but now apply the $k$-medoids method by using (i) extended Gowda–Diday dissimilarities and (ii) extended Ichino–Yaguchi distances. Compare your results with those for the $k$-means approach of Exercise 6.7.

**6.12** At the beginning of Chapter 6.3 it was suggested that by using dissimilarities/distances instead of means, more internal information is used in the partitioning procedures. Do you agree with this assertion? Why, why not?

**6.13** Refer to Example 6.8. Explain why the two partitions differ from each other.

# Appendix

Table 6.25 Statistics for $Y_3$ household union and intersection pairs (Example 6.10)

| House pair | Mean $\bar{Y}$ | Variance $S^2$ | Standard deviation $S$ | House pair | Mean $\bar{Y}$ | Variance $S^2$ | Standard deviation $S$ | House pair | Mean $\bar{Y}$ | Variance $S^2$ | Standard deviation $S$ | House pair | Mean $\bar{Y}$ | Variance $S^2$ | Standard deviation $S$ |
|---|---|---|---|---|---|---|---|---|---|---|---|---|---|---|---|
| | Union | | | | Intersection | | | | Union | | | | Intersection | | |
| 1∪2 | 50.623 | 2255.798 | 47.495 | 1∩2 | 43.925 | 1041.041 | 32.265 | 4∪8 | 58.625 | 3589.317 | 59.911 | 4∩8 | 48.646 | 1064.143 | 32.621 |
| 1∪3 | 48.012 | 2426.411 | 49.259 | 1∩3 | 45.831 | 1235.635 | 35.152 | 4∪9 | 48.618 | 2447.783 | 49.475 | 4∩9 | 43.705 | 1091.237 | 33.034 |
| 1∪4 | 49.583 | 2331.433 | 48.285 | 1∩4 | 45.336 | 1142.213 | 33.797 | 4∪10 | 48.632 | 2240.305 | 47.332 | 4∩10 | 45.749 | 1122.902 | 33.510 |
| 1∪5 | 47.585 | 2418.160 | 49.175 | 1∩5 | 40.845 | 898.821 | 29.980 | 4∪11 | 41.947 | 1601.890 | 40.024 | 4∩11 | 41.593 | 1246.425 | 35.305 |
| 1∪6 | 48.053 | 2397.758 | 48.967 | 1∩6 | 42.437 | 1086.782 | 32.966 | 4∪12 | 47.014 | 2080.877 | 45.617 | 4∩12 | 44.7946 | 1157.230 | 34.018 |
| 1∪7 | 51.799 | 2223.103 | 47.150 | 1∩7 | 49.220 | 1559.514 | 39.491 | 5∪6 | 37.291 | 1336.387 | 36.557 | 5∩6 | 33.429 | 960.794 | 30.997 |
| 1∪8 | 64.953 | 3200.431 | 56.572 | 1∩8 | 59.479 | 1923.470 | 43.857 | 5∪7 | 42.009 | 1814.826 | 42.601 | 5∩7 | 38.093 | 967.847 | 31.110 |
| 1∪9 | 56.743 | 2643.152 | 51.412 | 1∩9 | 50.965 | 1586.519 | 39.831 | 5∪8 | 56.199 | 3725.723 | 61.039 | 5∩8 | 43.795 | 842.391 | 29.024 |
| 1∪10 | 55.166 | 2139.030 | 46.250 | 1∩10 | 54.791 | 1896.283 | 43.546 | 5∪9 | 46.480 | 2496.443 | 49.964 | 5∩9 | 39.426 | 872.711 | 29.542 |
| 1∪11 | 50.776 | 2261.434 | 47.555 | 1∩11 | 47.488 | 1420.435 | 37.689 | 5∪10 | 46.796 | 2321.093 | 48.178 | 5∩10 | 41.222 | 883.356 | 29.721 |
| 1∪12 | 54.329 | 2104.066 | 45.870 | 1∩12 | 52.867 | 1844.666 | 42.950 | 5∪11 | 40.461 | 1652.591 | 40.652 | 5∩11 | 37.227 | 980.143 | 31.307 |
| 2∪3 | 38.437 | 1497.838 | 38.702 | 2∩3 | 41.094 | 1094.188 | 33.079 | 5∪12 | 45.332 | 2153.460 | 46.405 | 5∩12 | 40.243 | 911.486 | 30.191 |
| 2∪4 | 40.118 | 1339.819 | 36.604 | 2∩4 | 40.394 | 1093.637 | 33.070 | 6∪7 | 42.279 | 1801.495 | 42.444 | 6∩7 | 39.638 | 1161.794 | 34.085 |
| 2∪5 | 37.752 | 1270.097 | 35.638 | 2∩5 | 37.156 | 935.830 | 30.591 | 6∪8 | 56.820 | 3693.113 | 60.771 | 6∩8 | 45.539 | 1023.130 | 31.986 |
| 2∪6 | 38.509 | 1364.661 | 36.941 | 2∩6 | 38.036 | 1027.031 | 32.047 | 6∪9 | 46.925 | 2477.282 | 49.772 | 6∩9 | 41.047 | 1064.210 | 32.622 |
| 2∪7 | 44.403 | 1714.835 | 41.411 | 2∩7 | 41.624 | 1104.626 | 33.236 | 6∪10 | 47.231 | 2301.878 | 47.978 | 6∩10 | 42.797 | 1070.265 | 32.715 |
| 2∪8 | 59.451 | 3494.163 | 59.111 | 2∩8 | 47.085 | 970.224 | 31.148 | 6∪11 | 40.770 | 1639.244 | 40.488 | 6∩11 | 38.713 | 1177.292 | 34.312 |

(Continued)

**Table 6.25** (Continued)

| House pair | Mean $\bar{Y}$ | Variance $S^2$ | Standard deviation $S$ | House pair | Mean $\bar{Y}$ | Variance $S^2$ | Standard deviation $S$ | House pair | Mean $\bar{Y}$ | Variance $S^2$ | Standard deviation $S$ | House pair | Mean $\bar{Y}$ | Variance $S^2$ | Standard deviation $S$ |
|---|---|---|---|---|---|---|---|---|---|---|---|---|---|---|---|
| | Union | | | | Intersection | | | | Union | | | | Intersection | | |
| 2∪9 | 49.145 | 2316.091 | 48.126 | 2∩9 | 43.031 | 1052.911 | 32.449 | 6∪12 | 45.654 | 2137.295 | 46.231 | 6∩12 | 41.858 | 1099.959 | 33.166 |
| 2∪10 | 49.805 | 2164.540 | 46.525 | 2∩10 | 44.240 | 1023.886 | 31.998 | 7∪8 | 60.922 | 3432.942 | 58.591 | 7∩8 | 52.968 | 1457.622 | 38.179 |
| 2∪11 | 43.182 | 1550.351 | 39.374 | 2∩11 | 40.343 | 1131.187 | 33.633 | 7∪9 | 51.174 | 2390.586 | 48.894 | 7∩9 | 47.189 | 1470.373 | 38.345 |
| 2∪12 | 48.330 | 2007.286 | 44.803 | 2∩12 | 43.233 | 1056.936 | 32.511 | 7∪10 | 50.947 | 2134.032 | 46.196 | 7∩10 | 49.468 | 1539.795 | 39.240 |
| 3∪4 | 38.734 | 1498.483 | 38.710 | 3∩4 | 40.495 | 1257.932 | 35.467 | 7∪11 | 44.560 | 1712.071 | 41.377 | 7∩11 | 44.739 | 1508.552 | 38.840 |
| 3∪5 | 37.742 | 1486.969 | 38.561 | 3∩5 | 35.593 | 1033.569 | 32.149 | 7∪12 | 49.428 | 1980.501 | 44.503 | 7∩12 | 48.343 | 1582.661 | 39.783 |
| 3∪6 | 38.319 | 1570.578 | 39.631 | 3∩6 | 36.693 | 1138.911 | 33.748 | 8∪9 | 63.982 | 3273.500 | 57.215 | 8∩9 | 57.500 | 1988.002 | 44.587 |
| 3∪7 | 42.062 | 1829.452 | 42.772 | 3∩7 | 42.850 | 1326.030 | 36.415 | 8∪10 | 64.779 | 3193.570 | 56.512 | 8∩10 | 58.792 | 1833.926 | 42.824 |
| 3∪8 | 57.178 | 3714.156 | 60.944 | 3∩8 | 49.155 | 1156.849 | 34.012 | 8∪11 | 59.793 | 3494.363 | 59.113 | 8∩11 | 51.104 | 1329.367 | 36.460 |
| 3∪9 | 47.170 | 2555.841 | 50.555 | 3∩9 | 43.976 | 1168.593 | 34.185 | 8∪12 | 63.709 | 3258.050 | 57.079 | 8∩12 | 56.842 | 1717.616 | 41.444 |
| 3∪10 | 47.091 | 2331.881 | 48.290 | 3∩10 | 46.262 | 1215.970 | 34.871 | 9∪10 | 55.080 | 2560.235 | 50.599 | 9∩10 | 52.059 | 1575.548 | 39.693 |
| 3∪11 | 39.607 | 1689.646 | 41.105 | 3∩11 | 42.990 | 1319.115 | 36.320 | 9∪11 | 49.938 | 2395.862 | 48.948 | 9∩11 | 45.701 | 1358.837 | 36.862 |
| 3∪12 | 45.491 | 2166.727 | 46.548 | 3∩12 | 45.261 | 1251.102 | 35.371 | 9∪12 | 53.998 | 2478.618 | 49.786 | 9∩12 | 50.418 | 1562.848 | 39.533 |
| 4∪5 | 38.672 | 1348.464 | 36.721 | 4∩5 | 35.973 | 1016.153 | 31.877 | 10∪11 | 49.938 | 2170.446 | 46.588 | 10∩11 | 47.761 | 1401.505 | 37.437 |
| 4∪6 | 39.654 | 1474.698 | 38.402 | 4∩6 | 36.668 | 1074.922 | 32.786 | 10∪12 | 53.304 | 2024.885 | 44.999 | 10∩12 | 53.216 | 1818.281 | 42.641 |
| 4∪7 | 43.571 | 1755.286 | 41.896 | 4∩7 | 42.495 | 1230.467 | 35.078 | 11∪12 | 48.339 | 2015.114 | 44.890 | 11∩12 | 46.752 | 1439.732 | 37.944 |

**Table 6.26** Extended Ichino–Yaguchi distances for household data $\gamma = 0.25$ (Example 6.10)

| House pair | By variable $Y_j$ $d_1$ | $d_2$ | $d_3$ | $\sum \frac{d_j}{d}$ | Euclidean ED | Normalized NED | House pair | By variable $Y_j$ $d_1$ | $d_2$ | $d_3$ | $\sum \frac{d_j}{d}$ | Euclidean ED | Normalized NED |
|---|---|---|---|---|---|---|---|---|---|---|---|---|---|
| 1-2 | 0.186 | 0.024 | 11.565 | 11.775 | 11.566 | 6.678 | 4-8 | 0.91 | 0.162 | 20.980 | 22.054 | 21.000 | 12.124 |
| 1-3 | 0.402 | 0.591 | 10.760 | 11.753 | 10.784 | 6.226 | 4-9 | 1.203 | 0.341 | 12.296 | 13.840 | 12.360 | 7.136 |
| 1-4 | 1.044 | 0.128 | 11.006 | 12.177 | 11.056 | 6.383 | 4-10 | 0.811 | 0.033 | 10.487 | 11.331 | 10.518 | 6.073 |
| 1-5 | 1.229 | 0.230 | 14.691 | 16.149 | 14.744 | 8.513 | 4-11 | 0.873 | 0.128 | 3.536 | 4.537 | 3.645 | 2.104 |
| 1-6 | 0.881 | 0.108 | 12.277 | 13.265 | 12.309 | 7.107 | 4-12 | 0.783 | 0.014 | 8.778 | 9.574 | 8.813 | 5.088 |
| 1-7 | 0.593 | 0.149 | 5.809 | 6.550 | 5.841 | 3.372 | 5-6 | 0.363 | 0.122 | 4.105 | 4.590 | 4.123 | 2.381 |
| 1-8 | 0.254 | 0.224 | 9.656 | 10.133 | 9.662 | 5.579 | 5-7 | 0.650 | 0.081 | 8.704 | 9.434 | 8.728 | 5.039 |
| 1-9 | 1.131 | 0.422 | 8.468 | 10.020 | 8.553 | 4.938 | 5-8 | 1.091 | 0.189 | 24.793 | 26.073 | 24.818 | 14.329 |
| 1-10 | 0.744 | 0.150 | 2.020 | 2.914 | 2.158 | 1.246 | 5-9 | 1.241 | 0.368 | 15.419 | 17.027 | 15.473 | 8.934 |
| 1-11 | 0.338 | 0.242 | 0.000 | 0.579 | 0.415 | 0.240 | 5-10 | 0.980 | 0.080 | 14.114 | 15.174 | 14.148 | 8.169 |
| 1-12 | 0.225 | 0.218 | 0.000 | 0.443 | 0.313 | 0.181 | 5-11 | 1.043 | 0.093 | 7.051 | 8.186 | 7.128 | 4.115 |
| 2-3 | 0.455 | 0.567 | 4.189 | 5.211 | 4.252 | 2.455 | 5-12 | 0.953 | 0.089 | 12.367 | 13.408 | 12.404 | 7.162 |
| 2-4 | 1.085 | 0.126 | 2.637 | 3.848 | 2.854 | 1.648 | 6-7 | 0.293 | 0.041 | 6.346 | 6.679 | 6.353 | 3.668 |
| 2-5 | 1.254 | 0.228 | 3.799 | 5.281 | 4.007 | 2.313 | 6-8 | 0.740 | 0.168 | 22.331 | 23.238 | 22.344 | 12.900 |
| 2-6 | 0.938 | 0.107 | 3.660 | 4.704 | 3.780 | 2.182 | 6-9 | 1.239 | 0.347 | 12.973 | 14.559 | 13.037 | 7.527 |
| 2-7 | 0.662 | 0.147 | 6.146 | 6.955 | 6.184 | 3.570 | 6-10 | 0.655 | 0.059 | 11.704 | 12.417 | 11.723 | 6.768 |
| 2-8 | 0.396 | 0.200 | 21.499 | 22.095 | 21.504 | 12.415 | 6-11 | 0.716 | 0.153 | 4.671 | 5.539 | 4.728 | 2.729 |
| 2-9 | 0.947 | 0.398 | 11.823 | 13.167 | 11.867 | 6.852 | 6-12 | 0.786 | 0.033 | 9.992 | 10.811 | 10.023 | 5.787 |
| 2-10 | 0.628 | 0.149 | 11.018 | 11.794 | 11.037 | 6.372 | 7-8 | 0.447 | 0.171 | 15.666 | 16.284 | 15.673 | 9.049 |

(Continued)

Table 6.26 (Continued)

| House pair | By variable $Y_j$ | | | $\sum \frac{d_j}{d}$ | Euclidean ED | Normalized NED |
|---|---|---|---|---|---|---|
| | $d_1$ | $d_2$ | $d_3$ | | | |
| 2-11 | 0.225 | 0.218 | 4.306 | 4.748 | 4.317 | 2.493 |
| 2-12 | 0.876 | 0.140 | 9.301 | 10.316 | 9.343 | 5.394 |
| 3-4 | 1.061 | 0.477 | 2.420 | 3.958 | 2.685 | 1.550 |
| 3-5 | 1.235 | 0.402 | 4.817 | 6.454 | 4.989 | 2.881 |
| 3-6 | 0.909 | 0.503 | 4.374 | 5.786 | 4.496 | 2.596 |
| 3-7 | 0.641 | 0.465 | 4.793 | 5.899 | 4.858 | 2.805 |
| 3-8 | 0.647 | 0.368 | 20.775 | 21.789 | 20.788 | 12.002 |
| 3-9 | 1.095 | 0.311 | 12.259 | 13.664 | 12.312 | 7.108 |
| 3-10 | 0.764 | 0.444 | 10.222 | 11.430 | 10.260 | 5.923 |
| 3-11 | 0.624 | 0.350 | 3.568 | 4.542 | 3.639 | 2.101 |
| 3-12 | 1.010 | 0.473 | 8.491 | 9.973 | 8.563 | 4.944 |
| 4-5 | 0.185 | 0.102 | 3.656 | 3.943 | 3.662 | 2.114 |
| 4-6 | 0.179 | 0.026 | 4.169 | 4.373 | 4.172 | 2.409 |
| 4-7 | 0.465 | 0.021 | 5.128 | 5.614 | 5.149 | 2.973 |
| 7-9 | 1.245 | 0.350 | 7.844 | 9.439 | 7.950 | 4.590 |
| 7-10 | 0.533 | 0.021 | 5.270 | 5.824 | 5.297 | 3.058 |
| 7-11 | 0.437 | 0.116 | 1.907 | 2.459 | 1.960 | 1.131 |
| 7-12 | 0.794 | 0.008 | 3.567 | 4.368 | 3.654 | 2.110 |
| 8-9 | 1.251 | 0.198 | 9.653 | 11.102 | 9.735 | 5.621 |
| 8-10 | 0.607 | 0.152 | 10.403 | 11.161 | 10.421 | 6.017 |
| 8-11 | 0.195 | 0.114 | 17.441 | 17.750 | 17.443 | 10.071 |
| 8-12 | 0.882 | 0.171 | 11.921 | 12.974 | 11.955 | 6.902 |
| 9-10 | 0.716 | 0.330 | 8.014 | 9.061 | 8.053 | 4.649 |
| 9-11 | 1.067 | 0.275 | 9.034 | 10.375 | 9.101 | 5.254 |
| 9-12 | 0.453 | 0.350 | 7.542 | 8.344 | 7.564 | 4.367 |
| 10-11 | 0.412 | 0.095 | 6.959 | 7.465 | 6.971 | 4.025 |
| 10-12 | 0.278 | 0.029 | 1.772 | 2.079 | 1.794 | 1.036 |
| 11-12 | 0.689 | 0.123 | 5.267 | 6.079 | 5.314 | 3.068 |

# 7

# Divisive Hierarchical Clustering

A general overview of divisive hierarchical clustering was given in section 5.3. In the present chapter, we study this in detail as it pertains to symbolic data. Some basics relating to finding these clusters are first presented in section 7.1. Divisive clustering techniques are (broadly) either monothetic or polythetic methods. Monothetic methods involve one variable at a time considered successively across all variables. In contrast, polythetic methods consider all variables simultaneously. Each is considered in turn in sections 7.2 and 7.3, respectively. Typically, the underlying clustering criteria involve some form of distance or dissimilarity measure, and since for any set of data there can be many possible distance/dissimilarity measures, then the ensuing clusters are not necessarily unique across methods/measures. Recall, from Chapter 3, that distances are dissimilarities for which the triangle property holds (see Definition 3.2). Therefore, the divisive techniques covered in this chapter use the generic "dissimilarity" measure unless it is a distance measure specifically under discussion. A question behind any clustering procedure relates to how many clusters is optimal. While there is no definitive answer to that question, some aspects are considered in section 7.4. Agglomerative hierarchical methods are considered in Chapter 8.

## 7.1 Some Basics

### 7.1.1 Partitioning Criteria

In Chapters 3 and 4, we supposed our sample space $\Omega$ consisted of objects $A_u$, $u = 1, \ldots, m$. These "objects" may correspond to a single observation or they may represent a collection or aggregation of single observations. In this chapter, we refer to the object $A_u$ as an "observation". For ease of presentation, we sometimes will refer to these $A_u$ as observation $u = 1, \ldots, m$. We write the set of observations as $\{u_1, \ldots, u_m\} \equiv \{1, \ldots, m\} \in \Omega$. Each observation $u$ is described by the random variable $\mathbf{Y}_u = (Y_{u1}, \ldots, Y_{up})$, $u = 1, \ldots, m$, where the

*Clustering Methodology for Symbolic Data*, First Edition. Lynne Billard and Edwin Diday.
© 2020 John Wiley & Sons Ltd. Published 2020 by John Wiley & Sons Ltd.

variable $Y_j$, takes values in $\mathcal{Y}_j$, $j = 1, \ldots, p$; these values can be multi-valued (or, lists) either modal or non-modal values, interval values, or histogram values, i.e., the realizations are symbolic-valued (for which a classical value is a special case, e.g., the point $a$ is the interval $[a, a]$).

The goal is to execute a hierarchical methodology to obtain a partition of $\Omega$ into a number of clusters. Suppose the $r$th bi-partition gives the $r$ clusters $P_r = (C_1, \ldots, C_r)$ where the clusters $C_v$ are disjoint, i.e., $C_v \cap C_{v'} = \phi$, $v \neq v'$ for all $v, v' = 1, \ldots, r$, and where all observations are in a cluster so that $\cup_v C_v = \Omega$. Suppose the cluster $C_v$ contains $m_v$ observations, i.e., $C_v = \{A_1, \ldots, A_{m_v}\}$ (or, equivalently, $C_v = \{1, \ldots, m_v\}$). Clearly, $\sum_{v=1}^{r} m_v = m$. (Note that the focus here is to look at the hierarchy at a particular $r$th stage. When looking at the entire hierarchy construction, $C_v \cap C_{v'} = \{\phi, C_v, C_{v'}\}$; see Definition 5.2.)

Then, at the $(r+1)$th stage, we select one of the clusters in $P_r$, $C_v$ say, and divide it into two further sub-clusters $C_v = (C_v^{(1)}, C_v^{(2)})$, to give

$$P_{r+1} = (P_r \cup \{C_v^{(1)}, C_v^{(2)}\}) - \{C_v\}. \tag{7.1.1}$$

The question then is which particular cluster $C_v$ is to be divided (i.e., how do we choose $v$), and then how do we divide that selected cluster into two sub-clusters. There are clearly many choices. We want these clusters to be internally as homogeneous as possible and as heterogeneous as possible across clusters. This leads us to the concepts of within-cluster variations and between-cluster variations.

**Definition 7.1** Suppose we have the cluster $C_v = \{u_1, \ldots, u_{m_v}\} \in \Omega$. Then, the **within-cluster variation** $I(C_v)$ is given by

$$I(C_v) = \frac{1}{2\lambda} \sum_{u_1=1}^{m_v} \sum_{u_2=1}^{m_v} w_{u_1} w_{u_2} d^2(u_1, u_2) \equiv \frac{1}{\lambda} \sum_{u_2=1}^{m_k} \sum_{\substack{u_1=1 \\ u_1 > u_2}}^{m_k} w_{u_1} w_{u_2} d^2(u_1, u_2)$$

$$\tag{7.1.2}$$

where $d(u_1, u_2)$ is a distance or dissimilarity measure between the observations $u_1$ and $u_2$ in $C_v$, $u_1, u_2 = 1, \ldots, m_v$, and where $w_u$ is the weight associated with the observation $u$ and $\lambda = \sum_{u=1}^{m_v} w_u$. □

When all observations are equally weighted by $w_u = 1/m$, Eq. (7.1.2) simplifies to

$$I(C_v) = \frac{1}{2m_v m} \sum_{u_1=1}^{m_v} \sum_{u_2=1}^{m_v} w_{u_1} w_{u_2} d^2(u_1, u_2).$$

**Definition 7.2** For the partition $P_r = (C_1, \ldots, C_r)$, the **total within-cluster variation** $W(P_r)$ is the sum of the within-cluster variations over all clusters in $P_r$, i.e.,

$$W(P_r) = \sum_{v=1}^{r} I(C_v) \qquad (7.1.3)$$

where $I(C_v)$ is the within-cluster variation for cluster $C_v$ given in Definition 7.1.

□

**Definition 7.3** Suppose we have the complete set of observations $\Omega$, i.e., $\Omega = \{A_1, \ldots, A_m\}$. Then, the **total variation** $W(\Omega)$ is given by

$$W(\Omega) = \frac{1}{2\lambda} \sum_{u_1=1}^{m} \sum_{u_2=1}^{m} w_{u_1} w_{u_2} d^2(u_1, u_2) \equiv \frac{1}{\lambda} \sum_{u_1 < u_2 = 2}^{m} \sum_{u_2=1}^{m} w_{u_1} w_{u_2} d^2(u_1, u_2)$$

$$(7.1.4)$$

where $d^2(u_1, u_2)$ and $w_u$ for $u, u_1, u_2 = 1, \ldots, m$, and $\lambda$ are given in Definition 7.1.

□

Notice that the total variation $W(\Omega)$ of Eq. (7.1.4) is just the within-cluster variation of Eq. (7.1.2) for the special case that the total set of observations $\Omega \equiv C_1$ (or equivalently that $\Omega \equiv P_1$ of Eq. (7.1.3)).

**Definition 7.4** For the partition $P_r = (C_1, \ldots, C_r)$, the **between-cluster variation** $B(P_r)$ is given by, from Huygens theorem,

$$B(P_r) = W(\Omega) - W(P_r) \qquad (7.1.5)$$

where $W(\Omega)$ is the total variation of the entire set of observations (of Eq. (7.1.4)) and $W(P_r)$ is the total within-cluster variation of the clusters of the partition $P_r$ (of Eq. (7.1.3)).

□

It follows that at the $(r + 1)$th partition, from Eq. (7.1.1), we have

$$W(P_{r+1}) = W(P_r) - [I(C_{v^*}) - I(C_{v^*}^{(1)}) - I(C_{v^*}^{(2)})]. \qquad (7.1.6)$$

Therefore, given the partition $P_r$, to achieve our goal to obtain clusters with the smallest internal variation as possible, at each stage, we want to minimize the total within-cluster variation $W(P_{r+1})$, or, equivalently, from Eq. (7.1.5), maximize the between-cluster variation $B_r$. That is, from Eq. (7.1.6), we see that we select that cluster $C_{v^*}$ which maximizes the change in total within-cluster variation; this is written as

$$\Delta_{v^*} = \max_v \{\Delta_v\} \text{ with } \Delta_v = I(C_v) - I(C_v^{(1)}) - I(C_v^{(2)}), \quad v = 1, \dots, r.$$

$$(7.1.7)$$

Notice that these variations (total, total within-cluster, and between-cluster) and therefore this criterion of Eq. (7.1.7) all depend in some manner on dissimilarity measures. Since, as we saw in Chapters 3 and 4, distance and dissimilarity measures take many differing forms and values for any set of observations, it follows that resulting partitions will not necessarily be the same for all such measures.

The variations in Eqs. (7.1.2)–(7.1.7) are defined implicitly as being based on dissimilarities evaluated over all variables $Y_j, j = 1, \dots, p$ (needed when undertaking polythetic algorithms, see section 7.3), or they can be evaluated for a specific variable $Y_j$. In these cases, a $j$ superscript is added, e.g., in Eq. (7.1.7), $\Delta_{v^*}$ becomes $\Delta_{v^*}^j$. Monothetic algorithms use these entities by variable (see section 7.2).

The number of bi-partitions to be considered when seeking the sub-clusters, $C_v = (C_v^{(1)}, C_v^{(2)})$, can be quite large. To reduce the number of possibilities for interval data, Chavent (1998) ordered the intervals by their mean values and then based the algorithm on these means. In contrast, for modal-valued data, this number can be even larger, especially if the number of possible categories $s_j, j = 1, \dots, p$, is large. To reduce this number, the divisive algorithm uses a so-called association measure (described in section 7.1.2 for each variable) which identifies the most representative category to be used in the algorithm. An intuitive derivation of this measure is provided in Kim and Billard (2012).

## 7.1.2 Association Measures

When an observation takes values that are modal multi-valued or lists, an association measure is useful as part of the partitioning process. As in Chapter 4, let us suppose each observation contains modal list values for the $p$-dimensional random variable $\mathbf{Y} = (Y_1, \dots, Y_p)$ in the form (see Eq. (4.1.2)), i.e.,

$$\mathbf{Y}_u = (\{Y_{ujk_j}, p_{ujk_j}; \, k_j = 1, \dots, s_j\}, j = 1, \dots, p), \quad u = 1, \dots, m, \quad (7.1.8)$$

where $Y_{ujk_j}$ is the $k_j$th category observed from the set of possible categories $\mathcal{Y}_j = \{Y_{j1}, \dots, Y_{js_j}\}$ for the observation $u$ for the variable $Y_j$, and where $p_{ujk_j}$ is the probability or relative frequency associated with $Y_{ujk_j}$, with $\sum_k p_{ujk_j} = 1$. Again, by comparing Eq. (7.1.8) with Eq. (4.1.1), we are assuming without loss of generality that $s_{uj} \equiv s_j$ for all observations $Y_{uj}, u = 1, \dots, m$, by setting $p_{ujk_j} \equiv 0$ for those categories in $\mathcal{Y}_j$ that do not occur for that particular observation. For non-modal list values, we assume each observed $Y_{ujk_j}$ from $\mathcal{Y}_j$ has probability $p_{ujk_j} = 1/s_{uj}$ and non-observed values have probability zero. Also, unless necessary for clarification, let us write for notational simplicity $k_j \equiv k, j = 1, \ldots, p$.

**Definition 7.5**   Suppose we have the cluster $C_v = \{A_1, \ldots, A_{m_v}\}$ where each observation $A_u$ has realization $\mathbf{Y}_u$ taking the form of Eq. (7.1.8). Then, for a given variable $Y_j$, the **association** $\alpha_{jk}^v$ between the category $Y_{jk}$ and the other categories in $\mathcal{Y}_j$ (i.e., $Y_{jk'}, k' = 1, \ldots, s_j, k \neq k'$) is, for each $j = 1, \ldots, p$,

$$\alpha_{jk}^v = \sum_{k'=1, k' \neq k}^{s_j} |a_{j(k,k')}^v d_{j(k,k')}^v - b_{j(k,k')}^v c_{j(k,k')}^v|, \quad v = 1, \ldots, r, \ k = 1, \ldots, s_j,$$

(7.1.9)

where

$$a_{j(k,k')}^v = \sum_{u=1}^{m_v} h_u^v p_{ujk} p_{ujk'},$$

(7.1.10)

$$b_{j(k,k')}^v = \sum_{u=1}^{m_v} h_u^v p_{ujk}(1 - p_{ujk'}),$$

(7.1.11)

$$c_{j(k,k')}^v = \sum_{u=1}^{m_v} h_u^v (1 - p_{ujk}) p_{ujk'},$$

(7.1.12)

$$d_{j(k,k')}^v = \sum_{u=1}^{m_v} h_u^v (1 - p_{ujk})(1 - p_{ujk'}),$$

(7.1.13)

where $h_u^v$ is the weight of the observation $\mathbf{Y}_{uj}$ in the cluster $C_v$, i.e.,

$$h_u^v = w_u \Big/ \sum_{u=1}^{m_v} w_u,$$

(7.1.14)

and where $w_u$ is the weight of the observation $\mathbf{Y}_u$ in $\Omega$.   □

When two categories $k$ and $k'$, $k \neq k'$, have a high and positive relationship in the cluster $C_v$, then the first term on the right-hand side of Eq. (7.1.9) will be high, if there is a high but negative relationship, then the second term in Eq. (7.1.9) will be high, while if there is no relationship, then both terms in Eq. (7.1.9) will be similar. Hence, for each variable, when one particular category ($Y_{jk}$) is highly related to the other categories ($Y_{jk'}, k \neq k'$), then the association value is large. Thus, that category can be regarded as a representative category for that variable. This has a consequence that the number of possible partitions to be considered in the divisive algorithm can be greatly reduced.

**Example 7.1** Consider the household data of Table 4.3 (in Example 4.3), which we recall consist of two random variables, $Y_1$ = household makeup with $s_1 = 6$ categories from $\mathcal{Y}_1$ = {married couple with children under 18, married couple without children under 18, female head with children under 18, female head without children under 18, householder living alone aged under 65, householder living alone aged 65 or over} and $Y_2$ = household occupancy and tenure with $s_2 = 3$ categories from $\mathcal{Y}_2$ = {owner occupied, renter occupied, vacant}. There are $m = 6$ counties. Thus, for example, the proportion of households consisting of a married couple and children under 18 years old in the first county is 0.274 or 27.4%.

Take $Y_2$, and suppose for illustrative simplicity all weights $h_u \equiv 1$. Then, from Eq. (7.1.10) and Table 4.3, we have, for $k = 1$, $k' = 2$ (say),

$$a_{2(1,2)} = 0.530 \times 0.439 + \cdots + 0.587 \times 0.378 = 1.291$$

and, similarly, from Eqs. (7.1.11)–(7.1.13), we have

$$b_{2(1,2)} = 2.017, \quad c_{2(1,2)} = 1.061, \quad d_{2(1,2)} = 1.631.$$

Values for all the $\alpha_{jk}$ components, namely, $a_j(k, k')$, $b_j(k, k')$, $c_j(k, k')$, and $d_j(k, k')$, for all $(k, k')$, $k, k' = 1, 2, 3$, are displayed in Table 7.1.

Hence, substituting into Eq. (7.1.9), we obtain, for $j = 2$, $k = 1$,

$$\alpha_{21} = |1.291 \times 1.631 - 2.017 \times 1.061| + |0.187 \times 2.539 - 3.121 \times 0.153|$$
$$= 0.041.$$

Likewise, the complete set of $\alpha_{jk}$ for all observations is shown in Table 7.2 (first row).

When the observations are assumed to have equal weights $h_u = 1/m$, then the association measures $\alpha_{jk}$, $k = 1, \ldots s_j, j = 1, 2$, are as shown in Table 7.2

**Table 7.1** Components of association measures for $Y_2$ (Example 7.1)

| $k'$ | 1 | 2 | 3 | 1 | 2 | 3 |
|---|---|---|---|---|---|---|
| $k$ | | $a_2(k, k')$ | | | $b_2(k, k')$ | |
| 1 | 1.831 | 1.291 | 0.187 | 1.477 | 2.017 | 3.121 |
| 2 | 1.291 | 0.929 | 0.132 | 1.061 | 1.423 | 2.220 |
| 3 | 0.187 | 0.132 | 0.021 | 0.153 | 0.208 | 0.319 |
| $k$ | | $c_2(k, k')$ | | | $d_2(k, k')$ | |
| 1 | 1.477 | 1.061 | 0.153 | 1.215 | 1.631 | 2.539 |
| 2 | 2.017 | 1.423 | 0.208 | 1.631 | 2.225 | 3.440 |
| 3 | 3.121 | 2.220 | 0.319 | 2.539 | 3.440 | 5.341 |

**Table 7.2** Association measures $\alpha_{jk}^{v}$ (Example 7.1)

| Weight | Cluster | $Y_1$ | | | | | | $Y_2$ | | |
|---|---|---|---|---|---|---|---|---|---|---|
| $h_j$ | k | 1 | 2 | 3 | 4 | 5 | 6 | 1 | 2 | 3 |
| $h_u = 1$ | $C_1$ | 0.122 | 0.045 | 0.036 | 0.013 | 0.088 | 0.054 | 0.041 | 0.045 | 0.012 |
| $h_u = 1/m$ | $C_1$ | 0.00340 | 0.00125 | 0.00100 | 0.00035 | 0.00244 | 0.00150 | 0.00113 | 0.00125 | 0.00034 |
| $h_u = n_u/n$ | $C_1$ | 1.468 | 0.813 | 0.420 | 0.372 | 1.268 | 0.259 | 0.885 | 0.822 | 0.208 |

(second row). Table 7.2 (third row) also gives the association measures (to the order $10^{-3}$) for the case in which the weights are proportional to the population of each county, i.e., $h_{uj} = n_{uj}/n_j$ where $n_{uj}$ is the population size for county $u = 1, \ldots, m = 6$, and $n_j = \sum_u n_{uj}$, for variable $Y_j, j = 1, 2$ (see Exercise 7.1).

These measures tell us that, for these data for the $Y_1$ variable, the category married couple with children under 18 is the most representative category for household makeup since the association measure here is largest. For $Y_2$, the most representative category for occupancy and tenure is either owner or renter occupied depending on what weights are used. □

## 7.2 Monothetic Methods

In a monothetic method, one variable at a time is used to create the best division of the cluster $C_v$ into $(C_v^{(1)}, C_v^{(2)})$ at the $r$th stage. That is, by using variable $Y_j$, we obtain the best (maximum) $\Delta_{v^*}^{j}$ of Eq. (7.1.7) for each $j = 1, \ldots, p$, and then select the partition which corresponds to, for each $v = 1, \ldots, r$,

$$\Delta_{v^*}^{j^*} = \max_{j=1,\ldots,p} \{\Delta_{v^*}^{j}\} = \max_{j=1,\ldots,p} \{ \max_{v=1,\ldots,r} \{\Delta_{v}^{j}\}\} \text{ with}$$
$$\Delta_{v}^{j} = I(C_v^j) - I(C_v)^{(1)j} - I(C_v^{(2)j}). \tag{7.2.1}$$

Typically, some ordering property is established for the selected variable to facilitate the methodology. For multi-valued list data (modal or non-modal) an ordering is based on the probabilities for one of the possible categories for that variable via the association measures, while for interval or histogram realizations the ordering is based on observation means. This ordering has the effect of reducing the total number of possible bi-partitions needing to be considered. Without some clearly defined ordering criteria, when bi-partitioning a cluster $C$ of size $m$ into two sub-clusters (of any unspecified size $m_1$ and $m_2$ with the only restriction being that $m_1 + m_2 = m$), there are $M = \sum_{i=1}^{m} \binom{m}{i}$ possible bi-partitions for each variable. This can be quite challenging computationally. With the ordering component, the number of possibilities reduces to

$n \leq m - 1$ where $n$ is the number of distinct ordered values in the set $C$. This ordering feature also has the advantage that the bi-partitioning criteria can be easily calculated and interpreted.

For descriptive clarity, where necessary, an observation $A_u$ (or simply $u$) will be represented by its realization $\mathbf{Y}_u$, or by $Y_{uj}$ for variable $j = 1, \ldots, p$, $u = 1, \ldots, m$. The essential execution of a monothetic algorithm involves the following steps:

**Step 1**  Stage $r = 1$: Start with $\Omega = P_1 = C_1 = \{\mathbf{Y}_1, \ldots, \mathbf{Y}_m\}$.

**Step 2**  Stage $j = 1$: i.e., start with variable $Y_1$.

**Step 3**  Stage $r$ and variable $j$: $\Omega = P_r = (C_1, \ldots, C_r)$ with each cluster $C_v$ described for variable $Y_j$ by $C_v^j = \{Y_{uj}, u = 1, \ldots, m_v\}$.

**Step 4**  Order the observations according to the ordering criterion for the variable $Y_j$.

**Step 5**  For each ordered $C_v$ in $P_r$, $v = 1, \ldots, r$, set up the bi-partitions $(C_{v(q)}^{(1)j}, C_{v(q)}^{(2)j})$, $q = 1, \ldots, n_v$, with observations for variable $Y_j$ described by $C_{v(q)}^{(1)j} = \{Y_{(1)j}, \ldots, Y_{(q)j}\}$ and $C_{v(q)}^{(2)j} = \{Y_{(q+1)j}, \ldots, Y_{(m_v)j}\}$.

**Step 6**  For each pair $(C_{v(q)}^{(1)j}, C_{v(q)}^{(2)j})$, i.e., for each $v$ and each $q$, from Step 5, calculate the change in the total within-cluster variation $\Delta_{v(q)}^j$ (from Eq. (7.2.4) below).

**Step 7**  Determine the sub-cluster $v^*$ in $P_r$ to be partitioned and its bi-partition $q^*$ from the maximum change in the total within-cluster variation $\Delta_{v(q)}^j$ calculated in Step 6 (from Eq. (7.2.5) below), i.e., for this $j$ and this $r$,

$$\Delta_{v^*(q^*)}^j = \max_{v,q}\{\Delta_{v(q)}^j, v = 1, \ldots, r, \ q = 1, \ldots, n_v\}.$$

**Step 8**  Now $j = j + 1$. Repeat Steps 2–7 until $j = p$.

**Step 9**  When $j = p$, determine that $j = j^*$, which maximizes $\Delta_{v^*(q^*)}^j$, i.e.,

$$\Delta_{v^*(q^*)}^{j^*} = \max_{j}\{\Delta_{v^*(q^*)}^j, j = 1, \ldots, p\}.$$

**Step 10**  Now $r = r + 1$. Repeat Steps 2–9 until $r = m$, the total number of observations, or until $r = R$, a preassigned maximum number of stages.

At each division, $r$, in a monothetic algorithm, a cut value, $c_r$, emerges for the relevant variable driving the actual division at that stage. The nature of this cut variable will depend on the ordering mechanism tied to the type of data involved. For example, if the data are multi-valued (modal or non-modal), the observations are ordered according to one of the probabilities $p_{ujk}$, and so the cut value will be the average of the probabilities on either side of the observation at which the division occurred (see, e.g., Eq. (7.2.17) in Example 7.2). In contrast, for interval or histogram data, the observations are ordered according to the mean values for the relevant variable, so the cut value will be the average of the means on either side of where the division occurred (see, e.g., Example 7.5). This cut value defines a question which determines which observation in $C$ goes into which sub-cluster, $C^{(1)}$ or $C^{(2)}$. Thus, in Example 7.2, the question becomes "Is $Y_{22} < c_r$?" (or is the probability that $Y_{22} < c_r$?). If "Yes", then that observation takes the left sub-cluster $C^{(1)}$, and if "No", then the observation goes into $C^{(2)}$.

While there are some common elements to monothetic algorithms for symbolic data, the different types of realizations can induce special elements. Therefore, we consider each in turn, namely, modal multi-valued list observations in section 7.2.1, non-modal multi-valued observations in section 7.2.2 (as a special case of modal multi-valued list observations), interval-valued observations in section 7.2.3, and histogram-valued observations in section 7.2.4.

### 7.2.1 Modal Multi-valued Observations

Let us consider modal multi-valued observations of the form of Eq. (7.1.8). As for any divisive method, we need a binary question to effect the division of the data set into its respective clusters. When, as in this case, the divisive algorithm is based on a single $Y_{jk}$ category for multi-valued observations, this takes the form "Is the probability that the category $Y_{jk}$ occurs $\leq c_r$?", or more succinctly "Is $Y_{ujk} \leq c_r$?", where $c_r$ is the cut point at the $r$th stage, for the selected cutting variable $Y_j$.

Suppose we have the partition $P_r = (C_1, \dots, C_r)$. Then, for each variable $Y_j$, $j = 1, \dots, p$, and each cluster $C_v$, $v = 1, \dots, r$, in $P_r$, we calculate the association $\alpha_{jk}^v$ for each category $Y_{jk}$ in $\mathcal{Y}_j$, $k = 1, \dots, s_j$, from Eq. (7.1.9). We calculate the maximum association measure, $\alpha_{jk^*}^v$, namely,

$$\alpha_{jk^*}^v = \max_{k=1,\dots,s_j} \{\alpha_{jk}^v\}, \quad j = 1, \dots, p, \tag{7.2.2}$$

i.e., the maximum association occurs for the category $Y_{jk^*}$ in $\mathcal{Y}_j$.

The observations are then sorted in ascending order of the probabilities $p_{ujk^*}$, i.e., according to the category $Y_{jk^*}$ for which the association is greatest. For $Y_j$, this ordered set is written as the cluster $C_v^j$ described by $C_v^j = \{Y_{(1)j}, \dots, Y_{(m_v)j}\}$,

$v = 1, \ldots, r$, where $Y_{(1)j}$ describes the observation that has the smallest probability for the category $Y_{jk^*}$ and so on. Note there is a different sorted order for different $j$ values. It is this sorted order set of observations which is used in the calculations of the variations of Eqs. (7.1.2)–(7.1.7). Suppose the cluster $C_v^j$ is partitioned into the two non-overlapping clusters $C_v^j = (C_{v(q)}^{(1)j}, C_{v(q)}^{(2)j})$ with $C_v^j = C_{v(q)}^{(1)j} \cup C_{v(q)}^{(2)j}$ where the observations in the sub-clusters are described by, respectively, for the variable $j = 1, \ldots, p$,

$$C_{v(q)}^{(1)j} = \{Y_{(1)j}, \ldots, Y_{(q)j}\}, \quad C_{v(q)}^{(2)j} = \{Y_{(q+1)j}, \ldots, Y_{(m_v)j}\}, \quad q = 1, \ldots, n_v,$$

$$(7.2.3)$$

where $n_v \leq m_v - 1$ is the number of distinct values for $p_{ujk^*}$ in $C_v^j$. When all values are distinct, then $n_v = m_v - 1$.

For each of these $n_v$ possible bi-partitions of $C_v^j$, we calculate the change in the total within-cluster variation of Eq. (7.1.7), i.e., we calculate

$$\Delta_{v(q)}^j = I(C_{v(q)}^j) - I(C_{v(q)}^{(1)j}) - I(C_{v(q)}^{(2)j}), \quad q = 1, \ldots, n_v. \tag{7.2.4}$$

Note that the components of Eq. (7.2.4) are calculated from dissimilarities which may be based on all variables; the $j$ superscript serves to identify the variable $Y_j$ by which the observations were sorted. Then, at the $r$th stage, for all clusters $C_v^j, v = 1, \ldots, r$, in $P_r$, and for all variables $Y_j, j = 1, \ldots, p$, we select the category $Y_{jk}$ and the bi-partition $C_v^j = (C_{v(q)}^{(1)j}, C_{v(q)}^{(2)j}), q = 1, \ldots, n_v,$ which satisfies

$$\Delta_{v^*(q^*)}^{j^*} = \max_{j,v,q} \{\Delta_{v(q)}^j, q = 1, \ldots, n_v, v = 1, \ldots, r, j = 1, \ldots, p\} \tag{7.2.5}$$

where $\Delta_{v(q)}^j$ is defined in Eq. (7.2.4).

This maximum in Eq. (7.2.5) identifies which $C_v$ is divided into two sub-clusters (i.e., $v^*$); it identifies the two sub-clusters, i.e., which ordered observation (i.e., $q^*$) is selected for which variable $Y_j$ (i.e., $j^*$) to form the two sub-clusters (from Eq. (7.2.3)) and which category $Y_{jk}$ (i.e., $k^*$) in $\mathcal{Y}_j$ is the sorting category. Hence, the cut point value $c_r$ is obtained from

$$c_r = (p_{(q^*)jk^*} + p_{(q^*+1)jk^*})/2 \tag{7.2.6}$$

where $p_{(q^*)jk^*}$ and $p_{(q^*+1)jk^*}$ are the relative frequencies for the category $Y_{jk^*}$ of the $q^*$th ordered and $(q^* + 1)$th ordered observations described by $Y_{(q^*)jk^*}$ and $Y_{(q^*+1)jk^*}$ in $C_v^{(1)j}$ and $C_v^{(2)j}$, respectively.

The algorithm is repeated iteratively until $r = m$ or until $r = R$.

**Example 7.2**   Consider the household data of Table 4.3 used in Example 7.1. Let us assume all observations in $\Omega$ are equally weighted, i.e., in Eq. (7.1.2), $w_u = 1/m, u = 1, \ldots, m = 6$. This implies that the weight $h_u^v$ in the sub-cluster of size $m_v$ is, from Eq. (7.1.14), $h_u^v = 1/m_v$.

Initially, we have $P_1 \equiv C_1 = \Omega$. From Table 7.2 calculated in Example 7.1, we see that, for the variable $Y_1$, the largest association value is for the category $Y_{11}$, i.e., from Eq. (7.2.2), $\alpha^1_{1k^*} = \max_k\{\alpha^1_{1k}, k = 1, \dots, 6\} = 0.0034$. Hence, the observations are sorted according to their $Y_{11}$ probabilities. Thus, the original ordering of the observations described by $Y_{11}, \dots, Y_{16}$ now corresponds to $Y^1_{(3)}, Y^1_{(5)}, Y^1_{(1)}, Y^1_{(2)}, Y^1_{(6)}, Y^1_{(4)}$, respectively. The extended Ichino–Yaguchi dissimilarity matrix, from Eq. (4.1.15), when $\gamma = 0.25$, for $Y_1$ when sorted by $Y_{11}$, becomes

$$
\mathbf{D}_1 = \begin{bmatrix}
0 & 0.156 & 0.126 & 0.212 & 0.197 & 0.267 \\
0.156 & 0 & 0.161 & 0.176 & 0.117 & 0.231 \\
0.126 & 0.161 & 0 & 0.117 & 0.137 & 0.195 \\
0.212 & 0.176 & 0.117 & 0 & 0.089 & 0.084 \\
0.197 & 0.117 & 0.137 & 0.089 & 0 & 0.122 \\
0.267 & 0.231 & 0.195 & 0.084 & 0.122 & 0
\end{bmatrix}
\tag{7.2.7}
$$

and the corresponding dissimilarity matrix for $Y_2$, when sorted by $Y_{11}$, is

$$
\mathbf{D}_2 = \begin{bmatrix}
0 & 0.122 & 0.080 & 0.029 & 0.089 & 0.039 \\
0.122 & 0 & 0.170 & 0.123 & 0.078 & 0.132 \\
0.080 & 0.170 & 0 & 0.053 & 0.092 & 0.042 \\
0.029 & 0.123 & 0.053 & 0 & 0.090 & 0.011 \\
0.089 & 0.078 & 0.092 & 0.090 & 0 & 0.090 \\
0.039 & 0.132 & 0.042 & 0.011 & 0.090 & 0
\end{bmatrix},
\tag{7.2.8}
$$

hence, by adding Eqs. (7.2.7) and (7.2.8), the Ichino–Yaguchi city block dissimilarity, when sorted by $Y_{11}$, becomes

$$
\mathbf{D}^{(1)} = \begin{bmatrix}
0 & 0.278 & 0.206 & 0.240 & 0.285 & 0.306 \\
0.278 & 0 & 0.330 & 0.299 & 0.195 & 0.363 \\
0.206 & 0.330 & 0 & 0.170 & 0.228 & 0.237 \\
0.240 & 0.299 & 0.170 & 0 & 0.179 & 0.095 \\
0.285 & 0.195 & 0.228 & 0.179 & 0 & 0.212 \\
0.306 & 0.363 & 0.237 & 0.095 & 0.212 & 0
\end{bmatrix}.
\tag{7.2.9}
$$

Likewise, for the variable $Y_2$, from Table 7.2, from Eq. (7.2.2), $\alpha^1_{2k^*} = \max_k\{\alpha^1_{2k}, k = 1, 2, 3\} = 0.00125$. Hence, the sorted observations in this case are described by $Y^2_{(5)}, Y^2_{(6)}, Y^2_{(3)}, Y^2_{(2)}, Y^2_{(4)}$, and $Y^2_{(1)}$, and the Ichino–Yaguchi city block dissimilarity matrix, when sorted by $Y_{22}$, is

$$
\mathbf{D}^{(2)} =
\begin{bmatrix}
0 & 0.195 & 0.278 & 0.299 & 0.363 & 0.330 \\
0.195 & 0 & 0.285 & 0.179 & 0.212 & 0.228 \\
0.278 & 0.285 & 0 & 0.240 & 0.306 & 0.206 \\
0.299 & 0.179 & 0.240 & 0 & 0.095 & 0.170 \\
0.363 & 0.212 & 0.306 & 0.095 & 0 & 0.237 \\
0.330 & 0.228 & 0.206 & 0.170 & 0.237 & 0
\end{bmatrix}.
\tag{7.2.10}
$$

Notice that these ordered dissimilarity matrices $(\mathbf{D}^{(1)}, \mathbf{D}^{(2)})$ are just that, a reordering determined by the association measures $(\alpha_{jk}, k = 1, \dots, s_j; j = 1, 2,$ respectively) of the original observations with dissimilarity matrix

$$
\mathbf{D} =
\begin{bmatrix}
0 & 0.170 & 0.206 & 0.237 & 0.330 & 0.228 \\
0.170 & 0 & 0.240 & 0.095 & 0.299 & 0.179 \\
0.206 & 0.240 & 0 & 0.306 & 0.278 & 0.285 \\
0.237 & 0.095 & 0.306 & 0 & 0.363 & 0.212 \\
0.330 & 0.299 & 0.278 & 0.363 & 0 & 0.195 \\
0.228 & 0.179 & 0.285 & 0.212 & 0.195 & 0
\end{bmatrix}.
\tag{7.2.11}
$$

At the first stage $(r = 1)$, we have one cluster $P_1 \equiv C_1 \equiv \Omega$, which is to be divided into two sub-clusters $(C_{1(q)}^{(1)}, C_{1(q)}^{(2)})$ where $(q)$ represents the position of the bi-partition at which the division into the two sub-clusters occurs, $q = 1, \dots, m - 1 = 5$ (see Eq. (7.2.3)). We need to calculate the $\Delta_{v(q)}^{j}$, for each of $j = 1, 2$, from Eq. (7.2.4); at this first stage, $v = r = 1$.

To illustrate, take $j = 1$ and $q = 2$. Therefore, when sorted by $Y_j = Y_1$, we see (from Table 7.2) that the maximum association measure is $\alpha_{11}$, hence the observations are sorted in ascending order by their $p_{11}$ values to give us $C_1 = \{(3), (5), (1), (2), (6), (4)\}$. When this ordered $C_1$ is bi-partitioned, for $q = 2$, we have $C_{1(2)}^{(1)1} = \{(3), (5)\}$ and $C_{1(2)}^{(2)1} = \{(1), (2), (6), (4)\}$. Therefore, from Eq. (7.1.2), by substituting from Eq. (7.2.9),

$$
I(C_{1(2)}^{(1)1}) = \frac{1}{(2/6)} \left[ \left( \frac{1}{6} \right) \left( \frac{1}{6} \right) (.278^2) \right] = 0.0064
\tag{7.2.12}
$$

and

$$I(C_{1(2)}^{(2)1}) = \frac{1}{(4/6)} \left[ \left(\frac{1}{6}\right)\left(\frac{1}{6}\right)(0.170^2) + \left(\frac{1}{6}\right)\left(\frac{1}{6}\right)(0.228^2) \right.$$
$$+ \left(\frac{1}{6}\right)\left(\frac{1}{6}\right)(0.237^2) + \left(\frac{1}{6}\right)\left(\frac{1}{6}\right)(0.179^2)$$
$$\left. + \left(\frac{1}{6}\right)\left(\frac{1}{6}\right)(0.095^2) + \left(\frac{1}{6}\right)\left(\frac{1}{6}\right)(0.212^2) \right]$$
$$= 0.0093. \tag{7.2.13}$$

Also, from Eqs. (7.1.2) and (7.2.9), we have

$$I(C_1^1) \equiv I(\Omega^1) = \frac{1}{(6/6)} \left[ \left(\frac{1}{6}\right)\left(\frac{1}{6}\right)(0.278^2) + \cdots + \left(\frac{1}{6}\right)\left(\frac{1}{6}\right)(0.212^2) \right]$$
$$= 0.0262. \tag{7.2.14}$$

Hence, substituting into Eq. (7.2.4), we have

$$\Delta_{1(2)}^1 = I(C_{1(2)}^1) - I(C_{1(2)}^{(1)1}) - I(C_{1(2)}^{(2)1}) = 0.0262 - 0.0064 - 0.0093$$
$$= 0.0105. \tag{7.2.15}$$

Similarly, we can calculate all $\Delta_{1(q)}^1$, $q = 1, \ldots, m - 1 = 5$. These values along with those for $I(C_{1(q)}^{(1)1})$ and $I(C_{1(q)}^{(2)1})$ are shown in Table 7.3(i).

Likewise, for variable $Y_2$, we calculate $\Delta_{1(q)}^2$, $q = 1, \ldots, m - 1 = 5$, where now the observations are ordered according to ascending values of $p_{22}$ since the maximum association measure for $Y_2$ was $\alpha_{22} = 0.00125$ (see Table 7.2). The corresponding $\Delta_{1(q)}^2$, $I(C_{1(q)}^{(1)2})$ and $I(C_{1(q)}^{(2)2})$, for $q = 1, \ldots, m - 1 = 5$, are shown in Table 7.3(ii).

Then, by Eq. (7.2.5), we want the maximum of these $\Delta_{1(q)}^j$ values. In this case,

$$\Delta_{v^*(q^*)}^j = \max_{v,j,q} \{\Delta_{v(q)}^j\} = \Delta_{1(2)}^{(2)2} = 0.0110. \tag{7.2.16}$$

Therefore, by using the bi-partition made when sorting by $Y_{22}$, the sub-clusters after the first stage are now $P_2 = (C_1, C_2) = (\{(5), (6)\}, \{(3), (2), (4), (1)\})$. The cut variable is $Y_{22}$ and its value is, from Eq. (7.2.6),

$$c_r = (p_{(q^*)jk^*} + p_{(q^*+1)jk^*})/2 = (p_{622} + p_{322})/2 = (0.378 + 0.388)/2 = 0.383. \tag{7.2.17}$$

That is, the bi-partition occurs by asking the question "Is the probability for $Y_{22} \leq 0.383$?" or. more succinctly "Is $Y_{22} \leq 0.383$?". If "Yes", then the observation goes into the sub-cluster $C_1$ and if "No", the observation goes into $C_2$.

This procedure is then repeated for each of the stages $r = 2, 3, \ldots$ . At each stage, we first calculate the association measures $\alpha_{jk}$ for each of the sub-clusters $C_v$, $v = 1, \ldots, r$. These are given in Table 7.4, for all sub-clusters that will be needed at one or more stage (in this case, $r = 1, 2, 3$; in general,

**Table 7.3** Variation components for first division $C_1 = (C_1^{(1)}, C_1^{(2)})$: extended Ichino–Yaguchi city block distances (Example 7.2)

| | | (i) Sorted by $Y_{11}$ | | | |
|---|---|---|---|---|---|
| $q$ | Bi-partition: $(C_{1(q)}^{(1)1}, C_{1(q)}^{(2)1})$ | $\Delta_{v(q)}^1$ | $I(C_{v(q)}^1)$ | $I(C_{v(q)}^{(1)1})$ | $I(C_{v(q)}^{(2)1})$ |
| 1 | $(\{(3)\}, \{(5), (1), (2), (6), (4)\})$ | 0.0065 | 0.0262 | 0 | 0.0197 |
| 2 | $(\{(3), (5)\}, \{(1), (2), (6), (4)\})$ | 0.0105 | 0.0262 | 0.0064 | 0.0093 |
| 3 | $(\{(3), (5), (1)\}, \{(2), (6), (4)\})$ | 0.0087 | 0.0262 | 0.0127 | 0.0048 |
| 4 | $(\{(3), (5), (1), (2)\}, \{(6), (4)\})$ | 0.0056 | 0.0262 | 0.0168 | 0.0037 |
| 5 | $(\{(3), (5), (1), (2), (6)\}, \{(4)\})$ | 0.0059 | 0.0262 | 0.0202 | 0 |

| | | (ii) Sorted by $Y_{22}$ | | | |
|---|---|---|---|---|---|
| $q$ | Bi-partition: $(C_{1(q)}^{(1)2}, C_{1(q)}^{(2)2})$ | $\Delta_{v(q)}^2$ | $I(C_{v(q)}^2)$ | $I(C_{v(q)}^{(1)2})$ | $I(C_{v(q)}^{(2)2})$ |
| 1 | $(\{(5)\}, \{(6), (3), (2), (4), (1)\})$ | 0.0096 | 0.0262 | 0 | 0.0166 |
| 2 | $(\{(5), (6)\}, \{(3), (2), (4), (1)\})$ | **0.0110** | 0.0262 | 0.0032 | 0.0120 |
| 3 | $(\{(5), (6), (3)\}, \{(2), (4), (1)\})$ | 0.0100 | 0.0262 | 0.0109 | 0.0052 |
| 4 | $(\{(5), (6), (3), (2)\}, \{(4), (1)\})$ | 0.0059 | 0.0262 | 0.0156 | 0.0047 |
| 5 | $(\{(5), (6), (3), (2), (4)\}, \{(1)\})$ | 0.0044 | 0.0262 | 0.0218 | 0 |

**Table 7.4** Association measures for stage $r = 2,\ 3$: extended Ichino–Yaguchi city block distances (Example 7.2)

| $Y_{jk}$ | | $r = 2$ | | $r = 3$ | |
|---|---|---|---|---|---|
| $j$ | $k$ | $\{1, 2, 3, 4\}$ | $\{5, 6\}$ | $\{1, 3\}$ | $\{2, 4\}$ |
| 1 | 1 | **0.004459** | **0.0015** | **0.0017** | **0.0008** |
| 1 | 2 | 0.001009 | 0.0009 | 0.0011 | 0.0002 |
| 1 | 3 | 0.001130 | 0.0000 | 0.0004 | 0.0000 |
| 1 | 4 | 0.000244 | 0.0003 | 0.0008 | 0.0003 |
| 1 | 5 | 0.004113 | 0.0000 | 0.0007 | 0.0006 |
| 1 | 6 | 0.001397 | 0.0013 | 0.0009 | 0.0003 |
| 2 | 1 | 0.000030 | 0.0005 | 0.0001 | 0.0000 |
| 2 | 2 | 0.000362 | **0.0007** | 0.0007 | **0.0000** |
| 2 | 3 | 0.000364 | 0.0006 | **0.0007** | 0.0000 |

**Table 7.5** $\Delta^j_{v(q)}$ for bi-partitions of $(C_1, C_2)$ at Stage $r = 2, 3, 4$: extended Ichino–Yaguchi city block distances (Example 7.2)

| $r$ | $C^j_v$ | Sorted by $Y_{jk}$ | Bi-partition $q$ | Sub-clusters $(C^{(1)1}_{v(q)}, C^{(2)1}_{v(q)})$ | $\Delta^j_{v(q)}$ |
|---|---|---|---|---|---|
| | | | | $(C_1, C_2) = (\{1,2,3,4\}, \{5,6\})$ | |
| 2 | $C^1_1$ | $Y_{11}$ | 1 | $(\{(3)\}, \{(1), (2), (4)\})$ | 0.00676 |
| | | | 2 | $(\{(3), (1)\}, \{(2), (4)\})$ | **0.00771** |
| | | | 3 | $(\{(3), (1), (2)\}, \{(4)\})$ | 0.00483 |
| | $C^2_1$ | $Y_{23}$ | 1 | $(\{(1)\}, \{(4), (2), (3)\})$ | 0.00307 |
| | | | 2 | $(\{(1), (4)\}, \{(2), (3)\})$ | 0.00249 |
| | | | 3 | $(\{(1), (4), (2)\}, \{(3)\})$ | 0.00076 |
| | $C^1_2$ | $Y_{11}$ | 1 | $(\{(5)\}, \{(6)\})$ | 0.00317 |
| | $C^2_2$ | $Y_{22}$ | 1 | $(\{(5)\}, \{(6)\})$ | 0.00317 |
| | | | | $(C_1, C_2, C_3) = (\{1,3\}, \{2,4\}, \{5,6\})$ | |
| 3 | $C^1_1$ | $Y_{11}$ | 1 | $(\{(3)\}\{, (1)\})$ | **0.00352** |
| | $C^2_1$ | $Y_{23}$ | 1 | $(\{(3)\}\{, (1)\})$ | 0.00352 |
| | $C^1_2$ | $Y_{11}$ | 1 | $(\{(2)\}\{, (4)\})$ | 0.00074 |
| | $C^2_2$ | $Y_{22}$ | 1 | $(\{(2)\}\{, (4)\})$ | 0.00074 |
| | $C^1_3$ | $Y_{11}$ | 1 | $(\{(5), (6)\})$ | 0.00317 |
| | $C^2_3$ | $Y_{22}$ | 1 | $(\{(5), (6)\})$ | 0.00317 |
| | | | | $(C_1, C_2, C_3, C_4) = (\{2,4\}, \{5,6\}, \{3\}, \{1\}))$ | |
| 4 | $C^1_1$ | $Y_{11}$ | 1 | $(\{(2)\}, \{(4)\})$ | 0.00074 |
| | $C^2_1$ | $Y_{22}$ | 1 | $(\{(2)\}, \{(4)\})$ | 0.00074 |
| | $C^1_2$ | $Y_{11}$ | 1 | $(\{(5)\}, \{(6)\})$ | **0.00317** |
| | $C^2_2$ | $Y_{22}$ | 1 | $(\{(5)\}, \{(6)\})$ | 0.00317 |

$r = 1, \ldots, m - 1$). A summary of the values for $\Delta^j_{v(q)}$, $v = 1, \ldots, r$, $j = 1, 2$, $r = 2, 3, 4$, is shown in Table 7.5.

From Tables 7.4 and 7.5, by selecting in turn the maximum $\alpha_{jk}$ and maximum $\Delta^j_{v(q)}$ values, we see that at the $r = 2$ stage, the cluster $C_1 = \{1, 2, 3, 4\}$ is bi-partitioned into $C_1 = (C^{(1)}_1, C^{(2)}_1)$ with $C^{(1)}_1 = \{1, 3\}$ and $C^{(2)}_1 = \{2, 4\}$, with cutting variable "Is $Y_{11} \leq 0.2965$?" Then, at stage $r = 3$, the cluster $C_1 = \{1, 3\}$ becomes $C_1 = (C^{(1)}_1, C^{(2)}_1) = (\{3\}, \{1\})$ with cutting value "Is $Y_{11} \leq 0.2425$?" At stage $r = 4$, the sub-cluster $C_2 = \{5, 6\}$ is partitioned with cutting variable "Is $Y_{11} \leq 0.2975$?" Finally, at the last stage (here, $r = 5$), clearly

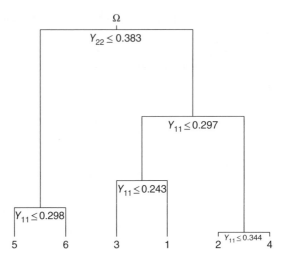

**Figure 7.1** Monothetic divisive hierarchy. Modal multi-valued data: extended Ichino–Yaguchi city block distances (Example 7.2).

the sub-cluster $C_{(\cdot)} = \{2, 4\}$ is bi-partitioned. Because the largest association measure is $\alpha_{11} = 0.0008$ (from Table 7.4), then the cutting question becomes "Is $Y_{11} \leq 0.3435$?" The final hierarchy is displayed in Figure 7.1.

The proportion of the total variation explained at each stage $r$ is discussed and illustrated for this hierarchy in Example 7.15 in section 7.4. Thus, if the vertical axis in Figure 7.1 is set on a $(0, 1)$ scale, then the respective heights at which a cluster is partitioned at stage $r$ is equivalent to $1 - E_r = e_r$, as indicated in Figure 7.17. □

**Example 7.3** Suppose, instead of the city block extended Ichino–Yaguchi dissimilarities used in Example 7.2, we had used the Euclidean extended Ichino–Yaguchi distances. Now, the distance matrix of Eq. (4.1.15), when $\gamma = 0.25$, is

$$
\mathbf{D}_1 = \begin{bmatrix}
0 & 0.128 & 0.149 & 0.199 & 0.233 & 0.164 \\
0.128 & 0 & 0.213 & 0.085 & 0.214 & 0.126 \\
0.149 & 0.213 & 0 & 0.270 & 0.198 & 0.216 \\
0.199 & 0.085 & 0.270 & 0 & 0.266 & 0.151 \\
0.233 & 0.214 & 0.198 & 0.266 & 0 & 0.141 \\
0.164 & 0.126 & 0.216 & 0.151 & 0.141 & 0
\end{bmatrix}.
$$

Then the hierarchy obtained is that shown in Figure 7.2. The relevant statistics (corresponding to Tables 7.3 and 7.5) are found in Tables 7.6 and 7.7, respectively. The details are left to the reader (see Exercise 7.3). □

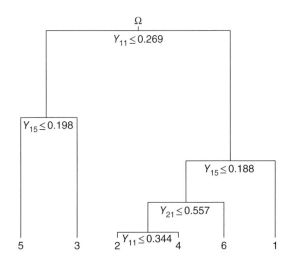

**Figure 7.2** Monothetic divisive hierarchy. Modal multi-valued data: extended Ichino–Yaguchi Euclidean distances (Example 7.3).

**Table 7.6** Variation components for first division $C_1 = (C_1^{(1)}, C_1^{(2)})$: extended Ichino–Yaguchi Euclidean distances (Example 7.3)

| | | (i) Sorted by $Y_{11}$ | | | |
|---|---|---|---|---|---|
| $q$ | Bi-partition: $(C_{1(q)}^{(1)1}, C_{1(q)}^{(2)1})$ | $\Delta_{v(q)}^1$ | $I(C_{v(q)}^1)$ | $I(C_{v(q)}^{(1)1})$ | $I(C_{v(q)}^{(2)1})$ |
| 1 | $(\{(3)\}, \{(5), (1), (2), (6), (4)\})$ | 0.0045 | 0.0152 | 0 | 0.0107 |
| 2 | $(\{(3), (5)\}, \{(1), (2), (6), (4)\})$ | 0.0065 | 0.0152 | 0.0033 | 0.0054 |
| 3 | $(\{(3), (5), (1)\}, \{(2), (6), (4)\})$ | 0.0062 | 0.0152 | 0.0064 | 0.0026 |
| 4 | $(\{(3), (5), (1), (2)\}, \{(6), (4)\})$ | 0.0039 | 0.0152 | 0.0093 | 0.0019 |
| 5 | $(\{(3), (5), (1), (2), (6)\}, \{(4)\})$ | 0.0041 | 0.0152 | 0.0111 | 0 |
| | | (ii) Sorted by $Y_{22}$ | | | |
| $q$ | Bi-partition: $(C_{1(q)}^{(1)2}, C_{1(q)}^{(2)2})$ | $\Delta_{v(q)}^2$ | $I(C_{v(q)}^2)$ | $I(C_{v(q)}^{(1)2})$ | $I(C_{v(q)}^{(2)2})$ |
| 1 | $(\{(5)\}, \{(6), (3), (2), (4), (1)\})$ | 0.0046 | 0.0152 | 0 | 0.0105 |
| 2 | $(\{(5), (6)\}, \{(3), (2), (4), (1)\})$ | 0.0050 | 0.0152 | 0.0016 | 0.0085 |
| 3 | $(\{(5), (6), (3)\}, \{(2), (4), (1)\})$ | 0.0058 | 0.0152 | 0.0059 | 0.0035 |
| 4 | $(\{(5), (6), (3), (2)\}, \{(4), (1)\})$ | 0.0030 | 0.0152 | 0.0089 | 0.0033 |
| 5 | $(\{(5), (6), (3), (2), (4)\}, \{(1)\})$ | 0.0023 | 0.0152 | 0.0129 | 0 |

**Table 7.7** $\Delta^j_{v(q)}$ for bi-partitions of $(C_1, C_2)$ at Stage $r = 2, 3, 4$: extended Ichino–Yaguchi Euclidean distances (Example 7.3)

| $r$ | $C^j_v$ | Sorted by $Y_{jk}$ | Bi-partition $q$ | Sub-clusters $(C^{(1)j}_{v(q)}, C^{(2)j}_{v(q)})$ | $\Delta^j_{v(q)}$ |
|---|---|---|---|---|---|
| | | | | $(C_1, C_2) = \{(1), (2), (6), (4)\}, \{(3),(5)\})$ | |
| 2 | $C^1_1$ | $Y_{11}$ | 1 | $(\{(1)\}, \{(2), (6), (4)\})$ | 0.00283 |
| | | | 2 | $(\{(1), (2)\}, \{(6), (4)\})$ | 0.00211 |
| | | | 3 | $(\{(1), (2), (6)\}, \{(4)\})$ | 0.00208 |
| | $C^2_1$ | $Y_{21}$ | 1 | $(\{(1)\}, \{(2), (6), (4)\})$ | 0.00040 |
| | | | 2 | $(\{(1), (2)\}, \{(6), (4)\})$ | 0.00254 |
| | | | 3 | $(\{(1), (2), (6)\}, \{(4)\})$ | 0.00186 |
| | $C^1_2$ | $Y_{15}$ | 1 | $(\{(5)\}, \{(3)\})$ | **0.00326** |
| | $C^2_2$ | $Y_{21}$ | 1 | $(\{(3)\}, \{(5)\})$ | 0.00326 |
| | | | | $(C_1, C_2, C_3) = (\{(1), (2), (4), (6)\}, \{(3)\}, \{(5)\})$ | |
| 3 | $C^1_1$ | $Y_{15}$ | 1 | $(\{(4)\}, \{(2), (6), (1)\})$ | 0.00208 |
| | | | 2 | $(\{(4), (2)\}, \{(6), (1)\})$ | 0.00254 |
| | | | 3 | $(\{(4), (2), (6)\}, \{(1)\})$ | **0.00283** |
| | $C^2_1$ | $Y_{21}$ | 1 | $(\{(4)\}, \{(2), (1), (6)\})$ | 0.00040 |
| | | | 2 | $(\{(4), (2)\}, \{(1), (6)\})$ | 0.00254 |
| | | | 3 | $(\{(4), (2), (1)\}, \{(6)\})$ | 0.00186 |
| | | | | $(C_1, C_2, C_3, C_4) = (\{(2), (4), (6)\}, \{(3)\}, \{(5)\}, \{(1)\})$ | |
| 4 | $C^1_1$ | $Y_{15}$ | 1 | $(\{(4)\}, \{(2), (6)\})$ | 0.00123 |
| | | | 2 | $(\{(4), (2)\}, \{(6)\})$ | **0.00196** |
| | $C^2_1$ | $Y_{21}$ | 1 | $(\{(6)\}, \{(2), (4)\})$ | 0.00065 |
| | | | 2 | $(\{(6), (2)\}, \{(4)\})$ | 0.00196 |

Chavent (2000) also introduced a monothetic divisive algorithm for ordered (only) modal-valued observations. Chavent (1997) considered the non-ordered case by exploring all possible bi-partitions to split a partition at each stage. However, the monothetic algorithms described herein can be used on either ordered or non-ordered modal-valued observations and also are more efficient than the Chavent (1997, 2000) methods, as described in Kim (2009) and Kim and Billard (2012).

### 7.2.2 Non-modal Multi-valued Observations

A set of non-modal multi-valued (or, simply, multi-valued) observations is a special case of modal multi-valued observations. In this case, the categories of $\mathcal{Y}_j$ that do occur are assumed to occur with equal probability, and those that do not occur clearly have probability $p_{ujk} \equiv 0$ (see Eq. (7.1.8)). The divisive method used on modal multi-valued observations in section 7.2.1 can be applied to this special case.

**Example 7.4**  Consider the data of Table 3.4. There are two variables $Y_1$ = energy and $Y_2$ = computer, taking values from $\mathcal{Y}_1$ = {gas, coal, electricity, wood, other} and $\mathcal{Y}_2$ = {laptop, personal computer (PC), other}, respectively. Reformulating these values in the format of Eq. (7.1.8) gives us the data in Table 7.8.

Suppose we base our dissimilarities on the Gowda–Diday dissimilarities of Eq. (3.2.6). Thus, we can show that the dissimilarity matrix is

$$\mathbf{D} = \begin{bmatrix} 0 & 2.000 & 0.833 & 3.167 & 4.000 \\ 2.000 & 0 & 1.167 & 2.633 & 3.167 \\ 0.833 & 1.167 & 0 & 3.467 & 4.000 \\ 3.167 & 2.633 & 3.467 & 0 & 1.167 \\ 4.000 & 3.167 & 4.000 & 1.167 & 0 \end{bmatrix}. \tag{7.2.18}$$

Then, using the methodology described in section 7.2.1, we obtain the hierarchy shown in Figure 7.3. The respective cutting variables and the questions asked are:

At stage $r = 1$ : "Is $Y_{11} \leq 0.25$?"
At stage $r = 2$ : "Is $Y_{21} \leq 0.75$?"
At stage $r = 3$ : "Is $Y_{13} \leq 0.25$?"
At stage $r = 4$ : "Is $Y_{14} \leq 0.167$?"

The details are left to the reader (see Exercise 7.4). ☐

**Table 7.8**  Multi-valued energy/computer data: relative frequencies (Example 7.4)

| Household | $Y_1$ = Energy | | | | | $Y_2$ = Computer | | |
|---|---|---|---|---|---|---|---|---|
| $u$ | Electricity | Gas | Wood | Coal | Other | Laptop | PC | Other |
| 1 | 0 | 0.5 | 0.5 | 0 | 0 | 1 | 0 | 0 |
| 2 | 0 | 0.333 | 0.333 | 0.333 | 0 | 0.5 | 0.5 | 0 |
| 3 | 0 | 0.333 | 0.333 | 0.333 | 0 | 1 | 0 | 0 |
| 4 | 0.5 | 0 | 0.5 | 0 | 0 | 0 | 0.5 | 0.5 |
| 5 | 0.5 | 0 | 0 | 0 | 0.5 | 0 | 0.5 | 0.5 |

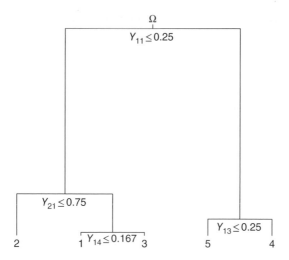

**Figure 7.3** Monothetic divisive hierarchy. Multi-valued data: Gowda–Diday dissimilarities (Example 7.4).

### 7.2.3 Interval-valued Observations

A divisive algorithm to build a hierarchy for interval-valued observations was first introduced by Chavent (1998, 2000). This is a monothetic methodology, therefore one variable at a time is considered. The partition is $P_r = (C_1, \ldots, C_r)$ where $C_v$ has observations $\{1, \ldots, m_v\}$, $v = 1, \ldots, r$. The partitioning criteria of section 7.1.1 are applied, and the Hausdorff (1937) distance (see Eq. (3.3.4)) is used as the dissimilarity measure $d(u_1, u_2)$ in Eqs. (7.1.2)–(7.1.4). The observations are ordered by increasing (or, equivalently, decreasing) observation mean values, and the cut value $c_q$, $q = 1, \ldots, n_v \leq m_v - 1$, is taken to be the midpoint between the observation means. The methodology is illustrated by the following examples.

**Example 7.5** Consider the data of Table 7.9 consisting of minimum and maximum monthly temperatures at 15 weather stations in China for 1988. The complete data set can be found at <http://dss.ucar.edu/datasets/ds578.5>. Each month is a variable with $Y_1$ = January, ..., $Y_{12}$ = December. The station elevation is the variable $Y_{13}$; note this is a classical value. Let us consider the first five stations for the months April, July, November, and elevation, and let us relabel these particular variables as $Z_1 \equiv Y_4$, $Z_2 \equiv Y_7$, $Z_3 \equiv Y_{11}$, $Z_4 \equiv Y_{13}$. These data are assembled into Table 7.10.

Let us focus on $Z_1$ as the partitioning variable (deferring the divisive procedure using all variables for Example 7.6). To reduce the number of possible sub-clusters to consider at each step (to $m_v - 1$ in cluster $C_v$, $v = 1, \ldots, r$), we first order the observations by increasing mean values (i.e., from Eq. (2.4.2),

**Table 7.9** China temperatures (Example 7.5)

| u | Station | $Y_1$ | $Y_2$ | $Y_3$ | $Y_4$ | $Y_5$ | $Y_6$ | $Y_7$ |
|---|---------|-------|-------|-------|-------|-------|-------|-------|
| 1 | BoKeTu | [−23.4, −15.5] | [−24.0, −14] | [−17.6, −4.7] | [−4.5, 9.5] | [1.9, 15.2] | [9.3, 23.0] | [12.9, 23.0] |
| 2 | Hailaer | [−28.4, −19.1] | [−29.6, −18.1] | [−20.2, −7.9] | [−2.9, 10.0] | [3.8, 16.5] | [12.5, 24.2] | [14.7, 25.4] |
| 3 | LaSa | [−8.4, 9.0] | [−3.5, 11.2] | [−1.2, 13.7] | [1.7, 16.4] | [5.9, 21.5] | [9.7, 24.8] | [10.8, 23.2] |
| 4 | KunMing | [2.8, 16.6] | [4.0, 19.4] | [6.7, 21.4] | [10.4, 23.4] | [15.9, 25.3] | [16.4, 25.3] | [16.9, 24.4] |
| 5 | TengChong | [2.3, 16.9] | [4.0, 18.9] | [6.4, 21.3] | [9.9, 24.3] | [15.0, 22.6] | [17.2, 24.1] | [17.4, 22.8] |
| 6 | WuZhou | [10.0, 17.7] | [9.1, 14.7] | [10.4, 16.2] | [15.8, 23.9] | [22.2, 30.3] | [23.9, 34.3] | [24.2, 33.8] |
| 7 | GuangZhou | [12.2, 20.6] | [11.4, 17.7] | [12.7, 17.9] | [17.4, 24.3] | [24.2, 30.3] | [25.1, 32.8] | [25.3, 33.6] |
| 8 | NanNing | [11.5, 17.7] | [10.3, 14.3] | [12.1, 17.3] | [17.8, 24.2] | [23.8, 30.7] | [24.9, 33.2] | [25.8, 33.5] |
| 9 | ShanTou | [11.8, 19.2] | [12.2, 17.7] | [12.7, 17.8] | [16.4, 22.7] | [22.5, 28.1] | [24.9, 31.5] | [25.6, 32.6] |
| 10 | HaiKou | [17.3, 23.0] | [17.3, 23.0] | [17.1, 23.7] | [19.9, 25.9] | [25.3, 33.4] | [25.5, 33.2] | [25.8, 34.0] |
| 11 | ZhanJiang | [15.8, 21.5] | [14.6, 19.4] | [14.4, 18.8] | [19.2, 24.2] | [26.0, 31.7] | [26.2, 32.3] | [27.1, 33.0] |
| 12 | MuDanJiang | [−18.4, −7.5] | [−19.2, −7.3] | [−10.7, 2.4] | [−0.1, 13.2] | [6.8, 18.5] | [13.7, 26.1] | [17.0, 26.5] |
| 13 | HaErBin | [−20.0, −9.6] | [−22.3, −10.0] | [−10.8, 1.7] | [0.2, 11.9] | [7.5, 19.0] | [14.6, 27.3] | [17.8, 27.2] |
| 14 | QiQiHaEr | [−22.7, −11.5] | [−21.1, −9.6] | [−11.5, 0.7] | [0.6, 13.8] | [7.8, 18.9] | [16.7, 27.4] | [19.1, 26.5] |
| 15 | NenJiang | [−27.9, −16.0] | [−27.7, −12.9] | [−16.5, −3.4] | [−1.5, 12.0] | [4.5, 17.5] | [13.7, 26.7] | [16.1, 25.0] |

(*Continued*)

**Table 7.9** (Continued)

| u | Station | $Y_8$ | $Y_9$ | $Y_{10}$ | $Y_{11}$ | $Y_{12}$ | $Y_{13}$ |
|---|---------|-------|-------|----------|----------|----------|----------|
| 1 | BoKeTu | [10.8, 23.6] | [4.1, 17.0] | [−4.0, 8.9] | [−13.5, −4.2] | [−13.1, 14.78] | 14.78 |
| 2 | Hailaer | [13.7, 24.8] | [5.3, 17.6] | [−3.2, 9.8] | [−13.8, −3.7] | [−26.0, −17.2] | 12.26 |
| 3 | LaSa | [10.2, 20.8] | [7.7, 19.9] | [1.4, 18.7] | [−4.9, 11.4] | [−7.5, 8.8] | 73.16 |
| 4 | KunMing | [16.6, 23.8] | [14.0, 21.7] | [12.4, 19.7] | [7.4, 16.3] | [3.8, 15.5] | 37.82 |
| 5 | TengChong | [17.4, 22.5] | [16.3, 23.2] | [14.5, 23.5] | [9.4, 20.0] | [4.7, 17.8] | 32.96 |
| 6 | WuZhou | [23.4, 32.1] | [22.3, 31.8] | [19.2, 27.6] | [12.4, 23.2] | [9.8, 21.6] | 2.38 |
| 7 | GuangZhou | [25.0, 31.9] | [24.1, 31.4] | [21.0, 28.5] | [14.2, 23.4] | [10.9, 22.2] | 0.14 |
| 8 | NanNing | [25.1, 31.9] | [24.2, 31.5] | [20.3, 26.9] | [14.8, 24.3] | [12.4, 22.4] | 1.44 |
| 9 | ShanTou | [24.9, 30.8] | [23.8, 30.0] | [20.4, 27.3] | [13.9, 22.6] | [10.0, 19.7] | 0.02 |
| 10 | HaiKou | [25.0, 32.6] | [25.1, 31.5] | [23.0, 26.7] | [18.6, 22.6] | [16.3, 21.6] | 0.28 |
| 11 | ZhanJiang | [25.6, 31.1] | [25.7, 31.9] | [22.0, 27.2] | [16.1, 23.2] | [14.5, 21.8] | 0.5 |
| 12 | MuDanJiang | [18.9, 28.4] | [10.3, 21.4] | [0.6, 13.1] | [−8.9, 1.5] | [−18.9, −7.8] | 4.82 |
| 13 | HaErBin | [17.7, 26.3] | [8.9, 20.8] | [−0.2, 12.5] | [−9.6, 0.4] | [−21.6, −10.1] | 3.44 |
| 14 | QiQiHaEr | [19.1, 26.9] | [10.5, 19.8] | [1.9, 12.2] | [−9.0, 0.2] | [−17.8, −9.1] | 2.92 |
| 15 | NenJiang | [14.5, 26.0] | [6.9, 19.2] | [−2.6, 10.9] | [−14.4, −3.6] | [−26.1, −13.8] | 4.84 |

**Table 7.10** China temperatures: five stations and four variables (Example 7.5)

| $u$ | Station | $Z_1$ April | $Z_2$ July | $Z_3$ November | $Z_4$ Elevation |
|---|---|---|---|---|---|
| 1 | BoKeTu | $[-4.5, 9.5]$ | $[12.9, 23.0]$ | $[-13.5, -4.2]$ | 14.78 |
| 2 | Hailaer | $[-2.9, 10.0]$ | $[14.7, 25.4]$ | $[-13.8, -3.7]$ | 12.26 |
| 3 | LaSa | $[1.7, 16.4]$ | $[10.8, 23.2]$ | $[-4.9, 11.4]$ | 73.16 |
| 4 | KunMing | $[10.4, 23.4]$ | $[16.9, 24.4]$ | $[7.4, 16.3]$ | 37.82 |
| 5 | TengChong | $[9.9, 24.3]$ | $[17.4, 22.8]$ | $[9.4, 20.0]$ | 32.96 |

$\bar{Z}_{u1} = (a_{u1} + b_{u1})/2)$. For $Z_1$, we have $\bar{Z}_{u1} = 2.50, 3.55, 9.05, 16.90, 17.10$, for stations $u = 1, \ldots, 5$, respectively. Thus, the ordered sequence is as presented in Table 7.11. (Note that this is not the same ordering if a different variable is used.)

Suppose the distance measure to be used in Eq. (7.1.2) is the Hausdorff distance of Eq. (3.3.4). Then, we obtain the distance matrix, $\mathbf{D}$, as

$$\mathbf{D} = \begin{bmatrix} 0 & 1.6 & 6.9 & 14.9 & 14.8 \\ 1.6 & 0 & 6.4 & 13.4 & 14.3 \\ 6.9 & 6.4 & 0 & 8.7 & 8.2 \\ 14.9 & 13.4 & 8.7 & 0 & 0.9 \\ 14.8 & 14.3 & 8.2 & 0.9 & 0 \end{bmatrix}. \qquad (7.2.19)$$

From these distances in Eq. (7.2.19) and by substituting into Eq. (7.1.4), we obtain the total variation $W(\Omega) = 42.399$, where we have assumed the observations have equal weights (i.e., for each $u = 1, \ldots, m$, $w_u = 1/m$ and hence $\lambda = 1$, or, equivalently, $w_u = 1$ and hence $\lambda = m$).

Table 7.11 shows the possible bi-partitions of $\Omega$ into $(C_{v(q)}^{(1)}, C_{v(q)}^{(2)})$ at the first step; here $v = r = 1$. Also shown are the respective cut-values for each potential $q$. Thus, for example, when $q = 2$, the bi-partition gives $\Omega \equiv P_2 = (C_{v(q)}^{(1)}, C_{v(q)}^{(2)}) = (\{1, 2\}, \{3, 4, 5\})$. The corresponding cut-value is $c_q = (3.55 + 9.05)/2 = 6.3$. The last line of Table 7.11 displays results for a so-called null bi-partition in which all observations are in $C_v^{(1)}$ and $C_v^{(2)}$ is the empty null cluster $\phi$; this bi-partition is denoted by $q = (0)$.

Then, for each possible bi-partition $q$, we calculate the respective within-cluster variations, from Eq. (7.1.2), namely, $I(C_{v(q)}^{(1)})$ and $I(C_{v(q)}^{(2)})$. Substituting into Eq. (7.2.4), we obtain the change, $\Delta_{v(q)}^1$, in total within-cluster variation for each bi-partition $q$. These values are summarized in Table 7.11. The actual bi-partition selected corresponds to that $q$ for which $\Delta_{v(q)}^1$ is greatest, according to Eq. (7.2.4). In this case, this corresponds to the $q = 3$

**Table 7.11** First bi-partitioning on $Z_1$ (Example 7.5)

| Station $u$ | Mean $\bar{Z}_{u1}$ | Bi-partition $(C^{(1)}_{v(q)}, C^{(2)}_{v(q)})$ | $q$ | Cut $c_q$ | (a) $I(C^{(1)}_{v(q)})$ | (b) $I(C^{(2)}_{v(q)})$ | (a)+(b) | $\Delta^1_{v(q)}$ |
|---|---|---|---|---|---|---|---|---|
| 1 | 2.50 | $(\{1\},\{2,3,4,5\})$ | 1 | 3.025 | 0 | 28.438 | 28.438 | 13.961 |
| 2 | 3.55 | $(\{1,2\},\{3,4,5\})$ | 2 | 6.3 | 0.256 | 9.583 | 9.839 | 32.560 |
| 3 | 9.05 | $(\{1,2,3\},\{4,5\})$ | 3 | 12.975 | 6.075 | 0.081 | 6.156 | **36.242** |
| 4 | 16.90 | $(\{1,2,3,4\},\{5\})$ | 4 | 17.0 | 28.420 | 0 | 28.420 | 13.979 |
| 5 | 17.10 | $(\{1,2,3,4,5\},\phi)$ | (0) | | 42.399 | – | 42.399 | |

**Table 7.12** Second bi-partitioning by $Z_1$ (Example 7.5)

| Station $u$ | Mean $\bar{Z}_{u1}$ | Bi-partition $(C^{(1)}_{v(q)}, C^{(2)}_{v(q)})$ | $q$ | Cut $c_q$ | (a) $I(C^{(1)}_{v(q)})$ | (b) $I(C^{(2)}_{v(q)})$ | (a)+(b) | $\Delta^1_{v(q)}$ |
|---|---|---|---|---|---|---|---|---|
| | | Partition of $C^{(1)}_1 = \{1,2,3\}$ | | | | | | |
| 1 | 2.50 | $(\{1\},\{2,3\})$ | 1 | 3.025 | 0 | 4.096 | 4.096 | 1.979 |
| 2 | 3.55 | $(\{1,2\},\{3\})$ | 2 | 6.3 | 0.256 | 0 | 0.256 | **5.819** |
| 3 | 9.05 | $(\{1,2,3\},\phi)$ | (0) | | 6.075 | – | 6.075 | |
| | | Partition of $C^{(2)}_1 = \{4,5\}$ | | | | | | |
| 4 | 16.90 | $(\{4\},\{5\})$ | 1 | 17.0 | 0 | 0 | 0 | 0.081 |
| 5 | 17.10 | $(\{4,5\},\phi)$ | (0) | | 0.081 | – | 0.081 | |

bi-partition with $\Delta^1_{v(q)} = 36.242$. This gives us two new clusters: $C_1 = \{1,2,3\}$ and $C_2 = \{4,5\}$; the cut value is, for this $r = 1$ step, $c_1 = 12.975$ at $q = 3$.

The procedure is repeated for both the $C^{(1)}_1$ and $C^{(2)}_1$ clusters obtained at the $r = 1$ step. The details for each cluster bi-partition possibilities are summarized in Table 7.12. If cluster $C^{(1)}_1$ is chosen to be bi-partitioned, then since $\Delta^1_{v(q)} = 5.819$ is the maximum $\Delta^1_{v(q)}$ value, the bi-partition $C^{(1)}_1 = (\{1,2\},\{3\})$ is selected, whereas for $C^{(2)}_1$, the best bi-partition (in fact, the only possible one) is $C^{(2)}_1 = (\{4\},\{5\})$. Which particular sub-cluster is chosen is found by selecting the sub-cluster that has the maximum of the maximum change of total within-cluster variation, i.e., $\max_{q,v}\{\Delta^1_{v(q)}\}$. Here, since $\max\{5.819, 0.081\} = 5.819$, the sub-cluster $C^{(1)}_1$ is selected. Hence, at the end of step $r = 2$, we have $\Omega = P_3 = (C_1, C_2, C_3) = (\{1,2\},\{3\},\{4,5\})$. If the divisive procedure is stopped at $R = 2$ steps, the resulting hierarchy is as shown in Figure 7.4, and the cut-value questions are

At stage $r = 1$ :"Is $Z_1 \leq 12.975$?"
At stage $r = 2$ :"Is $Z_1 \leq 6.3$?"

□

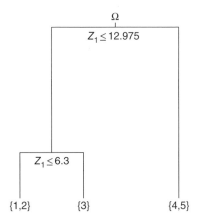

**Figure 7.4** Monothetic divisive hierarchy. Interval-valued data: Hausdorff distance on $Z_1$ (April), five stations (Example 7.5).

**Example 7.6** Consider the temperature data of Table 7.10 but now suppose we construct a hierarchy using all four variables. Therefore, we repeat the process described in Example 7.5 for each of the variables $Z_j$, $j = 1, \ldots, 4$, one by one. The possible bi-partitions are collectively those shown in Table 7.11 (for $Z_1$) and Table 7.13 (for $Z_2, Z_3, Z_4$). From these, it is seen that based on $Z_1, Z_2, Z_3$, and $Z_4$, the maximum changes in total within-cluster variation $\Delta_q^j$ are 36.243, 4.863, 90.330, and 379.548, for $j = 1, 2, 3$, and 4, respectively. Clearly, the maximum of these change values is the 379.548 relating to the $Z_4$ variable, with cut-value $c_1 = 55.49$ at $q = 2$. Hence, the optimal bi-partition at this first stage gives $P_2 = (C_1, C_2) = (\{1, 2, 4, 5\}, \{3\})$.

Then at the second stage, we consider the possible bi-partitions of $C_1$; in this case, there is no further bi-partition possible for $C_2$. Again, we obtain the maximum change in total within-cluster variations for each possible bi-partition of $C_1$, for each $Z_j$, $j = 1, \ldots, 4$, one by one. The resulting details are given in Table 7.14. As it happens, it is the second ($q = 2$) possible bi-partition that prevails for each variable. The maximum change in total within-cluster variation over all variables is $\max_j \{\Delta_{v(q)}^j\} = \Delta_{v(q)}^3 = 100.982$. Hence, at the end of the second stage, the new partition is $P_3 = (C_1, C_2, C_3) = (\{1, 2\}, \{4, 5\}, \{3\})$. The cut value is obtained from $Z_3$.

Suppose we stop at $R = 2$. Then Figure 7.5 displays the resulting hierarchy, and the cut-value questions are

At stage $r = 1$ : "Is $Z_4 \leq 55.49$?"

At stage $r = 2$ : "Is $Z_3 \leq 1.55$?"  $\quad\quad\quad\quad\quad\quad\quad\quad$ □

**Example 7.7** Consider now the data of Table 7.9 for all stations. Let us use the same four variables ($Z_1 = Y_4, Z_2 = Y_7, Z_3 = Y_{11}, Z_4 = Y_{13}$) and the Hausdorff distances obtained by using Eq. (3.3.4). The hierarchy of Figure 7.6 is

**Table 7.13** First bi-partitioning by $(Z_2, Z_3, Z_4)$ (Example 7.6)

| Station | Mean | Bi-partition | | Cut | | (a) | (b) | | |
|---|---|---|---|---|---|---|---|---|---|
| $u$ | $\bar{Z}_u$ | $(C^{(1)}_{v(q)}, C^{(2)}_{v(q)})$ | $q$ | $c_q$ | | $I(C^{(1)}_{v(q)})$ | $I(C^{(2)}_{v(q)})$ | (a)+(b) | $\Delta^j_{v(q)}$ |
| | | Bi-partition on $Z_2$, order $\{3, 1, 2, 5, 4\}$ | | | | | | | |
| 3 | 17.00 | $(\{3\}, \{1, 2, 5, 4\})$ | 1 | 17.475 | | 0 | 2.835 | 2.835 | 3.449 |
| 1 | 17.95 | $(\{3, 1\}, \{2, 5, 4\})$ | 2 | 19.0 | | 0.441 | 0.979 | 1.420 | **4.863** |
| 2 | 20.05 | $(\{3, 1, 2\}, \{5, 4\})$ | 3 | 20.075 | | 1.692 | 0.256 | 1.948 | 4.336 |
| 5 | 20.10 | $(\{3, 1, 2, 5\}, \{4\})$ | 4 | 20.375 | | 4.824 | 0 | 4.824 | 1.460 |
| 4 | 20.65 | $(\{1, 2, 3, 4, 5\}, \phi)$ | (0) | | | 6.284 | – | 6.284 | |
| | | Bi-partition on $Z_3$, order $\{1, 2, 3, 4, 5\}$ | | | | | | | |
| 1 | −8.85 | $(\{1\}, \{2, 3, 4, 5\})$ | 1 | −8.8 | | 0 | 80.814 | 80.814 | 34.173 |
| 2 | −8.75 | $(\{1, 2\}, \{3, 4, 5\})$ | 2 | −2.75 | | 0.025 | 24.631 | 24.656 | **90.330** |
| 3 | 3.25 | $(\{1, 2, 3\}, \{4, 5\})$ | 3 | 7.55 | | 31.441 | 1.369 | 32.810 | 82.176 |
| 4 | 11.85 | $(\{1, 2, 3, 4\}, \{5\})$ | 4 | 13.32 | | 75.458 | 0 | 75.458 | 39.529 |
| 5 | 14.79 | $(\{1, 2, 3, 4, 5\}, \phi)$ | (0) | | | 114.987 | – | 114.987 | |
| | | Bi-partition on $Z_4$, order $\{2, 1, 5, 4, 3\}$ | | | | | | | |
| 2 | 12.26 | $(\{2\}, \{1, 5, 4, 3\})$ | 1 | 13.52 | | 0 | 357.908 | 357.908 | 120.297 |
| 1 | 14.78 | $(\{2, 1\}, \{5, 4, 3\})$ | 2 | 23.87 | | 0.635 | 192.572 | 193.207 | 284.998 |
| 5 | 32.96 | $(\{2, 1, 5\}, \{4, 3\})$ | 3 | 35.385 | | 51.023 | 124.892 | 175.915 | 302.290 |
| 4 | 37.82 | $(\{2, 1, 5, 4\}, \{3\})$ | 4 | 55.49 | | 98.656 | 0 | 98.656 | **379.548** |
| 3 | 73.16 | $(\{1, 2, 3, 4, 5\}, \phi)$ | (0) | | | 478.205 | – | 478.205 | |

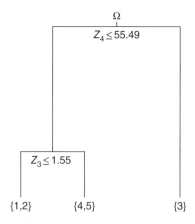

**Figure 7.5** Monothetic divisive hierarchy. Interval-valued data: Hausdorff distances on $Z_1, Z_2, Z_3$, and $Z_4$, five stations (Example 7.6).

**Table 7.14** Second bi-partitioning on $C^{(1)}$ by $(Z_1, Z_2, Z_3, Z_4)$ (Example 7.6)

| Station | Mean | Bi-partition | | Cut | (a) | (b) | | |
|---|---|---|---|---|---|---|---|---|
| $u$ | $\bar{Z}_u$ | $(C^{(1)}_{v(q)}, C^{(2)}_{v(q)})$ | $q$ | $c_q$ | $I(C^{(1)}_{v(q)})$ | $I(C^{(2)}_{v(q)})$ | (a)+(b) | $\Delta^j_{v(q)}$ |
| | | Bi-partition on $Z_1$, order $\{1, 2, 4, 5\}$ | | | | | | |
| 1 | 2.50 | $(\{1\}, \{2, 4, 5\})$ | 1 | 3.025 | 0 | 25.657 | 25.657 | 15.766 |
| 2 | 3.55 | $(\{1, 2\}, \{4, 5\})$ | 2 | 10.225 | 0.256 | 0.081 | 0.337 | **41.087** |
| 4 | 16.90 | $(\{1, 2, 4\}, \{5\})$ | 3 | 17.0 | 26.942 | 0 | 26.942 | 14.482 |
| 5 | 17.10 | $(\{1, 2, 4, 5\}, \phi)$ | (0) | | 41.424 | – | 41.424 | |
| | | Bi-partition on $Z_2$, order $\{1, 2, 5, 4\}$ | | | | | | |
| 1 | 17.95 | $(\{1\}, \{2, 5, 4\})$ | 1 | 19.0 | 0 | 0.979 | 0.979 | 1.856 |
| 2 | 20.05 | $(\{1, 2\}, \{5, 4\})$ | 2 | 20.075 | 0.576 | 0.256 | 0.832 | **2.003** |
| 5 | 20.10 | $(\{1, 2, 5\}, \{4\})$ | 3 | 20.375 | 2.220 | 0 | 2.220 | 0.615 |
| 4 | 20.65 | $(\{1, 2, 4, 5\}, \phi)$ | (0) | | 2.835 | – | 2.835 | |
| | | Bi-partition on $Z_3$, order $\{1, 2, 4, 5\}$ | | | | | | |
| 1 | −8.85 | $(\{1\}, \{2, 4, 5\})$ | 1 | −8.8 | 0 | 68.321 | 68.321 | 34.055 |
| 2 | −8.75 | $(\{1, 2\}, \{4, 5\})$ | 2 | 1.55 | 0.025 | 1.369 | 1.394 | **100.982** |
| 4 | 11.85 | $(\{1, 2, 4\}, \{5\})$ | 3 | 13.32 | 59.100 | 0 | 59.100 | 43.276 |
| 5 | 14.79 | $(\{1, 2, 4, 5\}, \phi)$ | (0) | | 102.376 | – | 102.376 | |
| | | Bi-partition on $Z_4$, order $\{2, 1, 5, 4\}$ | | | | | | |
| 2 | 12.26 | $(\{2\}, \{1, 5, 4\})$ | 1 | 13.52 | 0 | 58.998 | 58.998 | 36.658 |
| 1 | 14.78 | $(\{2, 1\}, \{5, 4\})$ | 2 | 23.87 | 0.635 | 2.362 | 2.997 | **95.659** |
| 5 | 32.96 | $(\{2, 1, 5\}, \{4\})$ | 3 | 35.385 | 51.023 | 0 | 51.023 | 47.633 |
| 4 | 37.82 | $(\{2, 1, 5, 4, 3\}, \phi)$ | (0) | | 98.656 | – | 98.656 | |

obtained. The details are left to the reader (see Exercise 7.5). If we stop at $R = 3$, the cut-value questions are

At stage $r = 1$ : "Is $Y_{13} \leq 23.87$?"

At stage $r = 2$ : "Is $Y_{11} \leq 7.05$?"

At stage $r = 3$ : "Is $Y_{13} \leq 55.49$?"

While each of the five stations of Example 7.6 are in the same distinct clusters as there are in Figure 7.6, the cut-values now differ. This is because of the influence of the other stations used in the current analysis. □

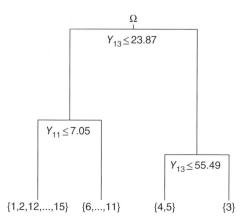

**Figure 7.6** Monothetic divisive hierarchy. Interval-valued data: Hausdorff distances on $Y_4, Y_7, Y_{11}$, and $Y_{13}$, 15 stations (Example 7.7).

**Example 7.8** When the temperature only variables $(Y_1, \ldots, Y_{12})$ are taken and the distance measures are the Hausdorff distances calculated by using Eq. (3.3.4) for all 15 stations, the hierarchy of Figure 7.7 is obtained. We stop at $R = 3$, with cut-value questions as

At stage $r = 1$ : "Is $Y_2 \leq -6.325$?"

At stage $r = 2$ : "Is $Y_1 \leq 11.775$?"

At stage $r = 3$ : "Is $Y_{11} \leq 7.55$?"

The details are left to the reader (see Exercise 7.7). Notice that the elevation variable $Y_{13}$ played a key role in the construction of the hierarchies of Figures 7.5

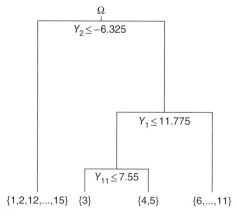

**Figure 7.7** Monothetic divisive hierarchy. Interval-valued data: Hausdorff distances on $Y_1, \ldots, Y_{12}$, 15 stations (Example 7.8).

and 7.6. However, this variable is not part of the analysis herein. Nevertheless, the five stations used in Example 7.6 are still in the same clusters as before. □

When the Hausdorff distance measures used are normalized using a span normalization, then these have to be recalculated at each step as the spans will change as the possible sub-clusters $(C_{v(q)}^{(1)}, C_{v(q)}^{(2)})$ change with $q$.

### 7.2.4 Histogram-valued Observations

A construction of a hierarchy for histogram-valued observations has been developed by Kim (2009). In one sense, this is a generalization of Chavent's divisive method for intervals (described in section 7.2.3) but it is more general in that it can use any dissimilarity measure $d(u_1, u_2)$ in Eq. (7.1.2) whereas Chavent's algorithm is limited to using Hausdorff distances.

Let the histogram observations be denoted by $\mathbf{Y}_u = (Y_{u1}, \ldots, Y_{up})$, $u = 1, \ldots, m$, where

$$Y_{uj} = (\{[a_{ujk}, b_{ujk}), p_{ujk}; k = 1, \ldots, s_j\}, j = 1, \ldots, p), \ u = 1, \ldots, m. \quad (7.2.20)$$

Although it is not necessary theoretically, it is computationally advantageous to transform the observations into a set of histogram values which have the same number of common sub-intervals, as described in section 4.2.1; we assume the observations of Eq. (7.2.20) have been so transformed. Therefore, we can write $s_{uj} \equiv s_j$, for all $u$.

Since this is a monothetic algorithm, it is run for each variable separately. Again, from Eqs. (7.1.2)–(7.1.4), we construct a hierarchy for histogram-valued observations based on dissimilarity measures. Then, for each observation, the algorithm first ranks observations in increasing (or, equivalently, decreasing) values of the means. At each stage, we want to partition a cluster $C_v$ into two sub-clusters $(C_v^{(1)}, C_v^{(2)})$ which maximize the change in the total within-cluster variation $\Delta_v^j$ of Eq. (7.1.7), for variable $j$.

This process is repeated for each variable $j = 1, \ldots, p$. The maximum of the maximum change in $\Delta_{v(q)}^j$ determines which cluster $C_v$ is being partitioned and for what variable $j$ at which split $q$. That is, we seek $\max_{j,v,q}\{\Delta_{v(q)}^j\}$. Once the division in $C_v$ is determined, at $q$, for variable $j$, the cutting criterion is determined by finding the mean value of the union, $\bar{Y}_{(q\cup q+1)j}$, of the two histograms $Y_{qj}$ and $Y_{q+1,j}$, and calculated from Eqs. (4.2.16) and (4.2.17). Notice that this mean is a type of centroid value between the two histograms.

**Example 7.9** To illustrate this algorithm, consider the histogram-valued data for $Y_1 =$ cholesterol level and $Y_2 =$ glucose level for the $m = 4$ regions, from Tables 4.8 and 4.14, respectively. These are transformed into histograms with common histogram sub-intervals and shown in Tables 4.9 and 4.15, respectively. Let us assume each region is equally weighted.

Suppose first we build the hierarchy based on $Y_1 =$ cholesterol only, and suppose we use the extended Ichino–Yaguchi dissimilarity calculated from Eq. (4.2.29) for $\gamma = 0.25$. After reordering the regions by increasing means, we obtain the dissimilarity matrix $D_1$, for the new ordered regions 1, 4, 2, and 3, as

$$
\mathbf{D}_1 = \begin{bmatrix}
0 & 3.995 & 3.697 & 4.625 \\
3.995 & 0 & 2.099 & 2.968 \\
3.697 & 2.099 & 0 & 4.298 \\
4.625 & 2.968 & 4.298 & 0
\end{bmatrix}. \tag{7.2.21}
$$

For the first step $(r = 1)$, $\Omega \equiv P_1 \equiv C_1 = \{1, 4, 2, 3\}$. From Eqs. (7.1.2)–(7.1.7), we can obtain the change $\Delta^1_{v(q)}$ for each split of $C_v$ into $(C^{(1)}_{v(q)}, C^{(2)}_{v(q)})$ at $q = 1, 2, 3$; here since $r = 1$, $v = 1$ only. These values are shown in Table 7.15(i). Since the maximum change in total within-cluster variation, $\max_{v(q)}\{\Delta^1_q\} = \Delta^1_{v(1)} = 2.528$, occurs at $q = 1$, the new sub-clusters are $P_2 = (C_1, C_2) = (\{1\}, \{4, 2, 3\})$. The cutting criterion is the mean of the union of the histograms for regions 1 and 4, namely, $\bar{Y}_{1 \cup 4} = 175.431$ calculated from Eqs. (4.2.16) and (4.2.17).

To find the next division, we repeat the procedure on the cluster $C_2$. At this step, we see, from Table 7.15(ii), that the maximum change in total within-cluster variation, $\max_q\{\Delta^1_{2(q)}\} = \Delta^1_{2(2)} = 2.090$, occurs at the $q = 2$ split to give $C_2 = (C^{(1)}_{2(2)}, C^{(2)}_{2(2)}) = (\{4, 2\}, \{3\})$. If we stop the construction at $R = 2$, then the complete data set is divided into clusters

**Table 7.15** Bi-partitioning on $Y_1 =$ cholesterol (Example 7.9)

| Region | Mean | Bi-partition | | Cut | (a) | (b) | | |
|---|---|---|---|---|---|---|---|---|
| $u$ | $\bar{Y}_u$ | $(C^{(1)}_{v(q)}, C^{(2)}_{v(q)})$ | $q$ | $c_q$ | $I(C^{(1)}_{v(q)})$ | $I(C^{(2)}_{v(q)})$ | (a)+(b) | $\Delta^1_{v(q)}$ |
| (i) First bi-partition of $C = \Omega$ | | | | | | | | |
| 1 | 172.520 | $(\{1\}, \{4, 2, 3\})$ | 1 | 175.431 | 0 | 2.641 | 2.641 | **2.528** |
| 4 | 175.800 | $(\{1, 4\}, \{2, 3\})$ | 2 | 176.159 | 1.995 | 2.309 | 4.304 | 0.865 |
| 2 | 176.160 | $(\{1, 4, 2\}, \{3\})$ | 3 | 177.459 | 2.836 | 0 | 2.836 | 2.333 |
| 3 | 177.800 | $(\{1, 4, 2, 3\}, \phi)$ | (0) | | 5.169 | – | 5.169 | |
| (ii) Second bi-partition of $C_2 = \{4, 2, 3\}$ | | | | | | | | |
| 4 | 175.800 | $(\{4\}, \{2, 3\})$ | 1 | 176.159 | 0 | 2.309 | 2.309 | 0.332 |
| 2 | 176.160 | $(\{4, 2\}, \{3\})$ | 2 | 177.459 | 0.551 | 0 | 0.551 | **2.090** |
| 3 | 177.800 | $(\{4, 2, 3\}, \phi)$ | (0) | | 2.641 | – | 2.641 | |

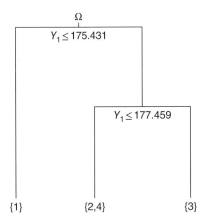

**Figure 7.8** Monothetic divisive hierarchy. Histogram-valued data: extended Ichino–Yaguchi distances on $Y_1$ = cholesterol (Example 7.9).

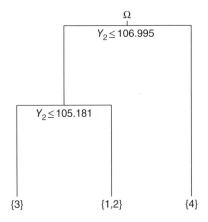

**Figure 7.9** Monothetic divisive hierarchy. Histogram-valued data: extended Ichino–Yaguchi distances on $Y_2$ = glucose (Example 7.9).

$(C_1, C_2, C_3) = (\{1\}, \{4, 2\}, \{3\})$ giving the hierarchy as shown in Figure 7.8, with cut criteria

At stage $r = 1$ : "Is $Y_1 \leq 175.431$?"

At stage $r = 2$ : "Is $Y_1 \leq 177.459$?"

If we build our hierarchy using $Y_2$ = glucose only, again based on the extended Ichino–Yaguchi dissimilarity calculated from Eq. (4.2.29) (where now the order of the regions is 3, 2, 1, 4), the hierarchy of Figure 7.9 is obtained (see Table 7.16

**Table 7.16** Bi-partitioning on $Y_2$ = glucose (Example 7.9)

| Region | Mean | Bi-partition | | Cut | (a) | (b) | | |
|---|---|---|---|---|---|---|---|---|
| $u$ | $\bar{Y}_u$ | $(C^{(1)}_{v(q)}, C^{(2)}_{v(q)})$ | $q$ | $c_q$ | $I(C^{(1)}_{v(q)})$ | $I(C^{(2)}_{v(q)})$ | (a)+(b) | $\Delta^2_{v(q)}$ |
| | | | (i) First bi-partition of $C = \Omega$ | | | | | |
| 3 | 104.400 | $(\{3\},\{2,1,4\})$ | 1 | 105.081 | 0 | 7.259 | 7.259 | 2.289 |
| 2 | 106.360 | $(\{3,2\},\{1,4\})$ | 2 | 106.248 | 3.240 | 1.810 | 5.050 | 4.498 |
| 1 | 106.780 | $(\{3,2,1\},\{4\})$ | 3 | 106.995 | 4.672 | 0 | 4.672 | **4.876** |
| 4 | 107.220 | $(\{1,4,2,3\}, \phi)$ | (0) | | 9.548 | – | 9.548 | |
| | | | (ii) Second bi-partition of $C_1 = \{3,2,1\}$ | | | | | |
| 3 | 104.400 | $(\{3\},\{2,1\})$ | 1 | 105.081 | 0 | 1.649 | 1.649 | **3.022** |
| 2 | 106.360 | $(\{3,2\},\{1\})$ | 2 | 106.248 | 3.240 | 0 | 3.240 | 1.431 |
| 1 | 106.780 | $(\{3,2,1\}, \phi)$ | 3 | | 4.671 | – | 4.671 | |

for the calculation details). In this case, the cut criteria are

At stage $r = 1$ : "Is $Y_2 \leq 106.995$?"

At stage $r = 2$ : "Is $Y_2 \leq 105.181$?"

It is immediately obvious that the hierarchies are different for the two variables. Recall, however, from Eq. (4.2.29), that the extended Ichino–Yaguchi dissimilarities include components relating to the standard deviations (which include the means) of the union and intersection of two histogram realizations. Take Figure 7.8 built from $Y_1$. If we look closely at these components, shown in Table 4.12 for the unordered observations, it is apparent that regions 2 and 4 in the sub-cluster $C_2$ have standard deviations that are different from that of region 3. This would have an impact on the construction. On the other hand, while clearly the standard deviations for $Y_2$ impact the construction of the hierarchy of Figure 7.9, the final construction, at both steps, reflects more the differences across the regions' means.                                                    □

When constructing a hierarchy by using a dissimilarity measure based on several variables, care is needed. For example, suppose we want to construct a hierarchy based on the Euclidean extended Ichino–Yaguchi distance obtained by substituting measures from Eq. (4.2.29) into Eq. (3.1.8). From Eq. (3.1.8), we see that this Euclidean distance is a sum involving distances for variables $Y_1, \ldots, Y_p$. However, as a monothetic algorithm, observations are ordered according to one variable at a time. Take $Y_1$. Then, the calculation of the distances for each of $D_1, \ldots, D_p$ (from Eq. (4.2.29)) used in Eq. (3.1.8) is based on the ordering of the $Y_1$ means. These are then used in Eqs. (7.1.2)–(7.1.7) to obtain the maximum change in total within-cluster variation, for $Y_1$, namely, $\max_q\{\Delta_q^1\}$. This is then repeated for each $Y_j$, where now the order of the observations is found from the means for $Y_j$, with a maximum change in total within-cluster variation, for $Y_j$, $\max_q\{\Delta_q^j\}$, $j = 1, \ldots, p$, obtained. Then, the selected variable to construct that branch of the hierarchy is the maximum of the $p$ maximums, i.e., $\max_j\{\max_q\{\Delta_q^j\}\}$.

**Example 7.10** Consider again the data used in Example 7.9. Let us mono-thetically construct the hierarchy using the Euclidean span-normalized extended Ichino–Yaguchi distances found by calculating the normalized extended Ichino–Yaguchi dissimilarities of Eq. (4.2.30) and substituting these distances into the (non-weighted) Euclidean distance of Eq. (3.1.9), taking the variable weights $w_j = 1$. When the observations are ordered by $Y_1 =$ cholesterol, the order of the regions is 1, 4, 2, 3, and when ordered by $Y_2 =$ glucose, the regions are ordered as 3, 2, 1, 4. For these orderings, the distance matrices, $D_{Y_j}$, ordered by $Y_j$, $j = 1, 2$, are, respectively, for $\gamma = 0.25$,

$$
\mathbf{D}_{Y_1} = \begin{bmatrix} 0 & 0.065 & 0.062 & 0.071 \\ 0.065 & 0 & 0.120 & 0.076 \\ 0.062 & 0.120 & 0 & 0.084 \\ 0.071 & 0.076 & 0.084 & 0 \end{bmatrix},
$$

$$
\mathbf{D}_{Y_2} = \begin{bmatrix} 0 & 0.084 & 0.071 & 0.076 \\ 0.084 & 0 & 0.062 & 0.120 \\ 0.071 & 0.062 & 0 & 0.065 \\ 0.076 & 0.120 & 0.065 & 0 \end{bmatrix}. \tag{7.2.22}
$$

Then, taking the distances between two observations given in $\mathbf{D}_{Y_j}$ for each $j = 1, 2$, in turn and substituting into Eqs. (7.1.2)–(7.1.7), we obtain the change in total within-cluster variation for each split $q = 1, 2, 3$. The calculations are shown in Table 7.17. Thus, when sorted by $Y_1$, we see from Table 7.17(i) that the maximum change in total within-cluster variation is $\Delta^1_{v(q)} = \Delta^1_{1(3)} = 0.00011$, and when sorted by $Y_2$, from Table 7.17(ii)), the maximum change in total within-cluster variation is $\Delta^2_{v(q)} = \Delta^2_{1(3)} = 0.00119$. Therefore, from Eq. (7.2.1), we have

$$
\Delta^{j^*}_{q^*} = \max_j\{\max_q\{\Delta_s^j\}\}
$$
$$
= \max\{\max\{.00025, .00011, .00066\}, \max\{.00066, .00111, .00119\}\}
$$
$$
= 0.00119.
$$

**Table 7.17** Bi-partitioning: Euclidean span-normalized extended Ichino–Yaguchi distance (Example 7.10)

| Region | Mean | Bi-partition | Cut | | (a) | (b) | | |
|---|---|---|---|---|---|---|---|---|
| $u$ | $\bar{Y}_u$ | $(C^{(1)}_{v(q)}, C^{(2)}_{v(q)})$ | $q$ | $c_q$ | $I(C^{(1)}_{v(q)})$ | $I(C^{(2)}_{v(q)})$ | (a)+(b) | $\Delta^2_{v(q)}$ |
| (i) First bi-partition of $C = \Omega$, sorted by $Y_1$ ||||||||||
| 1 | 172.520 | $(\{1\},\{4,2,3\})$ | 1 | 175.431 | 0 | 0.00227 | 0.00227 | 0.00025 |
| 4 | 175.800 | $(\{1,4\},\{2,3\})$ | 2 | 176.159 | 0.00053 | 0.00088 | 0.00141 | **0.00111** |
| 2 | 176.160 | $(\{1,4,2\},\{3\})$ | 3 | 177.459 | 0.00186 | 0 | 0.00186 | 0.00066 |
| 3 | 177.800 | $(\{1,4,2,3\}, \phi)$ | (0) | | 0.00252 | – | 0.00252 | |
| (ii) First bi-partition of $C = \Omega$, sorted by $Y_2$ ||||||||||
| 3 | 104.400 | $(\{3\},\{2,1,4\})$ | 1 | 105.081 | 0 | 0.00186 | 0.00186 | 0.00066 |
| 2 | 106.360 | $(\{3,2\},\{1,4\})$ | 2 | 106.248 | 0.00088 | 0.00053 | 0.00141 | 0.00111 |
| 1 | 106.780 | $(\{3,2,1\},\{4\})$ | 3 | 106.995 | 0.00133 | 0 | 0.00133 | **0.00119** |
| 4 | 107.220 | $(\{1,4,2,3\}, \phi)$ | (0) | | 0.00252 | – | 0.00252 | |
| (iii) Second bi-partition of $C_1 = \{3,2,1\}$, sorted by $Y_1$ ||||||||||
| 1 | 172.520 | $(\{1\},\{2,3\})$ | 1 | 175.431 | 0 | 0.01635 | 0.01635 | 0.01099 |
| 2 | 176.160 | $(\{1,2\},\{3\})$ | 2 | 177.459 | 0.01123 | 0 | 0.01123 | **0.01611** |
| 3 | 177.800 | $(\{1,2,3\}, \phi)$ | (0) | | 0.02734 | – | 0.02734 | |
| (iv) Second bi-partition of $C_1 = \{3,2,1\}$, sorted by $Y_2$ ||||||||||
| 3 | 104.400 | $(\{3\},\{2,1\})$ | 1 | 105.081 | 0 | 0.01123 | 0.01123 | **0.01611** |
| 2 | 106.360 | $(\{3,2\},\{1\})$ | 2 | 106.248 | 0.01635 | 0 | 0.01635 | 0.01099 |
| 1 | 106.780 | $(\{3,2,1\}, \phi)$ | (0) | | 0.02734 | – | 0.02734 | |

Thence, the first bi-partition gives the sub-clusters $C_1 = (C_1^{(1)}, C_1^{(2)}) \equiv (C_1, C_2) = (\{1,2,3\}, \{4\})$, with $Y_2$ being the cutting variable. Since the cutting value $c_r$ is the mean of the union between the relevant two observations, here between region 1 and region 4, the mean of their union $\bar{Y}_{(1\cup4)2}$, from Eq. (4.2.16), is $c_1 = 106.995$.

At the second step, $r = 2$, the process is repeated on the sub-cluster $C_1$; the details are shown in Table 7.22(iii) and Table 7.22(iv) when the sorted variables are $Y_1$ and $Y_2$, respectively. We see that the maximum change in total within-cluster variation is the same, $\Delta^j_{v(q)} = 0.01611$, for both variables though the split $q$ value differs; however, both produce the same sub-clusters, $(\{1,2\}, \{3\})$. Suppose we take the bi-partitioning produced by sorting on $Y_1$. In this case, the cutting value is $c_2 = 177.459 = \bar{Y}_{(2\cup3)1}$. The hierarchy is shown

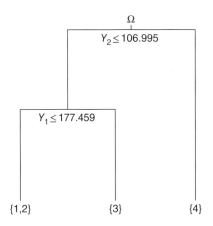

**Figure 7.10** Monothetic divisive hierarchy. Histogram-valued data: Euclidean span-normalized extended Ichino–Yaguchi distances, four regions (Example 7.10).

in Figure 7.10. The cut criteria become

$$\text{At stage } r = 1 : \text{ "Is } Y_2 \leq 106.995?"$$
$$\text{At stage } r = 2 : \text{ "Is } Y_1 \leq 177.459?"$$

$\square$

These algorithms depend only on the mean of each observation when undertaking the division of a cluster. This makes it difficult for the algorithm to distinguish between two histograms with similar means but different variances. One way to overcome that difficulty is to use the double monothetic algorithm introduced by Kim and Billard (2019). Now both sample means and sample variances are calculated at each step. Thus, the algorithm Steps 4–6 are done twice, once when the ordering variable is the mean and once when it is the standard deviation. At the end of Step 6, we have the total within-cluster variations $\Delta_{v(q)}^{M_j}$ and $\Delta_{v(q)}^{S_j}$ based on the means $M_j$ and on the standard deviations $S_j$, respectively.

Then, for the double monothethic algorithm, Step 7 for the monothethic algorithm is replaced by

**Step 7\*** Determine the sub-cluster $v^*$ in $P_r$ to be partitioned and its bi-partition $q^*$ from the maximum change in the total within-cluster variations $\Delta_{v(q)}^{M_j}$ and $\Delta_{v(q)}^{S_j}$ calculated in Step 6 (from Eq. (7.2.5)), i.e., for this $j$ and this $r$,

$$\Delta_{v^*(q^*)}^{j} = \max\{\Delta_{v^*(q^*)}^{M_j}, \Delta_{v^*(q^*)}^{S_j}\}$$

where

$$\Delta_{v^*(q^*)}^{M_j} = \max_{v,q}\{\Delta_{v(q)}^{M_j}, v = 1,\dots,r, \ q = 1,\dots,n_v\}$$

$$\Delta_{v^*(q^*)}^{S_j} = \max_{v,q}\{\Delta_{v(q)}^{S_j}, v = 1,\dots,r, \ q = 1,\dots,n_v\}.$$

The rest of the algorithm is unchanged.

**Example 7.11**  Consider the data set of house property values (in $1000s) as displayed in Table 7.18, which gives histogram observations for $p = 4$ variables, $Y_1$ = property value for a house with two or fewer bedrooms, $Y_2$ = property value for a three-bedroom house, $Y_3$ = property value for a four-bedroom house, and $Y_4$ = property value for a house with five or more bedrooms. There are $m = 6$ states, $\Omega = $ {Colorado (1), North Carolina (2), New Mexico (3), Nevada (4), Minnesota (5), Vermont (6)}. The complete data set can be downloaded from <http://www2.census.gov/acs2012/pums>.

Suppose we take the extended Ichino–Yaguchi dissimilarity of Eq. (4.2.29) with $\gamma = 0.25$. For these calculations, the sample means (from Eq. (4.2.14)) and sample standard deviations (from Eq. (4.2.15)) are needed for each variable for each observation. These are shown in Table 7.19. Then, the extended Ichino–Yaguchi dissimilarity matrix over all variables is

$$\mathbf{D} = \begin{bmatrix} 0 & 2.181 & 2.173 & 2.359 & 2.405 & 1.989 \\ 2.181 & 0 & 2.300 & 1.219 & 0.733 & 1.743 \\ 2.173 & 2.300 & 0 & 2.065 & 2.037 & 1.909 \\ 2.359 & 1.219 & 2.065 & 0 & 0.834 & 0.705 \\ 2.405 & 0.733 & 2.037 & 0.834 & 0 & 1.517 \\ 1.989 & 1.743 & 1.909 & 0.705 & 1.517 & 0 \end{bmatrix}.$$

At the first stage, the states are ordered by sample means by variable. Thus, for $Y_1$, we obtain in ascending order states 2, 4, 5, 6, 3, and 1 (see Table 7.19). The within-cluster variations $I(C_{v(q)}^{M_1(1)})$ and $I(C_{v(q)}^{M_1(2)})$ for each of the bi-partitions corresponding to the respective splits $q = 1,\dots,5$, calculated from Eq. (7.1.2), are shown in Table 7.20; here $v = r = 1$. Also shown is the change in total variation $\Delta_{v(q)}^{M_1}$ calculated from Eq. (7.1.7) for each split $q$. For the double algorithm, we repeat this step now using the standard deviations. In this case, from Table 7.19, the increasing ordered observations becomes 4, 6, 5, 2, 1, and 3. For $Y_1$, the respective within-cluster variations $I(C_{v(q)}^{S_1(1)})$ and $I(C_{v(q)}^{S_1(2)})$, and the change in total variation $\Delta_{v(q)}^{S_1}$, for each split $q = 1,\dots,5$, are given in Table 7.20.

**Table 7.18** Transformed histogram-valued data (in US$1000s) (Example) 7.11

| State | [0, 350) | [350, 700) | [700, 1050) | [1050, 1400) | [1400, 1750) | [1750, 2100) | [2100, 2450) | [2450, 2800) |
|---|---|---|---|---|---|---|---|---|
| | | | | $Y_{1u} = \{[a_{u1k}, b_{u1k}), p_{u1k}; k = 1, \ldots, s_1\}$ | | | | |
| 1 | 0.867 | 0.111 | 0.016 | 0.002 | 0.000 | 0.000 | 0.000 | 0.005 |
| 2 | 0.964 | 0.025 | 0.004 | 0.000 | 0.000 | 0.007 | 0.000 | 0.000 |
| 3 | 0.907 | 0.069 | 0.010 | 0.000 | 0.000 | 0.000 | 0.014 | 0.000 |
| 4 | 0.933 | 0.058 | 0.004 | 0.000 | 0.003 | 0.001 | 0.000 | 0.000 |
| 5 | 0.929 | 0.061 | 0.004 | 0.000 | 0.000 | 0.005 | 0.000 | 0.000 |
| 6 | 0.888 | 0.107 | 0.001 | 0.000 | 0.005 | 0.000 | 0.000 | 0.000 |
| | | | | $Y_{2u} = \{[a_{u2k}, b_{u2k}), p_{u2k}; k = 1, \ldots, s_2\}$ | | | | |
| 1 | 0.792 | 0.179 | 0.020 | 0.003 | 0.000 | 0.000 | 0.000 | 0.005 |
| 2 | 0.933 | 0.057 | 0.005 | 0.000 | 0.000 | 0.005 | 0.000 | 0.000 |
| 3 | 0.877 | 0.102 | 0.010 | 0.000 | 0.000 | 0.000 | 0.011 | 0.000 |
| 4 | 0.929 | 0.054 | 0.009 | 0.000 | 0.004 | 0.004 | 0.000 | 0.000 |
| 5 | 0.914 | 0.072 | 0.007 | 0.000 | 0.000 | 0.006 | 0.000 | 0.000 |
| 6 | 0.859 | 0.121 | 0.015 | 0.000 | 0.005 | 0.000 | 0.000 | 0.000 |

(*Continued*)

**Table 7.18** (Continued)

| | [0, 350) | [350, 700) | [700, 1050) | [1050, 1400) | [1400, 1750) | [1750, 2100) | [2100, 2450) | [2450, 2800) |
|---|---|---|---|---|---|---|---|---|
| | | | | $Y_{3u} = \{[a_{u3k}, b_{u3k}), p_{u3k}; k = 1, \ldots, s_3\}$ | | | | |
| 1 | 0.005 | 0.646 | 0.288 | 0.048 | 0.006 | 0.000 | 0.000 | 0.000 |
| 2 | 0.000 | 0.769 | 0.189 | 0.028 | 0.000 | 0.000 | 0.014 | 0.000 |
| 3 | 0.000 | 0.790 | 0.183 | 0.016 | 0.000 | 0.000 | 0.000 | 0.012 |
| 4 | 0.000 | 0.830 | 0.137 | 0.018 | 0.000 | 0.011 | 0.004 | 0.000 |
| 5 | 0.000 | 0.845 | 0.133 | 0.011 | 0.000 | 0.000 | 0.010 | 0.000 |
| 6 | 0.000 | 0.719 | 0.226 | 0.030 | 0.000 | 0.025 | 0.000 | 0.000 |
| | | | | $Y_{4u} = \{[a_{u4k}, b_{u4k}), p_{u4k}; k = 1, \ldots, s_4\}$ | | | | |
| 1 | 0.000 | 0.011 | 0.533 | 0.323 | 0.093 | 0.021 | 0.000 | 0.000 |
| 2 | 0.000 | 0.000 | 0.545 | 0.314 | 0.079 | 0.000 | 0.000 | 0.062 |
| 3 | 0.012 | 0.000 | 0.654 | 0.279 | 0.050 | 0.000 | 0.000 | 0.000 |
| 4 | 0.000 | 0.000 | 0.670 | 0.249 | 0.036 | 0.000 | 0.033 | 0.011 |
| 5 | 0.000 | 0.000 | 0.730 | 0.211 | 0.034 | 0.000 | 0.000 | 0.025 |
| 6 | 0.000 | 0.000 | 0.684 | 0.230 | 0.039 | 0.000 | 0.047 | 0.000 |

**Table 7.19** Mean and standard deviation (Example 7.11)

| State | Mean $\bar{Y}_{uj} \equiv M_{uj}$ | | | | Standard deviation $S_{uj}$ | | | |
|---|---|---|---|---|---|---|---|---|
| $u$ | $Y_1$ | $Y_2$ | $Y_3$ | $Y_4$ | $Y_1$ | $Y_2$ | $Y_3$ | $Y_4$ |
| 1 | 238516.0 | 266694.6 | 344275.7 | 449158.8 | 241072.7 | 254832.4 | 341357.4 | 474052.3 |
| 2 | 198156.2 | 206654.3 | 285626.3 | 448326.1 | 187745.4 | 181628.9 | 280342.0 | 449529.6 |
| 3 | 235931.5 | 240416.3 | 274344.2 | 342708.3 | 288045.4 | 265980.2 | 285279.6 | 338053.5 |
| 4 | 204902.5 | 211753.5 | 257949.6 | 353276.4 | 169205.2 | 195505.9 | 250049.0 | 353945.5 |
| 5 | 208891.7 | 216462.5 | 247313.7 | 315863.2 | 187407.8 | 200435.6 | 240766.5 | 329280.3 |
| 6 | 219667.8 | 235041.8 | 310207.3 | 348632.8 | 176027.1 | 197215.9 | 289087.3 | 345409.9 |

**Table 7.20** Bi-partition on $Y_1$ based on $M_1$ and $S_1$ (Example 7.11)

| State | First bi-partition, sorted by $Y_1$ means | | | | | | | |
|---|---|---|---|---|---|---|---|---|
| $u$ | $\bar{Y}_{u1}$ | $(C_{v(q)}^{M(1)}, C_{v(q)}^{M(2)})$ | $q$ | $q_c^{M_1}$ | $I(C_{v(q)}^{M_1(1)})$ | $I(C_{v(q)}^{M_1(2)})$ | Sum$(M_1)$[a] | $\Delta_{v(q)}^{M_1}$ |
| 2 | 198.16 | $(\{2\}, \{4,5,6,3,1\})$ | 1 | 201.53 | 0 | 0.5998 | 0.5998 | 0.1272 |
| 4 | 204.90 | $(\{2,4\}, \{5,6,3,1\})$ | 2 | 206.90 | 0.1016 | 0.5013 | 0.6029 | 0.1241 |
| 5 | 208.89 | $(\{2,4,5\}, \{6,3,1\})$ | 3 | 214.28 | 0.1548 | 0.3373 | 0.4921 | 0.2349 |
| 6 | 219.67 | $(\{2,4,5,6\}, \{3,1\})$ | 4 | 227.80 | 0.2813 | 0.1811 | 0.4224 | **0.2646** |
| 3 | 235.93 | $(\{2,4,5,6,3\}, \{1\})$ | 5 | 237.22 | 0.5021 | 0 | 0.5021 | 0.2249 |

| State | First bi-partition, sorted by $Y_1$ standard deviations | | | | | | | |
|---|---|---|---|---|---|---|---|---|
| $u$ | $S_{u1}$ | $(C_{v(q)}^{S(1)}, C_{v(q)}^{S(2)})$ | $q$ | $q_c^{S_1}$ | $I(C_{v(q)}^{S_1(1)})$ | $I(C_{v(q)}^{S_2(2)})$ | Sum$(S_1)$[a] | $\Delta_{v(q)}^{S_1}$ |
| 4 | 169.21 | $(\{4\}, \{6,5,2,1,3\})$ | 1 | 172.62 | 0 | 0.6330 | 0.6330 | 0.094 |
| 6 | 176.03 | $(\{4,6\}, \{5,2,1,3\})$ | 2 | 181.72 | 0.0587 | 0.4929 | 0.5516 | 0.1754 |
| 5 | 187.41 | $(\{4,6,5\}, \{2,1,3\})$ | 3 | 187.58 | 0.1698 | 0.3697 | 0.5395 | 0.1875 |
| 2 | 187.75 | $(\{4,6,5,2\}, \{1,3\})$ | 4 | 214.41 | 0.2813 | 0.1811 | 0.4724 | **0.2646** |
| 1 | 241.07 | $(\{4,6,5,2,1\}, \{3\})$ | 5 | 264.56 | 0.5229 | 0 | 0.5229 | 0.2041 |

a) Sum$(M_1) = I(C_{v(q)}^{M(1)}) + I(C_{v(q)}^{M(2)})$, sum$(S_1) = I(C_{v(q)}^{S(1)}) + I(C_{v(q)}^{S(2)})$.

These calculations are repeated for each of the other variables $Y_2, \ldots, Y_4$. Then, taking the maximum change in total variation, we find that this occurs for the variable $Y_1$ at the $q = 4$ cut, this maximum being the same for both the mean and standard deviation cuts, i.e.,

$$\Delta_{v^*(q^*)}^j = \max\{\Delta_{v^*(q^*)}^{M_j}, \Delta_{v^*(q^*)}^{S_j}\} = \Delta_{1(4)}^{M_1} = \Delta_{1(4)}^{S_1} = 0.2646.$$

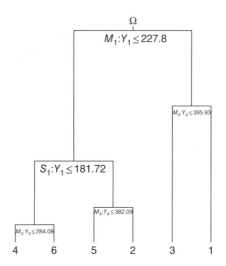

**Figure 7.11** Double monothetic divisive hierarchy. Histogram-valued data: extended Ichino–Yaguchi distances, six states (Example 7.11).

This gives the first bi-partition as $P_2 = (C_1, C_2)$ with $C_1 = \{2, 4, 5, 6\}$ and $C_2 = \{1, 3\}$; the cut question "Is $M_1 \leq 227.8$?", calculated as the average of the means $\bar{Y}_{u1}$ for states $u = 3$ and $u = 6$.

This process is now repeated on each of the clusters $C_1$ and $C_2$ in $P_2$. The resulting hierarchy is shown in Figure 7.11 where the statistic, mean or standard deviation for which variable $j$, that informs the division is indicated by "$M_j$" and "$S_j$", respectively.

Application of this double monothethic algorithm to the entire data set for all states can be found in Kim and Billard (2019). □

## 7.3 Polythethic Methods

A polythetic divisive algorithm for histogram data based on dissimilarity or distance measures was introduced by Kim (2009) and Kim and Billard (2011) (see also Billard and Kim (2017)). Unlike monothetic algorithms, which find sub-clusters for each variable separately, a polythetic algorithm is based on all variables simultaneously. This is advantageous compared to a monothetic method, which may perform poorly when trying to cluster over all variables using only one at a time, since a combination of variables may provide a better balance to the clustering approach. Since all variables are considered simultaneously, typically fewer possible sub-clusters need to be considered in order to select the optimal set.

Another feature is that the algorithm does not require the ordering of observations (by increasing means, association measures, etc.) that we have seen for monothetic methods. Instead, the polythetic algorithm involves finding a splinter observation (or cluster) which is farthest away from the main data set. Then the remaining observations are evaluated, through the dissimilarity measures, for their proximity to the splinter group, by moving them one at a time into the splinter group.

As before, at stage $r$, we have the partition $P_r = (C_1, \ldots, C_r)$. Thus a cluster $C_v$, $v = 1, \ldots, r$, is to be bi-partitioned into $C_v^{(1)}$ and $C_v^{(2)}$ by first finding the splinter sub-cluster for each, and then selecting which cluster $C_v$ (i.e., which $v$) is bi-partitioned at this $(r + 1)$th stage by seeking the maximum change in the within-cluster variation $\Delta_v$ (of Eq. (7.1.7)).

For reasons of descriptive clarity in this section (as in section 7.2), an observation $A_u$ (or, more simply, $u$) will sometimes be represented by its description $\mathbf{Y}_u$, $u = 1, \ldots, m$. Before outlining the steps for each iteration, we need the following entities.

**Definition 7.6**  Let $C^1$ and $C^2$ be a bi-partition of the cluster $C$. Let $TC^1$ and $TC^2$ be the clusters that occur when an observation $\mathbf{Y}_{(i)}$ in $C^1$ is moved into $C^2$, that is, $TC_{(i)}^1 = C^1 - \{\mathbf{Y}_{(i)}\}$ and $TC_{(i)}^2 = C^2 + \{\mathbf{Y}_{(i)}\}$. Then, the **difference**, $H_{(i)}$, of the sums of the within-cluster variations between the two partitions $(C^1, C^2)$ and $(TC^1, TC^2)$ that results when the observation $\mathbf{Y}_{(i)}$ is moved from $C^1$ into $C^2$ is

$$H_{(i)} = \{I(C^1) + I(C^2)\} - \{I(TC_{(i)}^1) + I(TC_{(i)}^2)\}, \quad i = 1, \ldots, t, \qquad (7.3.1)$$

where the within-cluster variation $I(C)$ is defined in Eq. (7.1.2) and $t$ is the number of observations in $C^1$. □

**Definition 7.7**  Let $(C^1, C^2)$ and $(TC^1, TC^2)$ be defined as in Definition 7.6. The **maximum difference**, $MH$, obtained by shifting each observation $\mathbf{Y}_{(i)}$, one at a time, from $C^1$ to $C^2$ is

$$MH = \max_i \{H_{(i)}, i = 1, \ldots, t\}. \qquad (7.3.2)$$
□

**Definition 7.8**  Let $\mathbf{Y}_u^v$ be a specific observation $u$ in the cluster $C_v$ of size $m_v$. The **average weighted dissimilarity**, $\bar{D}_v(\mathbf{Y}_u^v)$, between this $\mathbf{Y}_u^v$ and the other observations in $C_v$ is

$$\bar{D}_v(\mathbf{Y}_u^v) = \frac{1}{\lambda - w_u} \sum_{u_1 \neq u = 1}^{m_v} w_u w_{u_1} d(\mathbf{Y}_u^v, \mathbf{Y}_{u_1}^v), \quad u = 1, \ldots, m_v, \qquad (7.3.3)$$

where $d(\mathbf{Y}_{u_1}, \mathbf{Y}_{u_2})$ (i.e., equivalently, $d(u_1, u_2)$) is the dissimilarity or distance measure between two observations $u_1$ and $u_2$ with realizations $\mathbf{Y}_{u_1}$ and $\mathbf{Y}_{u_2}$, and $w_u$ is the weight for $\mathbf{Y}_u$ with $\lambda = \sum_{u=1}^{m_v} w_u$. □

When there are equal weights per observation with $w_u = 1/m$, Eq. (7.3.3) simplifies to

$$\bar{D}_v(\mathbf{Y}_u^v) = \frac{1}{m(m_v - 1)} \sum_{u_1 \neq u=1}^{m_v} d(\mathbf{Y}_u^v, \mathbf{Y}_{u_1}^v), \quad u = 1, \dots, m_v. \tag{7.3.4}$$

**Definition 7.9** Let the set of observations $\Omega$ be partitioned into $P_r = (C_1, \dots, C_r)$. Let $\bar{D}_v(\mathbf{Y}_u^v)$ be the average weighted dissimilarity between an observation $u$ with realization $\mathbf{Y}_u^v$ in $C_v$ and the other observations in $C_v$, as defined in Definition 7.8. Then, across all $C_v$, $v = 1, \dots, r$, in $P_r$, the **maximum average weighted dissimilarity**, *MAD*, for $P_r$ is

$$MAD = \max_{u,v} \{\bar{D}_v(\mathbf{Y}_u^v), u = 1, \dots, m_v, v = 1, \dots, r\}. \tag{7.3.5}$$

□

The polythetic algorithm for histogram data contains the following steps:

**Step 1**   Stage $r = 1$: Start with $\Omega = P_1 = C_1 = \{1, \dots, m\} \equiv \{\mathbf{Y}_1, \dots, \mathbf{Y}_m\}$ where observation $u$ has realization $\mathbf{Y}_u$, $u = 1, \dots, m$.

**Step 2**   Stage $r$: $\Omega = P_r = (C_1, \dots, C_r)$ with each cluster $C_v$ containing observations, $C_v = \{\mathbf{Y}_u, u = 1, \dots, m_v\}$, $v = 1, \dots, r$.

**Step 3**   For $P_r$, calculate the average weighted dissimilarity, $\bar{D}_v(\mathbf{Y}_u^v)$, from Eq. (7.3.3), and hence calculate the maximum average weighted dissimilarity, *MAD*, from Eq. (7.3.5).

**Step 4**   Let $\mathbf{Y}^*$ be the observation that gave the *MAD* value in $P_r$ at Step 3, and let $C^*$ be the cluster in $P_r$ that contains $\mathbf{Y}^*$. Then it is this cluster $C^*$ which is to be bi-partitioned into $C^1$ and $C^2$, and this $\mathbf{Y}^*$ is used as the seed observation to form $C^2$. Let us denote $C^1 = \{\mathbf{Y}_{(1)}, \dots, \mathbf{Y}_{(t)}\}$ where $t = m_v - 1$ and $C^2 = \{\mathbf{Y}^*\}$. The $C^1$ cluster is called the **main** cluster and $C^2$ is the **splinter cluster**.

**Step 5**   Set $TC^1 = C^1$ and $TC^2 = C^2$. Move one of the observations $\mathbf{Y}_{(i)}$ in $TC^1$ into $TC^2$; now $TC^1 = TC^1 - \{\mathbf{Y}_{(i)}\}$ and $TC^2 = TC^2 + \{\mathbf{Y}_{(i)}\}$. Calculate $H_{(i)}$ from Eq. (7.3.1). Return that observation $\mathbf{Y}_{(i)}$ to $TC^1$, i.e., $TC^1$ and $TC^2$ revert to their original $C^1$ and $C^2$ clusters, respectively.

**Step 6** Repeat Step 5 for all observations $\mathbf{Y}_{(i)}$, $i = 1, \ldots, t$, in $TC^1$.

**Step 7** From the $H_{(i)}$ difference values, $i = 1, \ldots, t$, calculated in Steps 5 and 6, calculate the maximum difference $MH$ from Eq. (7.3.2).

**Step 8** Let $\mathbf{Y}^{MH}$ be the observation in $TC^1$ which gives the $MH$ value. Then, if $MH > 0$, this observation $\mathbf{Y}^{MH}$ moves from $C^1$ into $C^2$. That is, now, $C^1 = C^1 - \{\mathbf{Y}^{MH}\}$ with $t = t - 1$ observations, and $C^2 = C^2 + \{\mathbf{Y}^{MH}\}$.

**Step 9** Repeat Steps 5–8 while $MH > 0$. If $MH < 0$, then this $r$th stage is ended. Now $r = r + 1$.

**Step 10** Repeat Steps 2–9 until $r = R$ or $r = m$, where $R$ is a prespecified number of clusters and $m$ is the number of observations in $\Omega$.

**Example 7.12** To illustrate this algorithm, consider the histogram-valued data obtained by aggregating 581,012 individual observations in a forestry cover-type data set (from the UCI Machine Learning Data Repository at <http://www.ics.uci.edu>). The aggregation is across $m = 7$ cover types (*spruce-fir, lodgepole pine, cottonwood/willow, aspen, douglas fir, krummholz* and *ponderosa pine*) for $p = 4$ variables, namely, $Y_1 =$ elevation, $Y_2 =$ horizontal distance to nearest roadway (simply, road), $Y_3 =$ relative measure of incident sunlight at 9.00 am on the summer solstice (simply, sunlight), and $Y_4 =$ aspect. The data (extracted from Kim, 2009) are shown in Table 7.21.

A summary of some extended Ichino–Yaguchi distances for these data is shown in Table 7.22, for $\gamma = 0.25$. Thus, column (a) gives the standard extended Ichino–Yaguchi distances obtained from summing over all $Y_j$, $j = 1, \ldots, 4$, the distance of Eq. (4.2.29), (b) gives the extended Ichino–Yaguchi normalized distances obtained from Eq. (4.2.30), (c) displays the Euclidean distances obtained by substituting the standard extended Ichino–Yaguchi distances from Eq. (4.2.29) into Eq. (3.1.8), (d) shows the Euclidean extended Ichino–Yaguchi normalized distances obtained by substituting the distances of Eq. (4.2.30) into Eq. (3.1.8), (e) provides the normalized (normed by $1/p$) standard extended Ichino–Yaguchi distances calculated by combining Eq. (4.2.29) with Eq. (3.1.9), and (f) shows the normalized (normed by $1/p$) Euclidean extended Ichino–Yaguchi normalized (by the $1/N_j$ of Eq. (4.2.31), $j = 1, \ldots, p$) distances obtained by substituting the distances of Eq. (4.2.30) into Eq. (3.1.9). To illustrate, let us take the normalized (normed by $1/p$) Euclidean extended Ichino–Yaguchi normalized distances of Table 7.23(f). Then, the distance matrix is

**Table 7.21** Cover-types histogram data (Example 7.12)

| Cover type $u$ | $Y_j$ | $Y_u = \{[a_{ujk}, b_{ujk}), p_{ujk}; k = 1,\ldots,s_{uj}, j = 1,\ldots,p, u = 1,\ldots,m\}$ |
|---|---|---|
| 1 | $Y_1$ | {[2450, 2600), 0.0024; [2600, 2800), 0.0312; [2800, 3000), 0.1614; [3000, 3200), 0.4547; [3200, 3400), 0.3251; [3400, 3600), 0.0239; [3600, 3700], 0.0013} |
| | $Y_2$ | {[0, 1000), 0.1433; [1000, 2000), 0.2617; [2000, 3000), 0.2251; [3000, 4000), 0.1695; [4000, 5000), 0.1123; [5000, 6000), 0.0762; [6000, 7000], 0.0120} |
| | $Y_3$ | {[0, 100), 0.0015; [100, 150), 0.0229; [150, 180), 0.0808; [180, 210), 0.2966; [210, 240), 0.5099; [240, 260], 0.0883} |
| | $Y_4$ | {[0, 40), 0.1924; [40, 80), 0.1765; [80, 120), 0.1284; [120, 160), 0.0875; [160, 200), 0.0613; [200, 240), 0.0486; [240, 280), 0.0602; [280, 320), 0.1090; [320, 360], 0.1362} |
| 2 | $Y_1$ | {[2100, 2300), 0.0009; [2300, 2500), 0.0085; [2500, 2700), 0.1242; [2700, 2900), 0.2887; [2900, 3100), 0.4084; [3100, 3300), 0.15932; [3300, 3450], 0.0100} |
| | $Y_2$ | {[0, 1000), 0.2077; [1000, 2000), 0.2838; [2000, 3000), 0.1960; [3000, 4000), 0.1227; [4000, 5000), 0.0836; [5000, 6000), 0.0846; [6000, 7500], 0.0217} |
| | $Y_3$ | {[0, 100), 0.0014; [100, 150), 0.0219; [150, 170), 0.0386; [170, 190), 0.0984; [190, 210), 0.2055; [210, 230), 0.3586; [230, 250), 0.2638; [250, 260], 0.0119} |
| | $Y_4$ | {[0, 40), 0.1631; [40, 80), 0.1749; [80, 120), 0.1428; [120, 160), 0.1095; [160, 200), 0.0836; [200, 240), 0.0746; [240, 280), 0.0691; [280, 320), 0.0742; [320, 360], 0.1083} |
| 3 | $Y_1$ | {[1950, 2050), 0.0532; [2050, 2150), 0.2155; [2150, 2250), 0.2927; [2250, 2350), 0.3386; [2350, 2450), 0.0917; [2450, 2550], 0.0084} |
| | $Y_2$ | {[0, 300), 0.0572; [300, 600), 0.1787; [600, 900), 0.2228; [900, 1200), 0.2763; [1200, 1500), 0.2428; [1500, 1800], 0.0222} |
| | $Y_3$ | {[120, 140), 0.0033; [140, 160), 0.0120; [160, 180), 0.0397; [180, 200), 0.0837; [200, 220), 0.1740; [220, 240), 0.2774; [240, 260], 0.4099} |
| | $Y_4$ | {[0, 40), 0.0914; [40, 80), 0.1445; [80, 120), 0.2799; [120, 160), 0.2377; [160, 200), 0.0630; [200, 240), 0.0390; [240, 280), 0.0299; [280, 320), 0.0379; [320, 360], 0.0768} |
| 4 | $Y_1$ | {[2450, 2550), 0.0199; [2550, 2650), 0.0665; [2650, 2750), 0.2413; [2750, 2850), 0.4034; [2850, 2950), 0.2474; [2950, 3050], 0.0216} |
| | $Y_2$ | {[0, 500), 0.2626; [500, 1000), 0.1699; [1000, 1500), 0.1419; [1500, 2000), 0.1974; [2000, 2500), 0.1428; [2500, 3500], 0.0488} |
| | $Y_3$ | {[120, 140), 0.0024; [140, 160), 0.0109; [160, 180), 0.0423; [180, 200), 0.1121; [200, 220), 0.1888; [220, 240), 0.3823; [240, 250), 0.2030; [250, 260], 0.0583} |
| | $Y_4$ | {[0, 40), 0.0863; [40, 80), 0.2261; [80, 120), 0.2335; [120, 160), 0.1255; [160, 200), 0.0944; [200, 240), 0.0560; [240, 280), 0.0618; [280, 320), 0.0537; [320, 360], 0.0627} |

(Continued)

**Table 7.21** (Continued)

| Cover type $u$ | $Y_j$ | $Y_u = \{[a_{ujk}, b_{ujk}), p_{ujk}; k = 1,\ldots, s_{uj}, j = 1,\ldots, p, u = 1,\ldots, m\}$ |
|---|---|---|
| 5 | $Y_1$ | {[1850, 2050), 0.0439; [2050, 2250), 0.1236; [2250, 2450), 0.3935; [2450, 2650), 0.3313; [2650, 2850), 0.1048; [2850, 2950], 0.0031} |
| | $Y_2$ | {[0, 400), 0.1282; [400, 800), 0.2692; [800, 1200), 0.2436; [1200, 1600), 0.1901; [1600, 2000), 0.0961; [2000, 2400), 0.0612; [2400, 2800), 0.0075; [2800, 3200], 0.0041} |
| | $Y_3$ | {[0, 60), 0.0004; [60, 100), 0.0014; [100, 140), 0.0701; [140, 180), 0.2871; [180, 220), 0.3928; [220, 240), 0.1918; [240, 260], 0.0565} |
| | $Y_4$ | {[0, 40),0.2391; [40, 80), 0.1446; [80, 120), 0.0749; [120, 160), 0.0338; [160, 200), 0.0241; [200, 240), 0.0251; [240, 280), 0.0579; [280, 320), 0.1613; [320, 360], 0.2391} |
| 6 | $Y_1$ | {[2850, 3050), 0.0041; [3050, 3250), 0.1138; [3250, 3350), 0.3220; [3350, 3450), 0.4460; [3450, 3650), 0.0878; [3650, 3850), 0.0260; [3850, 3950], 0.0004} |
| | $Y_2$ | {[0, 500), 0.0007; [500, 1000), 0.0714; [1000, 1500), 0.1948; [1500, 2000), 0.1392; [2000, 2500), 0.1373; [2500, 3000), 0.1422; [3000, 3500), 0.1073; [3500, 4000), 0.1136; [4000, 4500), 0.1102; [4500, 5500], 0.0734} |
| | $Y_3$ | {[80, 110), 0.0010; [110, 140), 0.0052; [140, 170), 0.0317; [170, 190), 0.0982; [190, 220), 0.3491; [220, 240), 0.3600; [240, 250), 0.1275; [250, 260], 0.0274} |
| | $Y_4$ | {[0, 40), 0.1603; [40, 80), 0.1533; [80, 120), 0.1792; [120, 160), 0.1068; [160, 200), 0.1084; [200, 240), 0.0425; [240, 280), 0.0282; [280, 320), 0.0830; [320, 360], 0.1382} |
| 7 | $Y_1$ | {[1850, 2050), 0.0503; [2050, 2200), 0.1882; [2200, 2400), 0.3496; [2400, 2600), 0.3239; [2600, 2800), 0.0863; [2800, 2900], 0.0017} |
| | $Y_2$ | {[0, 400), 0.2036; [400, 800), 0.2653; [800, 1200), 0.2374; [1200, 1800), 0.1973; [1800, 2400), 0.0660; [2400, 3000), 0.0258; [3000, 3600], 0.0045} |
| | $Y_3$ | {[40, 100), 0.0108; [100, 140), 0.0860; [140, 180), 0.1912; [180, 220), 0.2784; [220, 250), 0.2784; [250, 260], 0.3814; [250, 260], 0.0512} |
| | $Y_4$ | {[0, 40), 0.1160; [40, 80), 0.1059; [80, 120), 0.1340; [120, 160), 0.1462; [160, 200), 0.1031; [200, 240), 0.0788; [240, 280), 0.0596; [280, 320), 0.1152; [320, 360], 0.1413} |

**Table 7.22** Cover types distances (Example 7.12)

| Object pair | Extended Ichino–Yaguchi distances[a] | | | | | |
|---|---|---|---|---|---|---|
| | (a) | (b) | (c) | (d) | (e) | (f) |
| 1-2 | 588.669 | 0.327 | 402.949 | 0.206 | 201.475 | 0.103 |
| 1-3 | 3276.296 | 1.355 | 2289.094 | 0.891 | 1144.547 | 0.446 |
| 1-4 | 1646.722 | 0.675 | 1184.881 | 0.384 | 592.440 | 0.192 |
| 1-5 | 2747.628 | 1.136 | 1917.130 | 0.766 | 958.565 | 0.383 |
| 1-6 | 882.790 | 0.404 | 610.011 | 0.247 | 305.006 | 0.123 |
| 1-7 | 2811.853 | 1.144 | 1964.839 | 0.799 | 982.419 | 0.399 |
| 2-3 | 3061.092 | 1.232 | 2146.428 | 0.824 | 1073.214 | 0.412 |
| 2-4 | 1267.101 | 0.456 | 998.683 | 0.239 | 499.341 | 0.120 |
| 2-5 | 2411.706 | 1.040 | 1684.095 | 0.636 | 842.048 | 0.318 |
| 2-6 | 1274.766 | 0.535 | 887.570 | 0.374 | 443.785 | 0.187 |
| 2-7 | 2444.275 | 0.968 | 1719.758 | 0.646 | 859.879 | 0.323 |
| 3-4 | 2237.548 | 1.029 | 1590.154 | 0.792 | 795.077 | 0.396 |
| 3-5 | 1263.750 | 0.922 | 885.247 | 0.553 | 442.624 | 0.276 |
| 3-6 | 3328.911 | 1.356 | 2325.533 | 0.938 | 1162.766 | 0.469 |
| 3-7 | 1087.455 | 0.677 | 747.637 | 0.417 | 373.818 | 0.208 |
| 4-5 | 1622.908 | 1.037 | 1134.716 | 0.649 | 567.358 | 0.325 |
| 4-6 | 1849.394 | 0.768 | 1292.381 | 0.498 | 646.190 | 0.249 |
| 4-7 | 1627.615 | 0.915 | 1173.486 | 0.642 | 586.743 | 0.321 |
| 5-6 | 2897.104 | 1.282 | 2005.631 | 0.860 | 1002.816 | 0.430 |
| 5-7 | 393.345 | 0.348 | 250.089 | 0.191 | 125.044 | 0.096 |
| 6-7 | 2947.112 | 1.218 | 2057.394 | 0.880 | 1028.697 | 0.440 |

a) (a) Unweighted non-normalized extended Ichino–Yaguchi (eIY) distance; (b) normalized eIY distance; (c) Euclidean eIY distance; (d) Euclidean normalized (by $1/N_j$) eIY distance; (e) normalized (by $1/p$) Euclidean eIY distance; (f) normalized (by $1/p$) Euclidean normalized (by $1/N_j$) eIY distance.

$$\mathbf{D} = \begin{bmatrix} 0 & 0.103 & 0.446 & 0.192 & 0.383 & 0.123 & 0.399 \\ 0.103 & 0 & 0.412 & 0.120 & 0.318 & 0.187 & 0.323 \\ 0.446 & 0.412 & 0 & 0.396 & 0.276 & 0.469 & 0.208 \\ 0.192 & 0.120 & 0.396 & 0 & 0.325 & 0.249 & 0.321 \\ 0.383 & 0.318 & 0.276 & 0.325 & 0 & 0.430 & 0.096 \\ 0.123 & 0.187 & 0.469 & 0.249 & 0.430 & 0 & 0.440 \\ 0.399 & 0.323 & 0.208 & 0.321 & 0.096 & 0.440 & 0 \end{bmatrix}. \quad (7.3.6)$$

Initially, $\Omega \equiv P_1 \equiv C_1$. We first divide $C_1$ into a main cluster $C_1^{(1)}$ and a splinter cluster $C_1^{(2)}$ consisting of a single observation (Step 4). To obtain this splinter observation, we shift one observation $\mathbf{Y}_u$, in turn, from $C_1^{(1)}$ to form $C_1^{(2)}$, thus leaving $C_1^{(1)} = C_1 - \{\mathbf{Y}_u\}$, $u = 1, \ldots, m$. Let us denote the cluster $C_1 - \{\mathbf{Y}_u\}$ by $C_{1(u)}$. Then, for each $C_{1(u)}$, $u = 1, \ldots, m$, we calculate the average weighted dissimilarity $\bar{D}_1(\mathbf{Y}_u^1)$ from Eq. (7.3.4). Since $r = 1$, we have $v = 1$ and $m_1 = m$.

For example, when we shift the first observation $\mathbf{Y}_1$ (i.e., cover type *spruce-fir*) from $C_1$, the average weighted dissimilarity $\bar{D}_1(\mathbf{Y}_1^1)$ for equal weights per observation (i.e., $w_u = 1/m$), from Eq. (7.3.4) and the first row of Eq. (7.3.6), is

$$\bar{D}_1(\mathbf{Y}_1^1) = \frac{1}{7(7-1)}[0.103 + 0.446 + 0.192 + 0.383 + 0.123 + 0.399]$$
$$= 0.0392.$$

Similarly, we find

$$\bar{D}_1(\mathbf{Y}_2^1) = 0.0348, \ \bar{D}_1(\mathbf{Y}_3^1) = 0.0526, \ \bar{D}_1(\mathbf{Y}_4^1) = 0.0381,$$
$$\bar{D}_1(\mathbf{Y}_5^1) = 0.0435, \ \bar{D}_1(\mathbf{Y}_6^1) = 0.0452, \ \bar{D}_1(\mathbf{Y}_7^1) = 0.0426.$$

Then, from Eq. (7.3.5), we find the maximum of the $\bar{D}_1(\mathbf{Y}_u^1)$, $u = 1, \ldots, m$, values. Thus,

$$MAD = \max\{0.0392, 0.0348, 0.0526, 0.0381, 0.0435, 0.0452, 0.0426\}$$
$$= 0.0526.$$

This $MAD$ value occurs for observation $\mathbf{Y}_3$ (i.e., cover type *cottonwood/willow*), thus the first splinter cluster consists of this observation, $TC^2 = \{\mathbf{Y}_3\}$, and the main cluster is $TC^1 = C_1 - \{\mathbf{Y}_3\}$. We can now proceed to Step 5.

At Step 5, we shift one of the observations $(\mathbf{Y}_{(i)})$ in $TC^1$ into $TC^2$. Then, we calculate the difference $H_{(i)}$, given in Eq. (7.3.1), for each shift of the $(i)$th observation from $TC^1$. For example, when we shift the third observation in $TC^1$, i.e., $\mathbf{Y}_{(3)} \equiv \mathbf{Y}_4$, the new $TC^1 = \{\mathbf{Y}_1, \mathbf{Y}_2, \mathbf{Y}_5, \mathbf{Y}_6, \mathbf{Y}_7\}$ and the new $TC^2 = \{\mathbf{Y}_3, \mathbf{Y}_4\}$. That is, the new $TC$ clusters are $TC_{(3)}^1$ and $TC_{(3)}^2$, respectively. To calculate the difference $H_{(3)}$, we first calculate the within-cluster variations $I(C^1)$, $I(C^2)$, $I(TC_{(3)}^1)$, and $I(TC_{(3)}^2)$ from Eq. (7.1.2).

For $C^1 = \{\mathbf{Y}_1, \mathbf{Y}_2, \mathbf{Y}_4, \mathbf{Y}_5, \mathbf{Y}_6, \mathbf{Y}_7\}$, the within-cluster variation is

$$I(C^1) = \frac{1}{2 \times 6 \times 7}[0.103^2 + 0.192^2 + 0.383^2 + \cdots + 0.440^2] = 0.0305$$

where the distances $d(\mathbf{Y}_u, \mathbf{Y}_{u'})$ are taken from the full distance matrix $\mathbf{D}$ of Eq. (7.3.6) with the third row and column deleted. Since $C^2 = \{\mathbf{Y}_3\}$, one observation only, $I(C^2) = 0$. For $TC_{(3)}^1$, the within-cluster variation is

$$I(TC_{(3)}^1) = \frac{1}{2 \times 5 \times 7}[0.103^2 + 0.383^2 + 0.123^2 + \cdots + 0.440^2] = 0.0274$$

where now the third and fourth rows and columns of $\mathbf{D}$ in Eq. (7.3.6) are deleted. For $TC_{(3)}^2$, the within-cluster variation is found by taking the third and fourth components of $\mathbf{D}$ in Eq. (7.3.6), as

$$I(TC_{(3)}^2) = \frac{1}{2 \times 2 \times 7}[.396^2] = 0.0112.$$

Hence, from Eq. (7.3.1),

$$H_{(3)} = I(C^1) + I(C^2) - I(TC_{(3)}^1) - I(TC_{(3)}^2)$$
$$= 0.0305 + 0 - 0.0274 - 0.0112 = -0.0081.$$

Likewise, by shifting each of the other $(i)$th observations from $TC^1$ into $TC^2$, we can show that

$$H_{(1)} = -0.0098, H_{(2)} = -0.0106, H_{(4)} = 0.0041, H_{(5)} = -0.0078,$$
$$H_{(6)} = 0.0071.$$

Then, from Eq. (7.3.2), the maximum $H_{(i)}$ value is

$$MH = \max\{-0.0098, -0.0106, -0.0081, 0.0041, -0.0078, 0.0071\}$$
$$= 0.0071, \tag{7.3.7}$$

which corresponds to $(i) = 6$, i.e., to observation $u = 7$ (cover type *ponderosa pine*).

Since, from Eq. (7.3.7), $MH > 0$ , we have not yet completed the first bi-partition. Therefore, we move observation $\mathbf{Y}_7$ into the previous $TC^2$ to give a second new $TC^2 = \{\mathbf{Y}_3, \mathbf{Y}_7\}$ (as the current pending splinter cluster). The second new $TC^1 = \{\mathbf{Y}_1, \mathbf{Y}_2, \mathbf{Y}_4, \mathbf{Y}_5, \mathbf{Y}_6\}$. Shifting an observation $\mathbf{Y}_{(i)}$ from these new $TC^1$ and $TC^2$, we obtain the new $TC_{(i)}^1 = TC^1 - \{\mathbf{Y}_{(i)}\}$ and the new $TC_{(i)}^2 = \{TC_3^1, TC_7^1, TC_{(i)}^1\}$. Again, we calculate the $H_{(i)}, i = 1, \dots, 5$, for this new set of clusters, from Eq. (7.3.1). We obtain

$$H_{(1)} = -0.0136, H_{(2)} = -0.0113, H_{(3)} = -0.0086, H_{(4)} = 0.0111,$$
$$H_{(5)} = -0.0131.$$

Since the maximum of these $H_{(i)}$ values is $H_{(4)} = 0.0111$, we move the observation for $(i) = 4$, i.e., observation $u = 5$ into the next $TC^2$.

Again, since $MH > 0$, the process is repeated. Now, $TC^1 = \{\mathbf{Y}_1, \mathbf{Y}_2, \mathbf{Y}_4, \mathbf{Y}_6\}$ and $TC^2 = \{\mathbf{Y}_3, \mathbf{Y}_5, \mathbf{Y}_7\}$. When we shift one of the (four) observations in turn from this $TC^1$ into this $TC^2$, we obtain the $H_{(i)}$ values as

$$H_{(1)} = -0.0156, H_{(2)} = -0.0111, H_{(3)} = -0.0082, H_{(4)} = -0.0105;$$

the maximum $H_{(i)}$ value is $MH = -0.0082 < 0$. Hence, we stop with this $TC^1$ and $TC^2$, as the first bi-partition of $\Omega$ is completed.

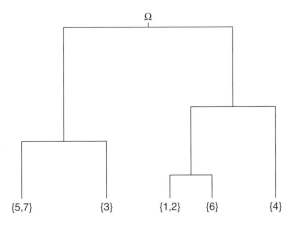

**Figure 7.12** Polythetic divisive hierarchy. Cover-types histogram-valued data: Euclidean span-normalized extended Ichino–Yaguchi distances (Example 7.12).

Now, $r = 2$ with $\Omega = P_2 = (C_1, C_2) = (\{\mathbf{Y}_1, \mathbf{Y}_2, \mathbf{Y}_4, \mathbf{Y}_6\}, \{\mathbf{Y}_3, \mathbf{Y}_5, \mathbf{Y}_7\})$. The distance matrices for these clusters $\mathbf{D}_v$, $v = 1, 2$, are, respectively,

$$\mathbf{D}_1 = \begin{bmatrix} 0 & 0.103 & 0.192 & 0.123 \\ 0.103 & 0 & 0.120 & 0.187 \\ 0.192 & 0.120 & 0 & 0.249 \\ 0.123 & 0.187 & 0.249 & 0 \end{bmatrix},$$

$$\mathbf{D}_2 = \begin{bmatrix} 0 & 0.276 & 0.208 \\ 0.276 & 0 & 0.096 \\ 0.208 & 0.0960 & 0 \end{bmatrix}. \tag{7.3.8}$$

The first step (Step 3 of the algorithm) is to determine which of $C_1$ or $C_2$ is to be partitioned. We use Eq. (7.3.5). The final hierarchy is shown in Figure 7.12.

Figure 7.13 shows the hierarchy obtained on these same data when the monothetic methodology of section 7.2.4 is used. While we observe that the first bi-partition gave the same sub-clusters $C_1 = \{1, 2, 4, 6\}$ and $C_2 = \{3, 5, 7\}$, the construction of the hierarchy changed at later $r$ stages. □

**Example 7.13** Instead of using the normalized (normed by $1/p$) Euclidean extended Ichino–Yaguchi normalized (by the $1/N_j$ of Eq. (4.2.31), $j = 1, \ldots, p$) distances for the data of Table 7.21, as in Example 7.12, suppose now we take the

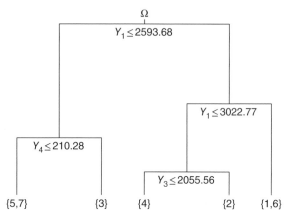

**Figure 7.13** Monothetic divisive hierarchy. Cover-types histogram-valued data: Euclidean span-normalized extended Ichino–Yaguchi distances (Example 7.12).

cumulative density function dissimilarity of Eqs. (4.2.44)–(4.2.46) (from Definition 4.16). We can show that the dissimilarity matrix becomes

$$\mathbf{D} = \begin{bmatrix} 0 & 0.170 & 0.831 & 0.463 & 0.706 & 0.240 & 0.705 \\ 0.170 & 0 & 0.674 & 0.310 & 0.617 & 0.333 & 0.593 \\ 0.831 & 0.674 & 0 & 0.393 & 0.443 & 0.898 & 0.311 \\ 0.463 & 0.310 & 0.393 & 0 & 0.515 & 0.536 & 0.463 \\ 0.706 & 0.617 & 0.443 & 0.515 & 0 & 0.867 & 0.190 \\ 0.240 & 0.333 & 0.898 & 0.536 & 0.867 & 0 & 0.836 \\ 0.705 & 0.593 & 0.311 & 0.463 & 0.190 & 0.836 & 0 \end{bmatrix}. \quad (7.3.9)$$

By following the polythetic algorithm steps, we obtain the hierarchy of Figure 7.14. The details are left to the reader (see Exercise 7.10).

For comparison, Figure 7.15 shows the hierarchy that emerges when the monothetic methodology, based on these cumulative density function dissimilarities of Eq. (7.3.9), is applied to these data (see Exercise 7.10). □

**Example 7.14** Consider now the mixed variable data set of Table 7.23. Data from Census (2002) consist of aggregated individual responses for those living in each of $m = 10$ regions, for $p = 6$ variables. The histogram variable $Y_1$ = age, takes values across the $s_1 = 12$ sub-intervals in $\mathcal{Y}_1 = \{[0, 4), [4, 17), [17, 20), [20, 24), [24, 34), [34, 44), [44, 54), [54, 64), [64, 74), [74, 84), [84, 94), [94, 120]\}$ years old, the histogram variable $Y_2$ = home value,

**Region**

$Y_1$ = Age, histogram sub-intervals

| u | [0,4), | [4,17), | [17,20), | [20,24), | [24,34), | [34,44), | [44,54), | [54,64), | [64,74), | [74,84), | [84,94), | [94,120] |
|---|---|---|---|---|---|---|---|---|---|---|---|---|
| 1 | 0.085 | 0.236 | 0.051 | 0.060 | 0.140 | 0.145 | 0.115 | 0.039 | 0.031 | 0.052 | 0.035 | 0.012 |
| 2 | 0.056 | 0.176 | 0.054 | 0.070 | 0.127 | 0.148 | 0.157 | 0.050 | 0.039 | 0.063 | 0.045 | 0.016 |
| 3 | 0.050 | 0.169 | 0.042 | 0.066 | 0.181 | 0.188 | 0.143 | 0.040 | 0.030 | 0.050 | 0.031 | 0.009 |
| 4 | 0.044 | 0.172 | 0.036 | 0.034 | 0.094 | 0.156 | 0.161 | 0.071 | 0.061 | 0.098 | 0.058 | 0.016 |
| 5 | 0.088 | 0.255 | 0.049 | 0.054 | 0.139 | 0.143 | 0.108 | 0.038 | 0.031 | 0.053 | 0.032 | 0.010 |
| 6 | 0.061 | 0.181 | 0.039 | 0.046 | 0.125 | 0.152 | 0.148 | 0.055 | 0.041 | 0.070 | 0.060 | 0.024 |
| 7 | 0.076 | 0.194 | 0.041 | 0.053 | 0.164 | 0.168 | 0.127 | 0.045 | 0.034 | 0.052 | 0.034 | 0.012 |
| 8 | 0.079 | 0.225 | 0.044 | 0.049 | 0.132 | 0.157 | 0.114 | 0.040 | 0.035 | 0.067 | 0.046 | 0.014 |
| 9 | 0.080 | 0.230 | 0.048 | 0.052 | 0.134 | 0.154 | 0.122 | 0.042 | 0.032 | 0.054 | 0.038 | 0.013 |
| 10 | 0.064 | 0.165 | 0.032 | 0.047 | 0.159 | 0.174 | 0.145 | 0.050 | 0.039 | 0.063 | 0.045 | 0.016 |

$Y_2$ = Home value in US$1000s, histogram sub-intervals

| u | [0,49), | [49,99), | [99,149), | [149,199), | [199,299), | [299,499), | [≥ 500] |
|---|---|---|---|---|---|---|---|
| 1 | 0.037 | 0.432 | 0.290 | 0.126 | 0.080 | 0.028 | 0.009 |
| 2 | 0.020 | 0.226 | 0.368 | 0.197 | 0.137 | 0.045 | 0.007 |
| 3 | 0.048 | 0.409 | 0.309 | 0.144 | 0.079 | 0.005 | 0.007 |
| 4 | 0.022 | 0.141 | 0.412 | 0.267 | 0.122 | 0.037 | 0.000 |
| 5 | 0.019 | 0.390 | 0.354 | 0.139 | 0.070 | 0.024 | 0.004 |
| 6 | 0.005 | 0.013 | 0.075 | 0.217 | 0.325 | 0.228 | 0.137 |
| 7 | 0.008 | 0.014 | 0.059 | 0.170 | 0.348 | 0.280 | 0.121 |
| 8 | 0.017 | 0.209 | 0.296 | 0.228 | 0.170 | 0.059 | 0.022 |
| 9 | 0.020 | 0.210 | 0.321 | 0.215 | 0.172 | 0.051 | 0.011 |
| 10 | 0.010 | 0.006 | 0.006 | 0.013 | 0.105 | 0.417 | 0.443 |

*(Continued)*

Table 7.23 (Continued)

| Region | $Y_3$ = Gender | | $Y_4$ = Fuel | | $Y_5$ = Tenure | | | $Y_6$ = Income | |
|---|---|---|---|---|---|---|---|---|---|
| $u$ | {male, | female} | | $\mathcal{Y}_4^a$ | {owner, | renter, | vacant} | [a, | b] |
| 1 | {0.499 | 0.501} | Gas | Electricity | 0.527 | 0.407 | 0.066 | [23.7, | 44.8] |
| 2 | {0.506 | 0.494} | Gas | Coal | 0.528 | 0.388 | 0.084 | [24.3, | 37.2] |
| 3 | {0.372 | 0.628} | Oil | Coal | 0.548 | 0.254 | 0.198 | [28.7, | 39.0] |
| 4 | {0.488 | 0.512} | Coal | Electricity | 0.523 | 0.226 | 0.251 | [31.5, | 41.8] |
| 5 | {0.502 | 0.498} | Gas | Electricity | 0.548 | 0.385 | 0.067 | [26.7, | 42.0] |
| 6 | {0.501 | 0.499} | Gas | | 0.609 | 0.326 | 0.065 | [41.7, | 62.8] |
| 7 | {0.502 | 0.498} | Gas | Electricity | 0.593 | 0.372 | 0.035 | [51.7, | 80.2] |
| 8 | {0.502 | 0.498} | Gas | Electricity | 0.596 | 0.270 | 0.134 | [28.1, | 63.4] |
| 9 | {0.500 | 0.500} | Gas | Electricity | 0.580 | 0.380 | 0.040 | [35.4, | 61.7] |
| 10 | {0.506 | 0.494} | Gas | Electricity | 0.599 | 0.376 | 0.025 | [64.7, | 83.7] |

a) $\mathcal{Y}_4$ = {gas, electricity, oil, coal, none}

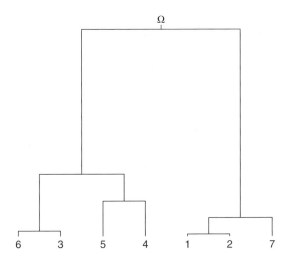

**Figure 7.14** Polythetic divisive hierarchy. Cover-types histogram-valued data: cumulative density function dissimilarities (Example 7.13).

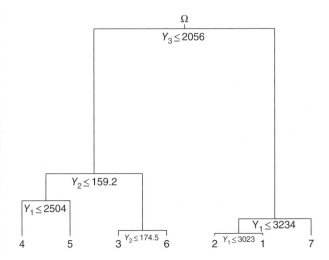

**Figure 7.15** Monothetic divisive hierarchy. Cover-types histogram-valued data: cumulative density function dissimilarities (Example 7.13).

takes values across the $s_2 = 7$ histogram sub-intervals in $\mathcal{Y}_2 = \{[0, 49), [49, 99), [99,149), [149,199), [199,299), [299,499), [\geq 500]\}$ in US\$1000s, $Y_3 =$ gender, with modal multi-valued values from $\mathcal{Y}_3 = \{$male, female$\}$, $Y_4 =$ fuel type used in the home, taking (non-modal) multi-valued values from $\mathcal{Y}_4 = \{$gas, electricity, oil, coal, none$\}$, $Y_5 =$ home tenure, taking modal multi-valued values from $\mathcal{Y}_5 = \{$owner, renter, vacant$\}$, and $Y_6 =$ income, with interval values $[a, b]$ from the real line expressed in US$1000s. For example, in region $u = 1$, the relative frequency of residents aged under 4 years is 0.085 (or 8.5%).

Suppose we want to construct a hierarchy using the polythetic divisive algorithm based on Gowda–Diday dissimilarities. Thus, for each region pair $(u_1, u_2)$, with $u_1, u_2, = 1, \ldots, 10$, we calculate the extended Gowda–Diday dissimilarities for the histogram variables $Y_1$ and $Y_2$ using Eq. (4.2.24), the extended Gowda–Diday dissimilarities for the modal multi-valued variables $Y_3$ and $Y_5$ using Eq. (4.1.7), the Gowda–Diday dissimilarities for the (non-modal) multi-valued variable $Y_4$ using Eq. (3.2.6), and the Gowda–Diday dissimilarities for the intervals of variable $Y_6$ from Eq. (3.3.15). These are then added to give the overall Gowda–Diday dissimilarity matrix for these data. This gives

$$D = \begin{bmatrix}
0 & 2.450 & 4.616 & 3.534 & 0.484 & 4.379 & 3.002 & 1.907 & 1.462 & 3.612 \\
2.450 & 0 & 3.462 & 3.237 & 2.472 & 4.565 & 4.980 & 3.734 & 3.488 & 5.237 \\
4.616 & 3.462 & 0 & 3.379 & 4.376 & 6.768 & 7.231 & 5.548 & 5.882 & 7.601 \\
3.534 & 3.237 & 3.379 & 0 & 3.297 & 6.404 & 5.876 & 4.033 & 4.425 & 6.138 \\
0.484 & 2.472 & 4.376 & 3.297 & 0 & 4.701 & 3.212 & 2.163 & 1.767 & 3.606 \\
4.379 & 4.565 & 6.768 & 6.404 & 4.701 & 0 & 3.192 & 3.558 & 3.384 & 4.095 \\
3.002 & 4.980 & 7.231 & 5.876 & 3.212 & 3.192 & 0 & 2.368 & 2.016 & 1.526 \\
1.907 & 3.734 & 5.548 & 4.033 & 2.163 & 3.558 & 2.368 & 0 & 1.048 & 3.505 \\
1.462 & 3.488 & 5.882 & 4.425 & 1.767 & 3.384 & 2.016 & 1.048 & 0 & 3.120 \\
3.612 & 5.237 & 7.601 & 6.138 & 3.606 & 4.095 & 1.526 & 3.505 & 3.120 & 0
\end{bmatrix}.$$
$$(7.3.10)$$

The resulting polythetic divisive hierarchy is shown in Figure 7.16. The details are left to the reader. □

## 7.4 Stopping Rule R

How the optimal stopping rule $R$ is established is an open question. In this section, we consider some possible approaches. One approach is to look at the changing value of the total within-cluster variation $W(P_r)$ as the number of clusters in $P_r = (C_1, \ldots, C_r)$, i.e., as $r$ increases. From Eqs. (7.1.3)–(7.1.5), it is clear that since the total variation $W(\Omega)$ is fixed, then as the total within-cluster variation $W(P_r)$ decreases, the between-cluster variation $B(P_r)$ must increase. A simple index is the following.

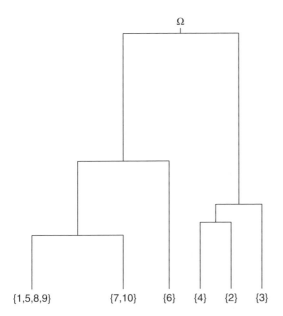

**Figure 7.16** Polythetic divisive hierarchy. Census mixed-valued data: Gowda–Diday dissimilarities (Example 7.14).

**Definition 7.10**   The **explained variation**, $E_r$, for the partition $P_r = (C_1, \ldots, C_r)$ is the proportion of the total variation explained by the between-cluster variation, i.e.,

$$E_r = \frac{B(P_r)}{W(\Omega)} = \frac{W(\Omega) - W(P_r)}{W(\Omega)}, \qquad r = 1, \ldots, R, \tag{7.4.1}$$

where $W(P_r)$, $W(\Omega)$, and $B(P_r)$ are defined in Eqs. (7.1.3)–(7.1.5), respectively, and $R \leq m$ is a preassigned stopping value. □

**Example 7.15**   Consider the modal multi-valued household occupancy and tenure data of Table 4.3 discussed in Example 7.2, with the city block extended Ichino–Yaguchi distance matrix of Eq. (7.2.11). There a hierarchy was built (using a monothetic algorithm). At each stage $r$, we can also calculate the total within-cluster variation $W(P_r)$ from Eq. (7.1.3), the total variation from Eq. (7.3.2), and the between-cluster variation $B(P_r)$ from Eq. (7.1.5), and hence the explained variation $E_r$, $r = 1, \ldots, R$, from Eq. (7.4.1). For example, take the stage $r = 3$ with $C_1 = \{1, 3\}$, $C_2 = \{2, 4\}$, and $C_3 = \{5, 6\}$ (see Example 7.2 and Table 7.5 for this bi-partitioning construction). Using the distances given in Eq. (7.2.11), we have, from Eq. (7.1.2),

$$I(C_1) = (6/2) \times [(1/6) \times (1/6) \times 0.206^2] = 0.00354,$$
$$I(C_2) = 0.00075, \qquad I(C_3) = 0.00317,$$

**Table 7.24** Within-, between-, explained-variation components based on extended Ichino–Yaguchi city block distances (Example 7.15)

| Clusters | Variations | | |
|---|---|---|---|
| | Within-clusters | Between clusters | Explained |
| $\{1,2,3,4\}, \{5,6\}$ | 0.0152 | 0.0110 | 0.4206 |
| $\{1,3\}, \{2,4\}, \{5,6\}$ | 0.0075 | 0.0187 | 0.7150 |
| $\{1\}, \{3\}, \{2,4\}, \{5,6\}$ | 0.0039 | 0.0222 | 0.8501 |
| $\{1\}, \{3\}, \{5\}, \{6\}\{2,4\}$ | 0.0008 | 0.0254 | 0.9713 |
| $\{1\}, \{2\}, \{3\}, \{4\}, \{5\}, \{6\}$ | 0.0000 | 0.02616 | 1.000 |

also, $W(\Omega) = 0.02616$. Hence, from Eqs. (7.1.3) and (7.4.1),

$$W(P_3) = I(C_1) + I(C_2) + I(C_3) = 0.00746,$$
$$B(P_3) = W(\Omega) - W(P_r) = 0.01871,$$
$$E_3 = B(P_3)/W(\Omega) = 0.71500.$$

The complete set of within-cluster, between-cluster, and explained variations for all stages is given in Table 7.24.

The tree of Figure 7.17 shows the tree heights at which the divisive cuts occur in the construction. These heights equal $(1 - E_r)$ for the $r$th cut. Thus, for example, the second cut occurs when the sub-cluster $\{1,2,3,4\}$ is divided

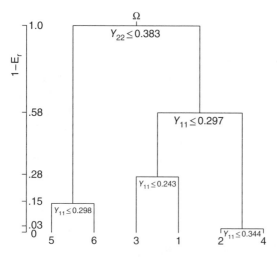

**Figure 7.17** Monothetic divisive hierarchy. Modal multi-valued data: extended Ichino–Yaguchi city block distances, tree heights (Example 7.15).

into $\{1,3\}$ and $\{2,4\}$ with explained rate $E_2 = 0.42$ and hence at a tree height of $1 - E_2 = .58$. □

As a guide for stopping at a specific $r^*$, we observe that if the difference in explained rate $E_{r+1} - E_r$ suddenly becomes small compared with $E_{r-1}$, then the effect of the bi-partition at stage $(r + 1)$ is weak. Therefore, it may be that $r^* = r$ is an optimal choice of the number of clusters. However, Milligan and Cooper (1985) performed a simulation study of many validity indexes (for classical data). That study suggested that the explained rate measure is not appropriate, despite its simplicity.

Kim and Billard (2011) proposed extensions of two indices developed by Dunn (1974) and Davis and Bouldin (1979) for classical data, which can be used on symbolic-valued realizations for both the monothetic algorithms and polythetic algorithms described in this chapter.

**Definition 7.11**    Suppose at stage $r$, the set of symbolic-valued observations $\Omega$ has been partitioned into $P_r = (C_1, \dots, C_r)$. Then, the **Dunn index for symbolic observations**, $DI_r^s$, for $P_r$ is given by

$$DI_r^s = \min_{v=1,\dots,r} \left\{ \min_{g=1,\dots,r, g \neq v} \left\{ \frac{I(C_g \cup C_v)) - I(C_g) - I(C_v)}{\max_{v=1,\dots,r} \{I(C_v)\}} \right\} \right\}, \quad r = 2, \dots, m - 1,$$

(7.4.2)

where $I(C_{(\cdot)})$ is the within-cluster variation of Eq. (7.1.2). □

**Definition 7.12**    Suppose at stage $r$, the set of symbolic-valued observations $\Omega$ has been partitioned into $P_r = (C_1, \dots, C_r)$. Then, the **Davis–Bouldin index for symbolic observations**, $DBI_r^s$, for $P_r$ is given by

$$DBI_r^s = \frac{1}{r} \sum_{v=1}^{r} \left[ \frac{\max_{h=1,\dots,r, h \neq v} \{I(C_h) + I(C_v)\}}{\min_{g=1,\dots,r, g \neq v} \{I(C_g \cup C_v)) - I(C_g) - I(C_v)\}} \right], \quad r = 2, \dots, m - 1,$$

(7.4.3)

where $I(C_{(\cdot)})$ is the within-cluster variation of Eq. (7.1.2). □

The between-cluster variation $I(C_g \cup C_v) - I(C_g) - I(C_v)$ in Eqs. (7.4.2) and (7.4.3) measures how far apart the two clusters $C_g$ and $C_v$, $g \neq v$, are in $P_r$, while the within-cluster variation $I(C_v)$ in the denominator of Eq. (7.4.2) measures how close the observations are in the cluster. The Dunn index looks at this ratio for all combinations of two clusters at a time in $P_r$. Since one aim of the clustering process is to minimize the within-cluster variation and to maximize the between-cluster variation, then the higher the value for the Dunn index $DI_r^s$, the better the clustering outcome.

In contrast, the Davis–Bouldin index of Eq. (7.4.3) takes an inverse ratio to that of the Dunn index, and it also takes the maximum of the average of the within-cluster variations for the two clusters $C_h$ and $C_v$, in the numerator of

Eq. (7.4.3) (rather than the maximum of the within-cluster variation of $C_v$ alone of the Dunn index in the denominator of Eq. (7.4.2)). As for the Dunn index, the Davis–Bouldin index also looks at this ratio for all combinations of two clusters at a time in $P_r$. Therefore, the lower the value for the Davis–Bouldin index $DBI_r^s$, the better the clustering result. These indices are illustrated in the following example (from Kim (2009) and Kim and Billard (2011)).

**Example 7.16**   Five sets, each of 200 classical observations, were simulated from the bivariate normal distributions $N_2(\boldsymbol{\mu}, \boldsymbol{\Sigma})$ shown in Table 7.25. Plots of these observations are shown in Figure 7.18, as $\mathbf{y}_u$, $u = 1, \dots, 5$, respectively. In each case, the classical observations were aggregated to give the histogram values of Table 7.26.

The normalized cumulative distribution function dissimilarity matrix, obtained from Eq. (4.2.47), is

$$\mathbf{D} = \begin{bmatrix} 0 & 0.016 & 0.101 & 0.106 & 0.500 \\ 0.016 & 0 & 0.094 & 0.106 & 0.511 \\ 0.101 & 0.094 & 0 & 0.028 & 0.573 \\ 0.106 & 0.106 & 0.028 & 0 & 0.566 \\ 0.500 & 0.511 & 0.573 & 0.566 & 0 \end{bmatrix}. \qquad (7.4.4)$$

Suppose we construct a tree for both the monothetic algorithm of sections 7.2 and 7.2.4, and the polythetic algorithm of section 7.3. At each stage $r$, we retain

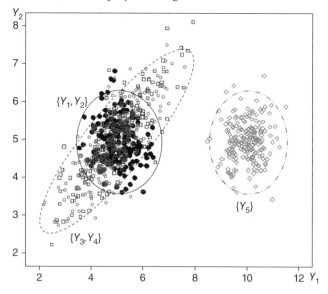

**Figure 7.18** Simulated histogram observations (Example 7.16).

**Table 7.25** Simulated bivariate normal distribution parameters (Example 7.16)

| Sample $u$ | $\mu$ | $\Sigma$ |
|---|---|---|
| 1, 2 | (5, 5) | $\begin{pmatrix} 0.3 & 0 \\ 0 & 0.3 \end{pmatrix}$ |
| ........ | ...... | ............ |
| 3, 4 | (5, 5) | $\begin{pmatrix} 1.0 & 0.8 \\ 0.8 & 1.0 \end{pmatrix}$ |
| ........ | ...... | ............ |
| 5 | (10, 5) | $\begin{pmatrix} 0.3 & 0 \\ 0 & 0.3 \end{pmatrix}$ |

the information for the $I(C_g \cup C_v)$, $I(C_g)$, and $I(C_v)$, $g = 1, \ldots, r$, $v = 1, \ldots, r$, $r = 1, \ldots, m - 1$, needed for each of the Dunn index and the Davis–Bouldin index of Eqs. (7.4.2) and (7.4.3), respectively. It is easy to show that the respective indices are as shown in Table 7.27(i).

Had we used the extended Gowda–Diday dissimilarity matrix calculated from Eq. (4.2.24), the normalized city block extended Ichino–Yaguchi distance matrix of Eq. (4.2.29), or the normalized Euclidean extended Ichino–Yaguchi distance matrix of Eq. (4.2.30), we obtain the Dunn indices and the Davis–Bouldin indices as shown in Table 7.27(ii)–(iv), respectively (see Kim (2009) for details). □

It is of interest to note that a polythetic algoritmic analysis will be more accurate in identifying the true number of clusters than a monothetic algorithm, particularly when the real clusters overlap.

**Example 7.17** Consider the situation described in Example 7.16. Kim and Billard (2011) repeated the analysis $B = 1000$ times. The polythetic algorithm correctly identified that there were three distinct clusters all 1000 times. In contrast, the monothetic algorithm correctly identified the three clusters only 21 times, and found two clusters the remaining 979 times. This is because, as we saw in section 7.2, the ordering process for the observations was based on the observation means, so that, for these data, this ordering was not able to distinguish between the observations $u = 1, 2, 3$, and 4, which were all generated from distributions with the same means (see Table 7.25). That is, the monothetic algorithm did not take into account the observation variations. In contrast, the polythetic algorithm was not limited to such an ordering of means;

**Table 7.26** Simulated histogram data (Example 7.16)

| u | $Y_1$ |
|---|-------|
| 1 | {[3.5, 4.0), 0.020; [4.0, 4.5), 0.165; [4.5, 5.0), 0.325; [5.0, 5.5), 0.290; [5.5, 6.0), 0.150; [6.0, 6.5], 0.050} |
| 2 | {[3.0, 3.5), 0.005; [3.5, 4.0), 0.020; [4.0, 4.5), 0.160; [4.5, 5.0), 0.300; [5.0, 5.5), 0.340; [5.5, 6.0), 0.150; [6.0, 6.5), 0.020; [6.5, 7.0], 0.005} |
| 3 | {[2.0, 3.0), 0.025; [3.0, 4.0), 0.135; [4.0, 5.0), 0.360; [5.0, 6.0), 0.345; [6.0, 7.0), 0.125; [7.0, 8.0], 0.010} |
| 4 | {[2.0, 3.0), 0.050; [3.0, 4.0), 0.130; [4.0, 5.0), 0.355; [5.0, 6.0), 0.290; [6.0, 7.0), 0.145; [7.0, 8.0], 0.030} |
| 5 | {[8.0, 8.5), 0.005; [8.5, 9.0), 0.020; [9.0, 9.5), 0.205; [9.5, 10.0), 0.255; [10.0, 10.5), 0.340; [10.5, 11.0), 0.155; [11.0, 11.5), 0.015; [11.5, 12.0], 0.005} |

| u | $Y_2$ |
|---|-------|
| 1 | {[3.5, 4.0), 0.035; [4.0, 4.5), 0.160; [4.5, 5.0), 0.345; [5.0, 5.5), 0.300; [5.5, 6.0), 0.120; [6.0, 6.5), 0.025; [6.5, 7.0], 0.015} |
| 2 | {[3.5, 4.0), 0.035; [4.0, 4.5), 0.165; [4.5, 5.0), 0.275; [5.0, 5.5), 0.305; [5.5, 6.0), 0.165; [6.0, 6.5), 0.050; [6.5, 7.0], 0.005} |
| 3 | {[2.0, 3.0), 0.020; [3.0, 4.0), 0.155; [4.0, 5.0), 0.295; [5.0, 6.0), 0.345; [6.0, 7.0), 0.170; [7.0, 8.0), 0.015} |
| 4 | {[2.0, 3.0), 0.030; [3.0, 4.0), 0.135; [4.0, 5.0), 0.360; [5.0, 6.0), 0.330; [6.0, 7.0), 0.115; [7.0, 8.0), 0.025; [8.0, 9.0], 0.005} |
| 5 | {[3.0, 3.5), 0.005; [3.5, 4.0), 0.020; [4.0, 4.5), 0.150; [4.5, 5.0), 0.365; [5.0, 5.5), 0.300; [5.5, 6.0), 0.135; [6.0, 6.5), 0.020; [6.5, 7.0], 0.005} |

**Table 7.27** Validity indices: histogram data (Example 7.16)

| Validity | Monothetic | | | Polythetic | | |
|---|---|---|---|---|---|---|
| | $r = 2$ | $r = 3$ | $r = 4$ | $r = 2$ | $r = 3$ | $r = 4$ |
| | (i) Normalized cumulative distribution function | | | | | |
| $D_r^s$ | 21.600 | **25.772** | 3.242 | 21.600 | **25.772** | 3.242 |
| $DB_r^s$ | 0.046 | **0.035** | 0.159 | 0.046 | **0.035** | 0.159 |
| | (ii) Extended Gowda–Diday | | | | | |
| $D_r^s$ | 2.023 | **53.387** | 1.698 | 2.023 | **53.387** | 1.698 |
| $DB_r^s$ | 0.494 | **0.025** | 0.301 | 0.494 | **0.025** | 0.301 |
| | (iii) Normalized city block extended Ichino–Yaguchi | | | | | |
| $D_r^s$ | 10.833 | **18.487** | 4.541 | 10.833 | **18.487** | 4.541 |
| $DB_r^s$ | 0.092 | **0.046** | 0.115 | 0.092 | **0.046** | 0.115 |
| | (iv) Normalized Euclidean extended Ichino–Yaguchi | | | | | |
| $D_r^s$ | 17.436 | **18.197** | 4.453 | 17.436 | **18.197** | 4.453 |
| $DB_r^s$ | 0.057 | **0.046** | 0.117 | 0.057 | **0.046** | 0.117 |

rather it was based on splinter observations using both the mean and variation information in the calculation of the dissimilarity values. □

Finally, while a definitive approach to determining the globally optimal number of clusters including determining whether or not a particular hierarchy is valid is not yet available, there are several other methods available in the literature for classical data, some of which may be adapted to symbolic data. One such measure of a global fit is the cophenetic correlation coefficient (Sneath and Sokal, 1973) which compares a distance matrix with a cophenetic matrix (with entries $c_{u_1 u_2}$ being the first level that observations first appear together in the same cluster). Other possible measures for model-based clustering can be found in, for example, Fraley and Raftery (1998).

## 7.5 Other Issues

There are a number of other aspects needing attention for divisive hierarchy trees. A different type of divisive procedure is the class of classification and regression tress (CART, originally developed for classical data by Breiman et al. (1984)). For these methods, the construction of the hierarchy tree is formed at each stage by either a classification method or by a regression analyses. A start on developing symbolic CART methods is provided in Seck (2012). This was successfully applied to the iris data in Seck et al. (2010) and to cardiac data in Quantin et al. (2011).

As suggested in Chapter 5, there are as yet no clear definitive answers as to which algorithms are best (however "best" may be defined) for any particular distance/dissimilarity measures. Chen (2014) and Chen and Billard (2019) have made an initial study, although it is limited to several versions of Hausdorff, Gowda–Diday, and Ichino–Yaguchi distances only. Broader investigations are still needed.

## Exercises

**7.1**  For the six counties of Table 4.3, suppose that the population sizes counted in producing those relative frequencies, for each variable, are as shown in Table 7.28.

Suppose the weights for each region are proportional to the population size of that county by variable. Show that the association measures $\alpha_{jk}$ for the complete set $\Omega$ evaluated in Example 7.1 are now as given in Table 7.2.

**7.2**  For the data of Table 4.3, show that the divisive method of section 7.2.1 when based on the extended Gowda–Diday dissimilarities produces a hierarchy that is the same as in Figure 7.1.

**7.3**  Refer to Example 7.3. Verify the details of Tables 7.6 and 7.7, and hence that the divisive algorithm produces Figure 7.2.

**7.4**  Refer to Example 7.4 for the data of Table 7.8. Construct the hierarchy of Figure 7.3.

**Table 7.28** Population sizes by county (Example 7.1 and Exercise 7.1)

|  |  | County |  |  |  |  |
|---|---|---|---|---|---|---|
| $Y_j$ | 1 | 2 | 3 | 4 | 5 | 6 |
| $Y_1$ | 449718 | 223543 | 42922 | 30720 | 40382 | 819617 |
| $Y_2$ | 540183 | 270767 | 55912 | 36563 | 48554 | 978484 |

**7.5** Take the data of Table 7.9 and the four variables $Y_4, Y_7, Y_{11}$, and $Y_{13}$ describing the temperature range for April, July, November and elevation at 15 stations. Complete the details of Example 7.7 to produce the hierarchy of Figure 7.6.

**7.6** For the data of Table 7.9, show that the hierarchy using the Hausdorff distances and all 13 variables is as in Figure 7.19. How does this hierarchy compare with the one you obtained in Exercise 7.5?

**7.7** Complete the details of Example 7.8 to build the hierarchy of Figure 7.7 using the temperature variables $(Y_1, \ldots, Y_{12})$ only and the Hausdorff distance.

**7.8** For the data of Table 7.9, build the hierarchy based on the normalized Euclidean Hausdorff distances for all 13 variables and all 15 stations.

**7.9** Take the histogram data for cholesterol and glucose used in Example 7.9. Instead of building the hierarchy by using only one variable, show that the hierarchy obtained by using both variables monothetically is comparable to that in Figure 7.9 when based on the basic Ichino–Yaguci distance $D_j, j = 1, 2$. How does this compare with your hierarchy if you used the combined distance $D = D_1 + D_2$ in your monothetic algorithm?

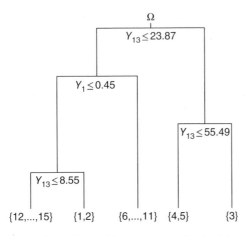

**Figure 7.19** Monothetic divisive hierarchy. Interval-valued data: Hausdorff distances on $Y_1, \ldots, Y_{13}$, 15 stations (Exercise 7.6).

**Table 7.29** Within-, between-, and explained-variation components
Gowda–Diday dissimilarities (Example 7.4 and Exercise 7.11)

| | Variations | | |
| Clusters | Within clusters | Between clusters | Explained |
|---|---|---|---|
| $\{1,2,3\}, \{4,5\}$ | 0.5398 | 2.5972 | 0.8279 |
| $\{1,3\}, \{2\}, \{4,5\}$ | 0.2056 | 2.9314 | 0.9345 |
| $\{1,3\}, \{2\}, \{4\}, \{5\}$ | 0.0694 | 3.0675 | 0.9779 |
| $\{1\}, \{2\}, \{3\}, \{4\}, \{5\}$ | 0.0000 | 3.1370 | 1.000 |

**7.10** Refer to Example 7.13. Complete the details of the polythetic algorithm of section 7.3 which produces the hierarchy of Figure 7.14.

**7.11** Refer to Example 7.4. Show that the explained variations, along with the respective within-cluster and between-cluster variations, are as shown in Table 7.29.

**7.12** Suppose a similarity matrix $\mathbf{D}$, extracted from Sokal (1963), is

$$\mathbf{D} = \begin{bmatrix} 1 & 0.033 & 0.995 & 0.034 & 0.927 & 0.331 \\ 0.033 & 1 & 0.028 & 0.973 & 0.090 & 0.450 \\ 0.995 & 0.028 & 1 & 0.030 & 0.941 & 0.394 \\ 0.034 & 0.973 & 0.030 & 1 & 0.058 & 0.400 \\ 0.927 & 0.090 & 0.941 & 0.058 & 1 & 0.452 \\ 0.331 & 0.450 & 0.394 & 0.400 & 0.452 & 1 \end{bmatrix}$$

where the entries are the Pearson's product-moment correlation coefficients for $p = 6$ taxonomies (A, ..., F) considered as variables. How might you construct a hierarchical tree to identify which taxonomies, if any, are clustered?

# 8

# Agglomerative Hierarchical Clustering

A general overview of hierarchical clustering was presented in Chapter 5. Divisive hierarchical clustering, in which the procedure starts with all observations in one cluster, was discussed in Chapter 7. In this chapter, the focus is on agglomerative hierarchical clustering. Agglomerative hierarchical clustering occurs when starting from the base of the hierarchical tree with each branch/node consisting of one observation only and where observations are progressively clustered into larger clusters until the top of the hierarchy tree is reached with all observations in a single cluster. In this chapter, we first consider the process where at any level of the tree, clusters are non-overlapping to produce hierarchical trees (see section 8.1). When clusters are allowed to overlap, we have pyramidal clustering (see section 8.2). Example 8.20 illustrates the distinction between these two trees.

## 8.1 Agglomerative Hierarchical Clustering

We start with describing some basic principles for agglomerative clustering in section 8.1.1. Then, agglomerative algorithms are applied to multi-valued list (modal and non-modal) observations in section 8.1.2, interval-valued observations in section 8.1.3, histogram-valued observations in section 8.1.4, and mixed-valued observations in section 8.1.5. In section 8.1.6, we consider an example where the data appear at first hand to be interval-valued, but where, after the application of logical rules, the data become histogram-valued in reality.

### 8.1.1 Some Basic Definitions

The basic principles for agglomerative hierarchical clustering for symbolic-valued observations are the same as those described briefly in Chapter 5 for classically valued observations. We have a set of $m$ observations $\Omega = \{1, \ldots, m\}$

*Clustering Methodology for Symbolic Data*, First Edition. Lynne Billard and Edwin Diday.
© 2020 John Wiley & Sons Ltd. Published 2020 by John Wiley & Sons Ltd.

with each described by the $p$-dimensional random variable $\mathbf{Y}_u = (Y_{u1}, \ldots, Y_{up})$, $u = 1, \ldots, m$. In contrast to the divisive clustering methods of Chapter 7, which start with all observations in a single cluster, agglomerative clustering starts with $m$ distinct non-overlapping clusters $(C_1, \ldots, C_m)$ each with a single observation with $\cup_v C_v = \Omega$.

The clustering criteria involve dissimilarity or distance measures between two clusters, and seek to merge clusters that are the most similar, i.e., that have the smallest dissimilarity between them. After merging, the dissimilarity matrix is re-calculated according to some criteria specific to the particular agglomerative method used. Thus, in effect, the major difference between agglomerative algorithms for classical and symbolically-valued data is in the calculation of the dissimilarity matrices. Once these matrices have been determined, the usual (i.e., classical) methods prevail. For completeness, some of the agglomerative methods widely used are re-defined herein. A coverage of the references from the classical literature can be found in Chapter 5.

Suppose at the $r$th step the clusters are $P_r = (C_1, \ldots, C_{m-r+1})$ and suppose cluster $C_v$ contains the observations $\{u_1^v, \ldots, u_{m_v}^v\}$ where the number of observations in $C_v$ is $m_v$, $v = 1, \ldots, m - r + 1$, $r = 1, \ldots, m$.

**Definition 8.1** Let $C_{v_1}$ and $C_{v_2}$ be two clusters in $P_r = (C_1, \ldots, C_{m-r+1})$, and let $d(C_{v_1}, C_{v_2})$ be the dissimilarity between these clusters, $v_1, v_2 = 1, \ldots, m - r + 1$. Suppose these particular clusters are merged into the cluster $C_{v_{12}}$. Then, for the **single-link**, also called **nearest neighbor**, criterion, the dissimilarity between this merged cluster and any other cluster $C_v \in P_{r+1}$, $v \neq v_1, v_2$, is

$$d(C_v, C_{v_{12}}) = \min_{v \neq v_1, v_2} \{d(C_v, C_{v_1}), d(C_v, C_{v_2})\} \tag{8.1.1}$$

where $d(C_v, C_{v'})$ is the dissimilarity or distance between any observation in $C_v$ and any observation in $C_{v'}$. □

**Definition 8.2** Let $C_{v_1}$ and $C_{v_2}$ be two clusters in $P_r = (C_1, \ldots, C_{m-r+1})$ and let $d(C_{v_1}, C_{v_2})$ be the dissimilarity between these clusters, $v_1, v_2 = 1, \ldots, m - r + 1$. Suppose these particular clusters are merged into the cluster $C_{v_{12}}$. Then, for the **complete-link**, also called **furthest neighbor**, criterion, the dissimilarity between this merged cluster and any other cluster $C_v \in P_{r+1}$, $v \neq v_1, v_2$, is

$$d(C_v, C_{v_{12}}) = \max_{v \neq v_1, v_2} \{d(C_v, C_{v_1}), d(C_v, C_{v_2})\} \tag{8.1.2}$$

where $d(C_v, C_{v'})$ is the dissimilarity or distance between any observation in $C_v$ and any observation in $C_{v'}$. □

**Definition 8.3** Let $C_{v_1}$ and $C_{v_2}$ be two clusters in $P_r = (C_1, \ldots, C_{m-r+1})$ and let $d(C_{v_1}, C_{v_2})$ be the dissimilarity between these clusters, $v_1, v_2 = 1, \ldots, m - r + 1$.

Suppose these particular clusters are merged into the cluster $C_{v_{12}}$. Then, for the **weighted average-link** criterion, the dissimilarity between this merged cluster and any other cluster $C_v \in P_{r+1}$, $v \neq v_1, v_2$, is

$$d(C_v, C_{v_{12}}) = [d(C_v, C_{v_1}) + d(C_v, C_{v_2})]/2 \tag{8.1.3}$$

where $d(C_v, C_{v'})$ is the dissimilarity or distance between any observation in $C_v$ and any observation in $C_{v'}$.  □

**Definition 8.4**  Let $C_{v_1}$ and $C_{v_2}$ be two clusters in $P_r = (C_1, \ldots, C_{m-r+1})$, and suppose $C_v$ contains the $m_v$ observations $\{u_1^{(v)}, \ldots, u_{m_v}^{(v)}\}$, $v = v_1, v_2$. Then, the **group average-link** criterion seeks those two clusters $C_{v_1}$ and $C_{v_2}$ which minimize the average between-cluster dissimilarities, i.e.,

$$d(C_{v_1}, C_{v_2}) = \min_{v,v'} \{ \frac{1}{m_v m_{v'}} \sum_{\substack{u^{(v)} \in C_v \\ u^{(v')} \in C_{v'}}} d(u^{(v)}, u^{(v')}) \} \tag{8.1.4}$$

where $d(u^{(v)}, u^{(v')})$ is the dissimilarity between an observation $u$ in $C_v$ and an observation $u'$ in $C_{v'}$, and $m_v m_{v'}$ is the total number of such pairs.  □

The weighted average-link criterion of Eq. (8.1.3) compares the average of all pairwise dissimilarities within one cluster with the corresponding average within the other cluster, while the grouped average-link criterion of Eq. (8.1.4) compares the average of all pairwise dissimilarities between one observation in one cluster and another observation in the second cluster.

Another important criterion for agglomerative clustering is that of Ward (1963). This criterion is based on Euclidean distances and it aims to reduce the variation within clusters. We first, in Definitions 8.5–8.7, present these as originally defined for classical observations, and then consider how they are adapted for symbolic data. Importantly, the key criterion (in Eq. (8.1.9)) is algebraically the same for both classical and symbolic data; however, the calculation of the sample means differs depending on the type of data.

**Definition 8.5**  Let $\mathbf{Y}_u = (Y_{u1}, \ldots, Y_{up})$, $u = 1, 2$, describe any two observations $A_u$, $u = 1, 2$, respectively. Then, the **squared Euclidean distance** between $A_1$ and $A_2$ is defined as

$$d^2(A_1, A_2) = \sum_{j=1}^{p} (Y_{1j} - Y_{2j})^2. \tag{8.1.5}$$

□

**Definition 8.6** Let the observation space $\Omega$ consist of the clusters $C_1, \ldots, C_K$. The **total within-cluster sum of squares** is

$$E = \sum_{k=1}^{K} E_k \tag{8.1.6}$$

where the **within-cluster sum of squares** for cluster $C_k$ with $m_k$ observations, $k = 1, \ldots, K$, is the squared Euclidean distance

$$E_k = \sum_{j=1}^{p} \sum_{u=1}^{m_k} (Y_{ujk} - \bar{Y}_{jk})^2 \tag{8.1.7}$$

where $\bar{Y}_{jk}$ is the midpoint or centroid value in cluster $C_k$ for the variable $Y_j$, $j = 1, \ldots, p$, and where the observation $u$ in $C_k$ is described by $\mathbf{Y}_{uk} = (Y_{u1k}, \ldots, Y_{upk})$, $u = 1, \ldots, m_k$, $k = 1, \ldots, K$. □

The within-cluster sum of squares is the cluster error sum of squares (also referred to as the inertia of the cluster in the French literature; see Bategelj (1988)).

**Definition 8.7** Let the observation space $\Omega$ consist of the clusters $C_1, \ldots, C_K$. Then, those two clusters $C_{v_1}$ and $C_{v_2}$, $v_1, v_2 = 1, \ldots, K$, for which the increase, $\Delta_{v_1 v_2}$, in the total within-cluster sum of squares of Eqs. (8.1.6) and (8.1.7) is minimized are said to be merged by the **Ward criterion**. That is, $v_1, v_2$ satisfy

$$\Delta_{v_1 v_2} = \min_{k_1, k_2} \{ E_{k_1 \cup k_2} - E_{k_1} - E_{k_2} \}, \tag{8.1.8}$$

where $C_{k_1 \cup k_2}$ is the new cluster when $C_{k_1}$ and $C_{k_2}$ are merged, $k_1, k_2 = 1, \ldots, K$. This is also called a **minimum variance** criterion. □

It follows, from Eqs. (8.1.7) and (8.1.8), that the increase in the total within-cluster variation by merging the clusters $C_{v_1}$ and $C_{v_2}$ satisfies

$$\Delta_{v_1 v_2} = \frac{m_{v_1} m_{v_2}}{m_{v_1} + m_{v_2}} \sum_{j=1}^{p} (\bar{Y}_{jv_1} - \bar{Y}_{jv_2})^2 \tag{8.1.9}$$

where $m_v$ is the number of observations in cluster $C_v$, $v = 1, \ldots, K$ (see, e.g., Ward (1963), Anderberg (1973) and Kaufman and Rousseeuw (1990) for details). The term $\Delta_{v_1 v_2}$ in Eq. (8.1.9) is also called the **Ward dissimilarity** between clusters $C_{v_1}$ and $C_{v_2}$.

While these concepts, as originally formulated by Ward (1963), dealt with classical quantitative observations, they are easily adapted to interval- and histogram-valued observations. Thus, we first note that the error sum of squares of Eq. (8.1.7) can be calculated for interval data from Eq. (2.4.5) and for histogram data from Eq. (2.4.11) with means given by Eqs. (2.4.2) and

Eq. (2.4.3), respectively. Thence, it can be easily shown that the Ward criterion of Eq. (8.1.9) still holds for interval and histogram data where now the centroid values $\bar{Y}_v$ are the corresponding sample means of the cluster observations obtained from Eqs. (2.4.2) and (2.4.3), respectively.

Another criterion for quantitative data is the centroid-link method which seeks the minimum distance between cluster centroids and so is applicable for quantitative data only. Formally, this is defined as follows.

**Definition 8.8**   Let $C_{v_1}$ and $C_{v_2}$ be two clusters in $P_r = (C_1, \ldots, C_{m-r+1})$ and let the Euclidean distance between their centroids be

$$d^2(C_{v_1}, C_{v_2}) = \sum_{j=1}^{p} (\bar{\mathbf{Y}}_{v_1 j} - \bar{\mathbf{Y}}_{v_2 j})^2 \tag{8.1.10}$$

where $\bar{\mathbf{Y}}_v$ is the centroid mean value of all observations in $C_v$. Then, the **centroid-link** criterion finds the two clusters $C_{v_1}$ and $C_{v_2}$ which minimize this distance, i.e.,

$$d^2(C_{v_1}, C_{v_2}) = \min_{v,v'} \left\{ \sum_{j=1}^{p} (\bar{\mathbf{Y}}_{vj} - \bar{\mathbf{Y}}_{v'j})^2 \right\}. \tag{8.1.11}$$

☐

When the data are interval- or histogram-valued, the centroids are calculated from Eqs. (2.4.2) and (2.4.3), respectively, as before.

Like the Ward criterion, the centroid-link method uses Euclidean distances. However, unlike the Ward criterion, we see, from Eqs. (8.1.9) and (8.1.11), that the Ward criterion uses a weighted centroid distance whereas the centroid-link criterion does not. One consequence of this is that the dissimilarities are not necessarily monotone so that reversals can occur when merging with the centroid-link criterion; these reversals do not occur for the Ward criterion.

Notice that the single-link, complete-link, and average-link criteria are expressed in terms of dissimilarity functions without specifically defining any specific dissimilarity function. Thus, any of the dissimilarity/distance functions provided in Chapters 3 and 4 can be used to build the agglomerative hierarchy. In contrast, Ward's criterion is based specifically on squared Euclidean distances, or equally the (non-squared) Euclidean distance, from Eq. (8.1.5),

$$d(A_1, A_2) = \left[ \sum_{j=1}^{p} (Y_{1j} - Y_{2j})^2 \right]^{1/2} \tag{8.1.12}$$

can be used instead.

The steps of the basic agglomerative algorithms are as follows, where it is assumed a particular criterion, such as the single-link method, had been adopted.

**Step 1**   Stage $r = 1$: Start with $\Omega = P_1 = (C_1, \ldots, C_m)$ with $C_v$ containing one observation $A_v$, $v = 1, \ldots, m$.

**Step 2**   Stage $r$:  $\Omega = P_r = (C_1, \ldots, C_{m-r+1})$ with dissimilarity matrix $\mathbf{D}_{(r)}$, $r = 1, \ldots, m$.

**Step 3**   Merge the two clusters in $\Omega = P_r = (C_1, \ldots, C_{m-r+1})$ for which the dissimilarity $d(C_v, C_{v'})$, $v \neq v' = 1, \ldots, m - r + 1$ in $\mathbf{D}_{(r)}$ is smallest.

**Step 4**   Calculate the dissimilarity matrix $\mathbf{D}_{(r+1)}$ for this new partition $\Omega = P_{r+1} = (C_1, \ldots, C_{m-r})$, $r = 1, \ldots, m$, using the criteria adopted for the algorithm, e.g., Eq. (8.1.1) for the single-link method.

**Step 5**   Now $r = r + 1$. Repeat Steps 2–4.

**Step 6**   When $r = m$, stop. Now $\Omega = P_m$ contains all the observations in a single cluster.

Since the algorithm is based on a dissimilarity matrix which is in turn based on all $p$ variables in $\mathbf{Y}$ under study (see Chapters 3 and 4), this algorithm is effectively a polythetic algorithm. The algorithm is essentially that of Kaufman and Rousseeuw (1990). It can be found in SAS ("proc cluster" procedure). In the R cluster library, there are two options, one using the "hclus" function, which is based on the squared Euclidean distance of Eq. (8.1.5), and one using the "agnes" function based on the Euclidean distance of Eq. (8.1.12) (both under the "method=ward" component). Murtagh and Legendre (2014) provide details of how to obtain one hierarchy from the other when using R. Their article includes case studies comparing the merits of using the squared Euclidean distance versus the Euclidean distance. Furthermore, beware that while the Ward criterion requires Euclidean dissimilarities, input matrices of other non-Euclidean dissimilarities would clearly run; however, the interpretation of the output could be problematic.

### 8.1.2   Multi-valued List Observations

Multi-valued list observations take values, as defined in Eqs. (2.2.1) and (2.2.2) for non-modal and modal multi–valued list observations, respectively. Since non-modal values are a special case of the modal valued case, let us restrict attention to this latter case. Thus, the observations are in the form

$$\mathbf{Y}_u = (\{Y_{ujk_j}, p_{ujk_j}; \, k_j = 1, \ldots, s_j\}, \, j = 1, \ldots, p), \sum_{k_j=1}^{s_j} p_{ujk_j} = 1, \, u = 1, \ldots, m,$$

$$(8.1.13)$$

where $Y_{ujk_j}$ is the $k_j$th category observed from the set of possible categories $\mathcal{Y}_j = \{Y_{j1}, \dots, Y_{js_j}\}$ for the variable $Y_j$, and where $p_{ujk_j}$ is the probability or relative frequency associated with $Y_{ujk_j}$.

Once dissimilarities and/or distances have been calculated, the agglomerative clustering methods used on classical data are the same as those to be used on symbolic valued data. Therefore, we illustrate briefly the method on modal multi-valued list data in the following example.

**Example 8.1** Take the multi-valued household data of Table 6.1. In particular, consider the two variables $Y_1 =$ fuel usage type with possible values $\mathcal{Y}_1 = \{$gas, electric, oil, coal/wood, none$\}$, and $Y_2 = \#$ rooms in the house with possible values $\mathcal{Y}_2 = \{(1,2), (3,4,5), \geq 6\}$. Let us perform an agglomerative clustering based on the extended multi-valued list Euclidean Ichino–Yaguchi dissimilarities calculated from Eq. (4.1.15), for $\gamma = 0.25$; these were provided in Table 6.2. (i) Suppose we perform a single-link nearest-neighbor clustering procedure. From the distance matrix of Table 6.2, the shortest distance between any two observations is the $d(4,6) = 0.180$ between $u = 4$ and $u = 6$. Hence, the first merging puts these two observations into one cluster. The next-nearest neighbors correspond to $u = 1$ and $u = 2$ with $d(1,2) = 0.188$, so these form the second merger of observations. Therefore, the dissimilarity matrix, after applying Eq. (8.1.1), becomes

$$
\mathbf{D}_{(3)} = \begin{bmatrix}
0 & 0.715 & 0.887 & 1.250 & 0.611 & 0.338 & 1.027 & 0.645 & 0.313 & 0.313 \\
0.715 & 0 & 1.039 & 1.298 & 0.791 & 0.744 & 1.138 & 0.884 & 0.715 & 1.115 \\
0.887 & 1.039 & 0 & 0.211 & 0.295 & 0.758 & 1.250 & 0.657 & 0.732 & 0.783 \\
1.250 & 1.298 & 0.211 & 0 & 0.655 & 1.107 & 1.294 & 0.983 & 1.047 & 0.957 \\
0.611 & 0.791 & 0.295 & 0.655 & 0 & 0.479 & 1.293 & 0.534 & 0.452 & 0.794 \\
0.338 & 0.744 & 0.758 & 1.107 & 0.479 & 0 & 1.267 & 0.625 & 0.226 & 0.898 \\
1.027 & 1.138 & 1.250 & 1.294 & 1.293 & 1.267 & 0 & 0.789 & 1.101 & 0.572 \\
0.645 & 0.884 & 0.657 & 0.983 & 0.534 & 0.625 & 0.789 & 0 & 0.422 & 0.280 \\
0.313 & 0.715 & 0.732 & 1.047 & 0.452 & 0.226 & 1.101 & 0.422 & 0 & 0.699 \\
0.313 & 1.115 & 0.783 & 0.957 & 0.794 & 0.898 & 0.572 & 0.280 & 0.699 & 0
\end{bmatrix}
\begin{matrix}
\{1,2\} \\ 3 \\ \{4,6\} \\ 5 \\ 7 \\ 8 \\ 9 \\ 10 \\ 11 \\ 12
\end{matrix}
$$

$$(8.1.14)$$

where, in Eq. (8.1.14), we apply Eq. (8.1.1) to all cluster pairs. Thus, for example, $d(\{4,6\}, 11) = \min\{0.883, 0.732\} = 0.732$, and $d(\{1,2\}, \{4,6\}) = \min\{1.052, 0.887, 1.092, 0.944\} = 0.887$. In Eq. (8.1.14), the entries down the right-hand side of $\mathbf{D}_{(3)}$ correspond to the observations in the respective clusters.

For the third merger, we have to recognize that we are seeking the shortest distance between any of the now-ten clusters, with one cluster containing observations $\{4,6\} \equiv C^1$ from the first merge and one cluster containing observations $\{1,2\} \equiv C^2$ from the second merge, and the other eight clusters being the as-yet-unmerged single observations. For the nearest-neighbor method, this means we seek the shortest distance between any observation in any cluster and an observation in any other cluster. Here, at this stage, the shortest remaining distance is the $d(4, 5) = 0.211$ between observation $u = 5$ and the $u = 4$ from inside $C^1$. Hence, $C^1 = \{4, 6\}$ becomes $C^1 = \{4, 5, 6\}$.

The same applies for all subsequent merges. For example, let us look at the ninth stage, where the current clusters now are $C^1 = \{1, 2, 8, 10, 11, 12\}$, $C^2 = \{4, 5, 6, 7\}$, and the single observations $u = 3$ and $u = 9$. Again, from Table 6.2 and Eq. (8.1.1), the dissimilarity matrix becomes

$$
\mathbf{D}_{(9)} = \begin{bmatrix}
0 & \mathbf{0.452} & 0.715 & 0.572 \\
\mathbf{0.452} & 0 & 0.791 & 1.250 \\
0.715 & 0.791 & 0 & 1.138 \\
0.572 & 1.250 & 1.138 & 0
\end{bmatrix}
\begin{matrix}
C^1 \\
C^2 \\
3 \\
9
\end{matrix}
\tag{8.1.15}
$$

where the shortest distance between any observation contained in this $C^1$ and $C^2$ is $d(7, 11) = 0.452$ between the observations $u = 7$ and $u = 11$. This distance is also shorter than any of the distances between $u = 3$, $u = 9$ and observations inside the current $C^1$ or $C^2$. Hence, these $C^1$ and $C^2$ are merged at this stage.

Notice also, for these distances, at the seventh stage, we have the shortest distance between the then-current clusters as being $d(1, 12) = 0.313$ between the clusters $C^1 = \{1, 2\}$ and $C^2 = \{10, 12\}$, but also $d(2, 11) = 0.313$ between clusters $C^1 = \{1, 2\}$ and $C^3 = \{8, 11\}$. Thus, at that stage, both the seventh and eighth stages occur simultaneously to form the new cluster $\{1, 2, 8, 10, 11, 12\}$.

The final agglomerative hierarchy is as shown in Figure 8.1(a).

(ii) Had the complete-link, or furthest-neighbor, criterion method been applied instead, then the hierarchy of Figure 8.1(b) emerges. The details are left to the reader. The differences in the two hierarchies are apparent, though the counties $\{1, 2\}$, $\{4, 6\}$, $\{8, 11\}$, and $\{10, 12\}$ are clustered together in both cases.

(iii) Alternatively, the dissimilarity of a cluster can be calculated to be the average of those dissimilarities of observations inside a cluster, giving the average-link method of Eq. (8.1.3). Suppose we are at the third stage of a group average-link agglomeration. In this case, the distance between the observation $u = 5$ and the cluster $C^1 = \{4, 6\}$ has a distance $d(5, C^1) = (0.211 + 0.383)/2 = 0.297$, while the distance between $u = 5$ and $C^2 = \{1, 2\}$ is $d(5, C^2) = (1.250 + 1.275)/2 = 1.2625$. Likewise, the distances between these clusters and all other observations can be calculated. Observe that in the pure single-link agglomeration, at this third stage, the distance between 5 and $\{4, 6\}$ at 0.297 is no longer the shortest available distance. Instead, the shortest distance is $d(8, 11) = 0.228$, so that observations $u = 8$ and $u = 11$ are merged. Continuing in this manner, we obtain the hierarchy of Figure 8.1(c). As before, the observations $\{1, 2\}$, $\{4, 6\}$, $\{8, 11\}$, and $\{10, 12\}$ are clustered together, but at different stages than when using single-link or complete-link methods; the observations $u = 3$ and $u = 9$, in particular, are merged at different stages. □

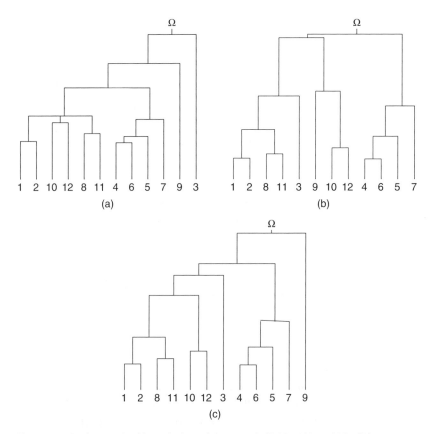

**Figure 8.1** Agglomerative hierarchy based on extended Ichino–Yaguchi Euclidean distances: household data (Example 8.1). (a) Single-link criterion (b) Complete-link criterion (c) Average-link criterion

### 8.1.3 Interval-valued Observations

Recall, from Chapter 2, that an interval-valued observation $u$ has realization

$$\mathbf{Y}_u = ([a_{u1}, b_{u1}], \ldots, [a_{up}, b_{up}]), \quad u = 1, \ldots, m. \tag{8.1.16}$$

We illustrate the agglomerative hierarchical procedure for such data.

**Example 8.2** Consider the *E. coli* species abundance data of Table 3.8, described in Example 3.10. There are $p = 4$ variables $Y_1 = CG$, $Y_2 = GC$, $Y_3 = TA$, and $Y_4 = AT$. Suppose the hierarchy is built by the nearest-neighbor, i.e., single-link, method, and suppose it is based on the Gowda–Diday dissimilarities, calculated in Example 3.11 and shown in Table 3.10(e). Therefore, the initial dissimilarity matrix, at the $r = 1$ stage, is

$$
\mathbf{D} \equiv \mathbf{D}_{(1)} = \begin{bmatrix}
0 & 7.137 & 6.917 & 5.565 & 5.847 & 4.622 \\
7.137 & 0 & 7.329 & 5.270 & 6.153 & 5.639 \\
6.917 & 7.329 & 0 & 6.567 & 6.039 & 7.357 \\
5.565 & 5.270 & 6.567 & 0 & \mathbf{3.624} & 3.988 \\
5.847 & 6.153 & 6.039 & \mathbf{3.624} & 0 & 5.789 \\
4.622 & 5.639 & 7.357 & 3.988 & 5.789 & 0
\end{bmatrix}.
$$
(8.1.17)

Then, from Eq. (8.1.17), since the smallest dissimilarity is between observations $u = 4$ and $u = 5$, these are merged into one cluster $\{4, 5\}$. Since we are using the nearest-neighbor method, the dissimilarity matrix, at the second stage $r = 2$, becomes

$$
\mathbf{D}_{(2)} = \begin{bmatrix}
0 & 7.137 & 6.917 & 5.565 & 4.622 \\
7.137 & 0 & 7.329 & 5.270 & 5.639 \\
6.917 & 7.329 & 0 & 6.039 & 7.357 \\
5.565 & 5.270 & 6.039 & 0 & \mathbf{3.988} \\
4.622 & 5.639 & 7.357 & \mathbf{3.988} & 0
\end{bmatrix}
\begin{matrix}
1 \\
2 \\
3 \\
\{4, 5\} \\
6
\end{matrix}
$$
(8.1.18)

Notice that the dissimilarities between the clusters $u$, $u \neq 4, 5$, and the cluster $\{4, 5\}$ are the minimum dissimilarities satisfying Eq. (8.1.1). Hence, for example, $d(1, \{4, 5\}) = d(1, 4) = 5.565$, whereas $d(3, \{4, 5\}) = d(3, 5) = 6.039$.

From Eq. (8.1.18), we see that the minimum dissimilarity is $d(\{4, 5\}, 6) = 3.988$. Hence, these two clusters are merged. Again, applying Eq. (8.1.1), we obtain the new dissimilarity matrix $\mathbf{D}_{(3)}$ as

$$
\mathbf{D}_{(3)} = \begin{bmatrix}
0 & 7.137 & 6.917 & \mathbf{4.622} \\
7.137 & 0 & 7.329 & 5.270 \\
6.917 & 7.329 & 0 & 6.039 \\
\mathbf{4.622} & 5.270 & 6.039 & 0
\end{bmatrix}
\begin{matrix}
1 \\
2 \\
3 \\
\{4, 5, 6\}
\end{matrix}
$$

and the iteration continues until all observations are in one cluster at the top of the tree. This final hierarchy is displayed in Figure 8.2. Notice how the clusters are strung out, as can happen when the single-link criterion is used. □

**Example 8.3** Let us now construct the agglomerative tree for the same situation of Example 8.2, except that we use the furthest-neighbor, or complete-link, criterion. This time, we seek the maximum dissimilarities as we merge observations at each stage. As in Example 8.2, at stage $r = 1$, from

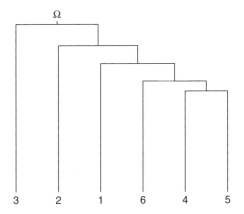

**Figure 8.2** Single-link agglomerative hierarchy based on Gowda–Diday dissimilarity: *E. coli* data (Example 8.2).

Eq. (8.1.17), the smallest dissimilarity is $d(4,5) = 3.624$. Thus, from Eqs. (8.1.17) and (8.1.2), the dissimilarity matrix, for stage $r = 2$, is now

$$
\mathbf{D}_{(2)} =
\begin{bmatrix}
0 & 7.137 & 6.917 & 5.847 & \mathbf{4.622} \\
7.137 & 0 & 7.329 & 6.153 & 5.639 \\
6.917 & 7.329 & 0 & 6.567 & 7.357 \\
5.847 & 6.153 & 6.567 & 0 & 5.789 \\
\mathbf{4.622} & 5.639 & 7.357 & 5.789 & 0
\end{bmatrix}
\begin{matrix}
1 \\ 2 \\ 3 \\ \{4,5\} \\ 6
\end{matrix}
\qquad (8.1.19)
$$

where now, for example, from Eq. (8.1.2), the new dissimilarity between observation $u = 1$ and $\{4,5\}$ is $\max\{5.565, 5.847\} = 5.847$. Then, for $r = 2$, since the smallest dissimilarity is $d(1,6) = 4.622$, clusters $\{1\}$ and $\{6\}$ are merged. Proceeding as before, from Eq. (8.1.2), gives the new dissimilarity matrix

$$
\mathbf{D}_{(3)} =
\begin{bmatrix}
0 & 7.137 & 7.357 & \mathbf{5.847} \\
7.137 & 0 & 7.329 & 6.153 \\
7.356 & 7.329 & 0 & 6.567 \\
\mathbf{5.847} & 6.153 & 6.567 & 0
\end{bmatrix}
\begin{matrix}
\{1,6\} \\ 2 \\ 3 \\ \{4,5\}
\end{matrix}
$$

Eventually, the tree of Figure 8.3 is built. Not surprisingly, especially in light of the discussion in Chapter 5, this tree is different from the nearest-neighbor tree of Figure 8.2. □

**Example 8.4** Suppose now we construct a complete-link tree on the *E. coli* data of Table 3.8 based on the Ichino–Yaguchi dissimilarities calculated in Example 3.13 and displayed in Table 3.13(e). At stage $r = 1$, the

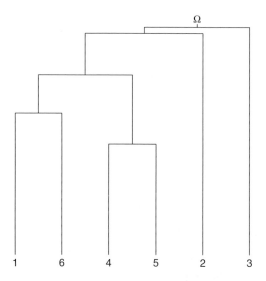

**Figure 8.3** Complete-link agglomerative hierarchy based on Gowda–Diday dissimilarity: *E. coli* data (Example 8.3).

dissimilarity matrix is

$$
\mathbf{D} \equiv \mathbf{D}_{(1)} =
\begin{bmatrix}
0 & 67.5 & 66.5 & 34.5 & 39.5 & 48.0 \\
67.5 & 0 & 76.0 & 53.0 & 54.0 & 40.5 \\
66.5 & 76.0 & 0 & 60.0 & 57.0 & 64.5 \\
34.5 & 53.0 & 60.0 & 0 & \mathbf{15.0} & 49.5 \\
39.5 & 54.0 & 57.0 & \mathbf{15.0} & 0 & 49.5 \\
48.0 & 40.5 & 64.5 & 49.5 & 49.5 & 0
\end{bmatrix}.
\tag{8.1.20}
$$

Since the smallest dissimilarity in $\mathbf{D}_{(1)}$ is $d(4,5) = 15.0$, these two observations are merged. The dissimilarity matrix becomes, from Eqs. (8.1.2) and (8.1.20),

$$
\mathbf{D}_{(2)} =
\begin{bmatrix}
0 & 67.5 & 66.5 & \mathbf{39.5} & 48.0 \\
67.5 & 0 & 76.0 & 54.0 & 40.5 \\
66.5 & 76.0 & 0 & 60.0 & 64.5 \\
\mathbf{39.5} & 54.0 & 60.0 & 0 & 49.5 \\
48.0 & 40.5 & 64.5 & 49.5 & 0
\end{bmatrix}
\begin{matrix}
1 \\ 2 \\ 3 \\ \{4,5\} \\ 6
\end{matrix}
\tag{8.1.21}
$$

Now, $r = 2$, and from Eq. (8.1.21), the observation $u = 1$ is merged into the cluster $\{4,5\}$. Then, again from Eq. (8.1.2), the dissimilarity matrix after this merger becomes

$$
\mathbf{D}_{(3)} =
\begin{bmatrix}
0 & 67.5 & 66.5 & 49.5 \\
67.5 & 0 & 76.0 & \mathbf{40.5} \\
66.5 & 76.0 & 0 & 64.5 \\
49.5 & \mathbf{40.5} & 64.5 & 0
\end{bmatrix}
\begin{matrix}
\{1,4,5\} \\ 2 \\ 3 \\ 6
\end{matrix}
\tag{8.1.22}
$$

By continuing in this manner until all observations are in the one cluster $\Omega$, the final hierarchy tree is obtained, as shown in Figure 8.4(a).

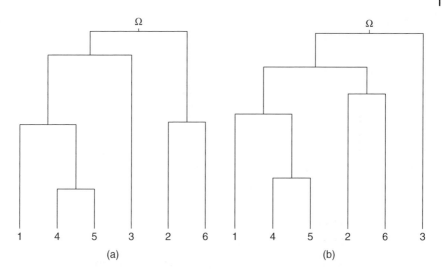

**Figure 8.4** Agglomerative hierarchy based on Ichino–Yaguchi dissimilarity: *E. coli* data (Example 8.4). (a) Complete-link (b) Single-link

However, had we used the single-link criterion of Eq. (8.1.1) or the average-link criterion of Eq. (8.1.3), the hierarchy trees would have a different structure from that in Figure 8.4(a) from the complete-link criterion, though both are similar to each other; see Figure 8.4(b) for the single-link hierarchy.                                                                □

**Example 8.5**  Suppose we want to build an agglomerative hierarchy on the *E. coli* data of Table 3.8, based on the span Euclidean non-normalized Ichino–Yaguchi dissimilarites of Table 3.13(i), calculated from Eqs. (3.1.9) and (3.3.9) in Example 3.13. Let us use the average-link criterion of Eq. (8.1.3). Thus, the initial dissimilarity matrix is

$$\mathbf{D} \equiv \mathbf{D}_{(1)} = \begin{bmatrix} 0 & 4.782 & 4.884 & 2.750 & 3.023 & 3.750 \\ 4.782 & 0 & 5.351 & 3.977 & 3.913 & 3.455 \\ 4.884 & 5.351 & 0 & 4.391 & 4.274 & 4.810 \\ 2.750 & 3.977 & 4.391 & 0 & \mathbf{1.202} & 3.947 \\ 3.023 & 3.913 & 4.274 & \mathbf{1.202} & 0 & 3.608 \\ 3.750 & 3.455 & 4.810 & 3.947 & 3.608 & 0 \end{bmatrix}. \quad (8.1.23)$$

The fourth and fifth observations are the closest; thus, the new dissimilarity matrix is

$$\mathbf{D}_{(2)} = \begin{bmatrix} 0 & 4.782 & 4.884 & \mathbf{2.887} & 3.750 \\ 4.782 & 0 & 5.351 & 3.945 & 3.455 \\ 4.884 & 5.351 & 0 & 4.333 & 4.810 \\ \mathbf{2.887} & 3.945 & 4.333 & 0 & 3.778 \\ 3.750 & 3.455 & 4.810 & 3.778 & 0 \end{bmatrix} \begin{array}{l} 1 \\ 2 \\ 3 \\ \{4,5\} \\ 6 \end{array} \quad (8.1.24)$$

where the dissimilarities between the cluster $\{4, 5\}$ and the other observations are obtained from Eq. (8.1.3), e.g., $d(2, \{4, 5\}) = (3.977 + 3.913)/2 = 3.945$.

From Eq. (8.1.24), we see that at the $r = 2$ stage, the clusters $u = 1$ and $\{4, 5\}$ are merged. The dissimilarity matrix becomes

$$\mathbf{D}_{(3)} = \begin{bmatrix} 0 & 5.351 & 4.224 & \mathbf{3.455} \\ 5.351 & 0 & 4.516 & 4.810 \\ 4.224 & 4.516 & 0 & 3.772 \\ \mathbf{3.455} & 4.810 & 3.772 & 0 \end{bmatrix} \begin{matrix} 2 \\ 3 \\ \{1, 4, 5\} \\ 6 \end{matrix} \qquad (8.1.25)$$

At the $r = 3$ stage, from Eq. (8.1.25), we see that the clusters are now $\{2, 6\}$, $\{3\}$, and $\{1, 4, 5\}$. The dissimilarity matrix becomes

$$\mathbf{D}_{(4)} = \begin{bmatrix} 0 & 5.081 & \mathbf{3.996} \\ 5.081 & 0 & 4.516 \\ \mathbf{3.996} & 4.516 & 0 \end{bmatrix} \begin{matrix} \{2, 6\} \\ 3 \\ \{1, 4, 5\} \end{matrix} \qquad (8.1.26)$$

where again from Eqs. (8.1.3) and (8.1.23), the respective dissimilarities can be obtained, e.g.,

$$d(\{2, 6\}, \{1, 4, 5\})$$

$$= \frac{1}{2 \times 3}(4.782 + 3.977 + 3.913 + 3.750 + 3.947 + 3.608) = 3.996.$$

Merging $\{2, 6\}$ with $\{1, 4, 5\}$ produces the dissimilarity matrix

$$\mathbf{D}_{(5)} = \begin{bmatrix} 0 & \mathbf{4.742} \\ \mathbf{4.742} & 0 \end{bmatrix} \begin{matrix} \{1, 2, 4, 5, 6\} \\ 3 \end{matrix} \qquad (8.1.27)$$

The final hierarchy is shown in Figure 8.5. Notice the height (along a $y$-axis, not shown) at which each merge occurs corresponds to the dissimilarity value that existed at that merger stage, highlighted in bold in Eqs. (8.1.23)–(8.1.27), respectively. □

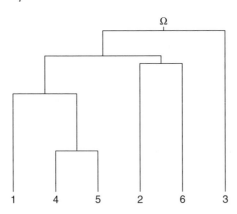

**Figure 8.5** Average-link agglomerative hierarchy based on span Euclidean non-normalized Ichino–Yaguchi dissimilarity: *E. coli* data (Example 8.5).

**Example 8.6** Consider again the *E. coli* interval data of Table 3.8. Suppose the tree is constructed based on the Euclidean dissimilarities for one variable $Y_1 = CG$ only (see Example 3.13 and Table 3.13). From Eq. (8.1.5), with $p = 1$, this initial dissimilarity matrix is

$$
\mathbf{D} \equiv \mathbf{D}_{(1)} = \begin{bmatrix}
0 & 380.25 & 12.25 & 156.25 & 169.00 & 30.25 \\
380.25 & 0 & 529.00 & 49.00 & 42.25 & 625.00 \\
12.25 & 529.00 & 0 & 256.00 & 272.25 & 4.00 \\
156.25 & 49.00 & 256.00 & 0 & \mathbf{0.25} & 324.00 \\
169.00 & 42.25 & 272.25 & \mathbf{0.25} & 0 & 342.25 \\
30.25 & 625.00 & 4.00 & 324.00 & 342.25 & 0
\end{bmatrix}.
$$

$$(8.1.28)$$

Thus, at the $r = 1$ stage, observations $u = 4$ and $u = 5$ are merged. This merged cluster has a centroid value of $\bar{\mathbf{Y}}_{4\cup5} = (90.25, 121.75, 79.25, 94.25)$. The new dissimilarity matrix becomes

$$
\mathbf{D}_{(2)} = \begin{bmatrix}
0 & 380.250 & 12.250 & 162.563 & 30.250 \\
380.25 & 0 & 529.00 & 45.563 & 625.00 \\
12.25 & 529.00 & 0 & 264.063 & \mathbf{4.00} \\
162.563 & 45.563 & 264.063 & 0 & 333.063 \\
30.25 & 625.00 & \mathbf{4.00} & 333.063 & 0
\end{bmatrix}
\begin{matrix}
1 \\ 2 \\ 3 \\ \{4,5\} \\ 6
\end{matrix}
$$

$$(8.1.29)$$

Now, the shortest dissimilarity is for $d(3, 6) = 4.0$. Hence, these observations are merged. The new dissimilarity matrix becomes

$$
\mathbf{D}_{(3)} = \begin{bmatrix}
0 & 380.250 & \mathbf{20.250} & 162.563 \\
380.25 & 0 & 576.00 & 45.563 \\
\mathbf{20.25} & 576.00 & 0 & 297.563 \\
162.563 & 45.563 & 297.563 & 0
\end{bmatrix}
\begin{matrix}
1 \\ 2 \\ \{3,6\} \\ \{4,5\}
\end{matrix}
$$

$$(8.1.30)$$

Then for this $r = 3$ stage, we merge $u = 1$ with $\{3, 6\}$, so that we now have the clusters $\{1, 3, 6\}$, $\{2\}$, and $\{4, 5\}$, with the cluster $\{1, 3, 6\}$ having the centroid $\bar{\mathbf{Y}}_{1\cup3\cup6} = (106, 114.3, 80, 99.83)$. The dissimilarity matrix becomes

$$
\mathbf{D}_{(4)} = \begin{bmatrix}
0 & 506.250 & 248.063 \\
506.250 & 0 & \mathbf{45.563} \\
248.063 & \mathbf{45.563} & 0
\end{bmatrix}
\begin{matrix}
\{1,3,6\} \\ 2 \\ \{4,5\}
\end{matrix}
$$

$$(8.1.31)$$

Therefore, $u = 2$ is merged with $\{4, 5\}$. The final tree is as shown in Figure 8.6(a). There are two distinct groups of observations here, one consisting of observations $\{1, 3, 6\}$ and the other with observations $\{2, 4, 5\}$. Indeed, the Euclidean distance between them is $d(\{1, 3, 6\}, \{2, 4, 5\}) = 676$. Yet, from Eqs. (8.1.28)–(8.1.31), in the construction of the tree, it is clear that

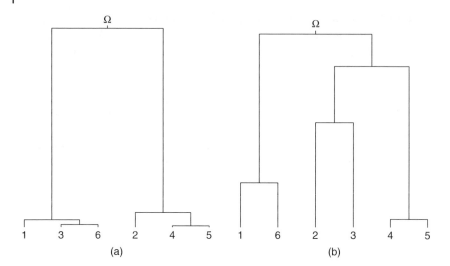

**Figure 8.6** Agglomerative hierarchy based on Ward criterion for $Y_1 = CG$: *E. coli* data (Example 8.6). (a) Euclidean distance (b) Ichino-Yaguchi dissimilarity

the observations within each of these two groups are relatively much closer together.

The increase in the total within-cluster variation can be found at each stage from Eq. (8.1.9). For example, the increase when merging the clusters $\{1, 3, 6\}$ and $\{2, 4, 5\}$ is

$$\Delta_{\{1,3,6\},\{2,4,5\}} = \frac{3 \times 3}{3 + 3}(106 - 88)^2 = 486,$$

where the centroid of the cluster $\{1, 3, 6\}$ is $\bar{Y}_1 = [(98 + 108)/2 + (88 + 125)/2 + (104 + 113)/2]/3 = 106$, and the centroid for $\{2, 4, 5\}$ is $\bar{Y}_1 = 88$, from Eq. (2.4.2). Similarly, the increase of total within-cluster variation at each stage can be found as

$$r = 1: \ \Delta_{4,5} = 0.125, \quad r = 2: \ \Delta_{3,6} = 2.000, \quad r = 3: \ \Delta_{1,\{3,6\}} = 13.500,$$
$$r = 4: \ \Delta_{2,\{4,5\}} = 30.375, \quad r = 5: \ \Delta_{\{1,3,6\},\{2,4,5\}} = 486.$$

As noted in section 8.1.1, the Ward criterion can be applied to any dissimilarity matrix. The plot in Figure 8.6(b) was obtained when this was done on the dissimilarity matrix obtained by using the Ichino–Yaguchi measure, on $Y_1$ only, and given in Table 3.13(a). It is immediately clear that the results are quite different, and that (as suggested by many authors) the interpretation is problematical. □

**Example 8.7** Consider the same *E. coli* species abundance interval observations of Example 8.6, and suppose now the agglomerative hierarchy uses the

**Table 8.1** Euclidean distances: *E. coli* data (Example 8.7)

| Object pair | By variable $Y_j$ $E_1$ | $E_2$ | $E_3$ | $E_4$ | Squared distance E | Distance $E^{1/2}$ |
|---|---|---|---|---|---|---|
| | (a) | (b) | (c) | (d) | (e) | (f) |
| 1-2 | 380.25 | 576.00 | 1.00 | 49.00 | 1006.25 | 31.721 |
| 1-3 | 12.25 | 4.00 | 196.00 | 36.00 | 248.25 | 15.756 |
| 1-4 | 156.25 | 4.00 | 36.00 | 100.00 | 296.25 | 17.212 |
| 1-5 | 169.00 | 20.25 | 72.25 | 110.25 | 371.75 | 19.281 |
| 1-6 | 30.25 | 576.00 | 72.25 | 64.00 | 742.50 | 27.249 |
| 2-3 | 529.00 | 484.00 | 225.00 | 1.00 | 1239.00 | 35.199 |
| 2-4 | 49.00 | 676.00 | 49.00 | 9.00 | 783.00 | 27.982 |
| 2-5 | 42.25 | 380.25 | 90.25 | 12.25 | 525.00 | 22.913 |
| 2-6 | 625.00 | 0.00 | 90.25 | 1.00 | 716.25 | 26.763 |
| 3-4 | 256.00 | 16.00 | 64.00 | 16.00 | 352.00 | 18.762 |
| 3-5 | 272.25 | 6.25 | 30.25 | 20.25 | 329.00 | 18.138 |
| 3-6 | 4.00 | 484.00 | 30.25 | 4.00 | 522.25 | 22.853 |
| 4-5 | 0.25 | 42.25 | 6.25 | 0.25 | 49.00 | 7.000 |
| 4-6 | 324.00 | 676.00 | 6.25 | 4.00 | 1010.25 | 31.784 |
| 5-6 | 342.25 | 380.25 | 0.00 | 6.25 | 728.75 | 26.995 |

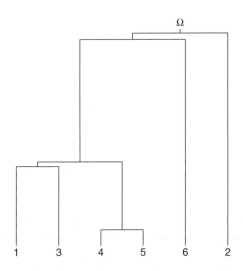

**Figure 8.7** Ward agglomerative hierarchy based on Euclidean distances: *E. coli* data (Example 8.7).

Ward criterion but on all four variables $Y_j, j = 1, \ldots, 4$. The squared Euclidean distances obtained from Eq. (8.1.5) are shown in Table 8.1 for each variable, $E_j$, in columns (a)–(d), respectively, and also the sum over all variables, $E = \sum_{j=1}^{4} E_j$, in column (e). The (non-squared) Euclidean distance of Eq. (8.1.12) is shown in column (f). We take the distances $E$ of column (e).

Then, following the steps taken in Example 8.6, we can obtain the agglomerative tree using the Ward criterion; the details are omitted. The resulting hierarchy is shown in Figure 8.7. □

### 8.1.4 Histogram-valued Observations

As for multi-valued list and interval-valued observations, when the data are histogram-valued, the agglomerative algorithm is based on the dissimilarity matrices. We illustrate briefly in the following examples, all of which relate to the airline data of Table 5.1.

**Example 8.8**  Consider the flight diagnostics data of Table 5.1 of Example 5.1. Recall that the variables are briefly $Y_1$ = air time, $Y_2$ = taxi-in time, $Y_3$ = arrival delay time, $Y_4$ = taxi-out time, and $Y_5$ = departure delay time. Let us take the first five and the last three airlines. We seek the agglomerative hierarchy obtained by applying the Ward criterion to the Euclidean distances between the cluster centroids. Calculation of the Euclidean distances between these $m = 8$ airlines, from Eq. (8.1.7) with $p = 5$, produces the distance matrix

$$\mathbf{D} \equiv \mathbf{D}_{(1)}$$

$$= \begin{bmatrix}
0 & 8824.59 & 16304.23 & 278.81 & 983.66 & \mathbf{63.34} & 16532.62 & 7010.14 \\
8824.59 & 0 & 1161.82 & 7228.07 & 4709.54 & 7744.56 & 1305.38 & 288.67 \\
16304.23 & 1161.82 & 0 & 14095.38 & 10377.03 & 14813.59 & 183.21 & 2252.45 \\
278.81 & 7228.07 & 14095.38 & 0 & 358.56 & 126.12 & 13963.67 & 5341.38 \\
983.66 & 4709.54 & 10377.03 & 358.56 & 0 & 587.92 & 10202.84 & 3280.88 \\
\mathbf{63.34} & 7744.56 & 14813.59 & 126.12 & 587.92 & 0 & 14950.95 & 6021.25 \\
16532.62 & 1305.38 & 183.21 & 13963.67 & 10202.84 & 14950.95 & 0 & 2137.07 \\
7010.14 & 288.67 & 2252.45 & 5341.38 & 3280.88 & 6021.25 & 2137.07 & 0
\end{bmatrix}.$$

$$(8.1.32)$$

From Eq. (8.1.32), the first merge occurs between the observations $u = 1$ and $u = 6$; the height of the hierarchy tree is $h = 63.34$. Proceeding with the analogous detail described in Example 8.6, the hierarchical tree of Figure 8.8 emerges. The heights at each merge are as shown in Table 8.2, where, for example, when the clusters $\{1, 6\}$ and $\{4\}$ are merged at the $r = 3$ stage, the tree height is $h = 186.63$. It is clear, from Figure 8.8, that these eight airlines fall into two distinct classes. Closer examination reveals that the airlines $\{1, 4, 5, 6\}$ (which are the airlines $\{1, 4, 5, 14\}$ of Table 5.1) are those

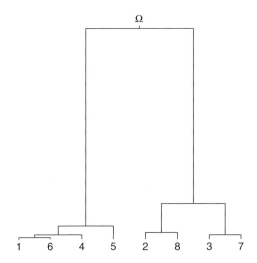

**Figure 8.8** Ward agglomerative hierarchy based on Euclidean distances: airlines data (Example 8.8).

**Table 8.2** Tree construction: airline data (Example 8.8)

| Stage $r$ | Clusters | Height $h$ |
|---|---|---|
| 1 | $\{1,6\},\{2\},\{3\},\{4\},\{5\},\{7\},\{8\}$ | 63.34 |
| 2 | $\{1,6\},\{2\},\{3,7\},\{4\},\{5\},\{8\}$ | 183.21 |
| 3 | $\{1,4,6\},\{2\},\{3,7\},\{5\},\{8\}$ | 186.63 |
| 4 | $\{1,4,6\},\{2,8\},\{3,7\},\{5\}$ | 288.67 |
| 5 | $\{1,4,5,6\},\{2,8\},\{3,7\}$ | 591.35 |
| 6 | $\{1,4,5,6\},\{2,3,7,8\}$ | 1596.21 |
| 7 | $\Omega = \{1,2,3,4,5,6,7,8\}$ | 9479.61 |

involved with long-distance flights, in contrast with the shorter distance flights of airlines $\{2,3,7,8\}$ (which correspond with the airlines $\{2,3,15,16\}$ of Table 5.1), although the other variables ($Y_j, j = 2,3,4,5$) also play a role in this construction. The histogram means for each variable and each airline given in Table 5.4 also help in the interpretation of the tree formation (see also Exercise 8.8). □

**Example 8.9** Suppose we take the same $m = 8$ airlines considered in Example 8.8, and suppose we seek the agglomerative hierarchy built from the normalized cumulative density dissimilarities. These were calculated from Eq. (4.2.47) and are displayed in Table 5.3 (for Example 5.1). Extracting the relevant dissimilarities for these eight airlines produces the dissimilarity matrix

$$\mathbf{D} \equiv \mathbf{D}_{(1)}$$

$$= \begin{bmatrix}
0 & 0.261 & 0.318 & 0.210 & 0.166 & 0.111 & 0.392 & 0.385 \\
0.261 & 0 & 0.115 & 0.241 & 0.201 & 0.258 & 0.141 & 0.206 \\
0.318 & 0.115 & 0 & 0.304 & 0.270 & 0.309 & \mathbf{0.093} & 0.237 \\
0.210 & 0.241 & 0.304 & 0 & 0.207 & 0.206 & 0.292 & 0.270 \\
0.166 & 0.201 & 0.270 & 0.207 & 0 & 0.179 & 0.295 & 0.306 \\
0.111 & 0.258 & 0.309 & 0.206 & 0.179 & 0 & 0.378 & 0.369 \\
0.392 & 0.141 & \mathbf{0.093} & 0.292 & 0.295 & 0.378 & 0 & 0.185 \\
0.385 & 0.206 & 0.237 & 0.270 & 0.306 & 0.369 & 0.185 & 0
\end{bmatrix}.$$

$$(8.1.33)$$

With these dissimilarities, the first merger is now between observations $u = 3$ and $u = 7$. Using the single-link criterion of Eq. (8.1.1), we eventually obtain the tree of Figure 8.9(a). The average-link criterion of Eq. (8.1.3) gives the tree of Figure 8.9(b). The details are omitted (see Exercise 8.8).

Comparing Figure 8.9(a) with Figure 8.8, except for the airline $u = 4$, we again observe two separate classes although the merging sequence differs depending on the clustering criterion. Airline $u = 4$ is, however, quite distinct. From the standard deviation values for each airline, given in Table 5.4, it is clear that the internal variation for the flight time variable ($Y_1$) for this airline is very much larger than for the other airlines. This phenomenon reflects the fact that both these internal variations along with the histogram means enter into the normalized cumulative density dissimilarities. From Figure 8.9(b), it is seen that

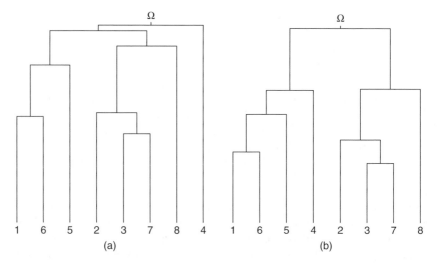

**Figure 8.9** Agglomerative hierarchy based on CDF distance: airlines data (Example 8.9). (a) Single-link (b) Average-link.

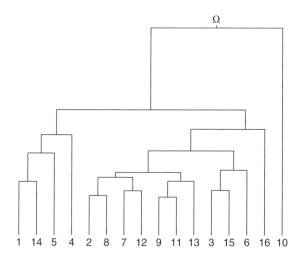

**Figure 8.10** Average-link agglomerative hierarchy based on CDF distance: airlines data (Example 8.10).

this separation of airline $u = 4$ does not occur when the hierarchy was built from the average-link criterion. □

**Example 8.10**  Consider all $m = 16$ airlines in Table 5.1, and the complete normalized cumulative density dissimilarities of Table 5.3. The hierarchy that pertains when using the average-link criterion Eq. (8.1.3) produces the tree of Figure 8.10. We omit the details (see Exercise 8.9).

When we compare Figures 8.9(b) and 8.10, we again see the impact of the $Y_1$ variable, with the separation of the longer flight airlines ({1, 4, 5, 14}) from the other airlines except for the airline $u = 10$, which is quite different again. This latter airline (see the sample diagnostics of Table 5.4) flies extremely long flight routes in comparison to the other 15 airlines. □

### 8.1.5  Mixed-valued Observations

The preceding cases were all distinguished by the fact that all variables were described by the same type of symbolic data. However, the agglomerative algorithm also works for data of different types across variables, i.e., for mixed-valued variables. This is illustrated briefly in the following example.

**Example 8.11**  Consider the mixed-valued variable census data of Table 7.23 and Example 7.14. Recall there are $p = 6$ variables, with $Y_1 =$ age and $Y_2 =$ home value being histogram-valued variables, $Y_3 =$ gender and $Y_5 =$ tenure being modal multi-valued variables, $Y_4 =$ fuel being a non-modal multi-valued variable, and $Y_6 =$ income being an interval-valued variable. Let us construct an agglomerative hierarchy tree based on the Gowda–Diday dissimilarity matrix of Eq. (7.3.10), calculated over all variables.

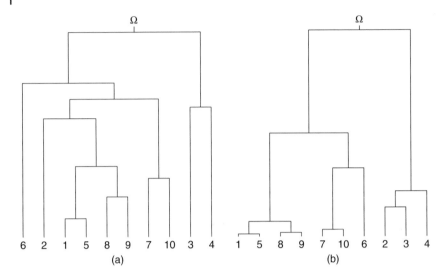

**Figure 8.11** Mixed variable census data: Gowda–Diday dissimilarity (Example 8.11). (a) Average-link (b) Ward criterion.

Then the tree built by the average-link agglomerative algorithm produces the tree of Figure 8.11(a). When the Ward criterion is used, the hierarchy tree becomes that of Figure 8.11(b). The details are left to the reader. □

### 8.1.6 Interval Observations with Rules

It is often the case that rules have to be applied to maintain the integrity of the data. This is particularly relevant when the symbolic-valued data are obtained by aggregating other observations. In these circumstances, the apparent data have to be transformed into virtual observations which adhere to the underlying rules. One consequence is that even when the apparent data are interval-valued, the virtual data can become histogram-valued (at least for some, not necessarily for all, observations). Subsequent analyses are performed on the virtual data set. This is evident in Examples 8.12 and 8.13.

**Example 8.12** This methodology is illustrated in the baseball data set of Table 8.3 (adapted from Billard and Diday (2006a,b); the original data are from Vanessa and Vanessa (2004)). Team observations result from aggregating individual statistics. The goal is to execute an agglomerative clustering procedure to these data.

There are two variables, $Y_1 = $ # at-bats and $Y_2 = $ # hits. Clearly, $Y_1 \geq Y_2$. Under this rule, not all apparent observations in the data rectangles in Table 8.3 are applicable. There are no problems with the first team. However, for the second team, values in the triangular region bounded by the points

**Table 8.3** At-bats and hits by team (Example 8.12)

| $u$ Team | $Y_1$ #At-bats | $Y_2$ #Hits | $u$ Team | $Y_1$ # At-bats | $Y_2$ #Hits |
|---|---|---|---|---|---|
| 1 | [289, 538] | [75, 162] | 11 | [177, 245] | [189, 238] |
| 2 | [88, 422] | [49, 149] | 12 | [342, 614] | [121, 206] |
| 3 | [189, 223] | [201, 254] | 13 | [120, 439] | [35, 102] |
| 4 | [184, 476] | [46, 148] | 14 | [80, 468] | [55, 115] |
| 5 | [283, 447] | [86, 115] | 15 | [75, 110] | [75, 110] |
| 6 | [168, 445] | [37, 135] | 16 | [116, 557] | [95, 163] |
| 7 | [123, 148] | [137, 148] | 17 | [197, 507] | [52, 53] |
| 8 | [256, 510] | [78, 124] | 18 | [167, 203] | [48, 232] |
| 9 | [101, 126] | [101, 132] | 19 | [24, 26] | [133, 141] |
| 10 | [212, 492] | [57, 151] | | | |

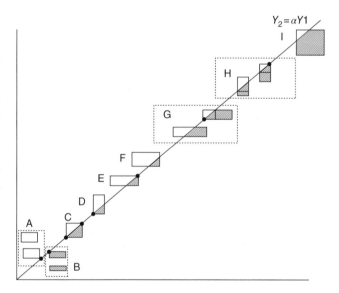

**Figure 8.12** Virtual patterns: baseball data (Example 8.12).

$(Y_1, Y_2) = (88, 88)$, $(88,149)$, $(149,149)$, are not feasible. Therefore, for this team, the virtual observation is the hypercube that results when this triangular region is omitted. There are nine possible patterns for the virtual regions, shown in Figure 8.12. The first team fits pattern B, while the second team fits pattern I. From these patterns, it is easy to calculate the corresponding

**Table 8.4** Virtual $\xi_1'$ and $\xi_2'$ under rule $v : Y_1 \geq Y_2$ (Example 8.12)

| Team $u$ | Pattern | Virtual $\xi_1'$ : $Y_1 = $ # At-bats | Virtual $\xi_2'$ : $Y_2 = $ # Hits |
|---|---|---|---|
| 1 | B | [289, 538] | [75, 162] |
| 2 | I | {[88, 149), 0.134; [149, 422], 0.866} | {[49, 88), 0.413; [88, 149], 0.587} |
| 3 | F | [201, 223] | [201, 223] |
| 4 | B | [184, 476] | [46, 148] |
| 5 | B | [283, 447] | [86, 115] |
| 6 | B | [168, 445] | [37, 135] |
| 7 | E | [137, 148] | [137, 148] |
| 8 | B | [256, 510] | [78, 124] |
| 9 | D | [101, 126] | [101, 126] |
| 10 | B | [212, 492] | [57, 151] |
| 11 | G | {[189, 238), 0.778; [238, 245], 0.222} | [189, 238] |
| 12 | B | [342, 614] | [121, 206] |
| 13 | B | [120, 439] | [35, 102] |
| 14 | I | {[80, 115), 0.066; [115, 468], 0.934} | {[55, 80), 0.428; [80, 115] 0.572} |
| 15 | C | [75, 110] | [75, 110] |
| 16 | I | {[116, 163), 0.072; [163, 557], 0.928} | {[95, 116), 0.321; [116, 163] 0.679} |
| 17 | B | [197, 507] | [52, 53] |
| 18 | H | [167, 203] | {[48, 167) 0.869; [167, 232], 0.131} |
| 19 | A | $\phi$ | $\phi$ |

virtual values for the observations. More details of these virtual derivations can be found in Billard and Diday (2006b). These are displayed in Table 8.4. Notice that some of these are now histogram-valued observations. Also notice that the virtual values for the last team form an empty set; thus, we omit this observation in the sequel. This leaves us with $m = 18$ teams.

Suppose we want to perform an agglomerative algorithm, based on the normalized Euclidean extended Ichino–Yaguchi dissimilarities. We know from section 4.2 that since the data are histogram-valued, we first need to convert them into transformed histograms where the histogram subintervals are the same for all observations, achieved by application of Eqs. (4.2.2)–(4.2.6) (see section 4.2.1).

Then, from Eqs. (2.4.3) and (2.4.11), we calculate the mean, variance, and standard deviations for each of $u_1$ and $u_2$, and from Eqs. (4.2.16), (4.2.18), (4.2.21), and (4.2.22), we calculate the mean, variance, and standard deviations for the union $u_1 \cup u_2$ and the intersection $u_1 \cap u_2$, for each pair of observations

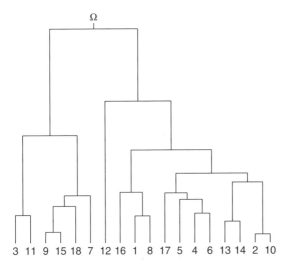

**Figure 8.13** Complete-link agglomerative hierarchy based on normalized extended Ichina–Yaguchi dissimilarity: baseball data (Example 8.12).

$u_1, u_2 = 1, \ldots, m = 18$. The union and intersection observations are found from Eqs. (4.2.10) and (4.2.12), respectively.

The basic extended Ichino–Yaguchi dissimilarity is found from Eq. (4.2.29). This can be used to calculate a variety of other dissimilarities. The normalized extended Ichino–Yaguchi dissimilarity matrix, with elements obtained from substituting the basic extended extended Ichino–Yaguchi dissimilarity into Eqs. (4.2.30)–(4.2.34), for $\gamma = 0.25$, is displayed in Table 8.5.

The agglomerative algorithm using the complete-link method is shown in Figure 8.13. Other methods, such as the single-link, average-link, and Ward's criterion, produce different hierarchical trees, though the teams $\{3, 7, 9, 11, 15, 18\}$ form a larger cluster in all cases. These teams are characterized by the fact that their ratio of hits to at-bats is quite high. □

**Example 8.13** The analysis described in Example 8.12 is essentially a non-adaptive clustering analysis. Suppose now after each iteration, prototype "observations" are determined for the newly formed cluster, with new dissimilarities calculated based on the new set of cluster observations. In the baseball example, the observations $u = 2$ and $u = 10$ were merged from the first $r = 1$ iteration. If these two observations are averaged, the resulting prototype corresponds to the transformed histogram observation $\{2, 10\}$ with realization $(Y_1, Y_2) = (\{[75, 200), 0.148; [200, 325), 0.400; [325, 450), 0.377; [450, 575], 0.075\}, \{[35, 85), 0.340; [85, 135), 0.508; [138, 185], 0.152\})$. The normalized extended Ichino–Yaguchi dissimilarity matrix for the effective $m = 17$ clusters is shown in Table 8.6, for $\gamma = 0.25$. By comparing Tables 8.5 and 8.6, we see that dissimilarities between any of the other observations have

**Table 8.5** Normalized Euclidean extended Ichino–Yaguchi dissimilarities: baseball data (Example 8.12)

| u | 1 | 2 | 3 | 4 | 5 | 6 | 7 | 8 | 9 | 10 | 11 | 12 | 13 | 14 | 15 | 16 | 17 | 18 |
|---|---|---|---|---|---|---|---|---|---|----|----|----|----|----|----|----|----|----|
| 1 | 0 | 68.435 | 84.754 | 36.221 | 31.735 | 53.619 | 133.555 | 15.150 | 132.680 | 23.641 | 93.967 | 38.672 | 66.402 | 61.769 | 132.588 | 33.866 | 36.211 | 123.403 |
| 2 | 68.435 | 0 | 68.865 | 31.435 | 42.925 | 16.978 | 70.558 | 55.649 | 65.619 | 0.000 | 61.621 | 107.904 | 11.668 | 10.485 | 64.269 | 38.572 | 49.025 | 54.149 |
| 3 | 84.754 | 68.865 | 0 | 70.658 | 63.283 | 69.948 | 63.988 | 78.627 | 75.716 | 70.965 | 15.617 | 109.563 | 77.758 | 78.265 | 80.133 | 76.609 | 88.454 | 68.021 |
| 4 | 36.221 | 31.435 | 70.658 | 0 | 28.163 | 17.111 | 98.703 | 24.291 | 95.155 | 12.544 | 71.452 | 75.125 | 29.361 | 25.407 | 94.243 | 29.224 | 23.575 | 83.888 |
| 5 | 31.735 | 42.925 | 63.283 | 28.163 | 0 | 26.958 | 105.879 | 20.118 | 103.178 | 24.577 | 72.656 | 68.750 | 39.726 | 44.193 | 103.543 | 44.743 | 33.741 | 97.100 |
| 6 | 53.619 | 16.978 | 69.948 | 17.111 | 26.958 | 0 | 86.371 | 39.325 | 79.488 | 29.718 | 66.849 | 94.289 | 12.452 | 20.579 | 78.575 | 37.142 | 32.612 | 69.551 |
| 7 | 133.555 | 70.558 | 63.988 | 98.703 | 105.879 | 86.371 | 0 | 122.499 | 23.764 | 110.846 | 52.998 | 163.307 | 79.998 | 81.810 | 31.396 | 103.862 | 117.697 | 27.890 |
| 8 | 15.150 | 55.649 | 78.627 | 24.291 | 20.118 | 39.325 | 122.499 | 0 | 119.384 | 13.392 | 87.260 | 54.452 | 51.544 | 47.171 | 119.291 | 26.447 | 21.321 | 111.768 |
| 9 | 132.680 | 65.619 | 75.716 | 95.155 | 103.178 | 79.488 | 23.764 | 119.384 | 0 | 108.363 | 67.026 | 164.875 | 70.585 | 74.358 | 8.687 | 103.454 | 110.518 | 23.307 |
| 10 | 23.641 | 0.000 | 70.965 | 12.544 | 24.577 | 29.718 | 110.846 | 13.392 | 108.363 | 0 | 79.059 | 62.794 | 42.117 | 38.602 | 107.609 | 29.660 | 22.225 | 99.424 |
| 11 | 93.967 | 61.621 | 15.617 | 71.452 | 72.656 | 66.849 | 52.998 | 87.260 | 67.026 | 79.059 | 0 | 118.770 | 71.593 | 70.377 | 72.007 | 65.482 | 94.118 | 57.621 |
| 12 | 38.672 | 107.904 | 109.563 | 75.125 | 68.750 | 94.289 | 163.307 | 54.452 | 164.875 | 62.794 | 118.770 | 0 | 107.400 | 101.713 | 166.042 | 69.927 | 75.040 | 158.339 |
| 13 | 66.402 | 11.668 | 77.758 | 29.361 | 39.726 | 12.452 | 79.998 | 51.544 | 70.585 | 42.117 | 71.593 | 107.400 | 0 | 10.374 | 68.947 | 42.372 | 41.506 | 61.578 |
| 14 | 61.769 | 10.485 | 78.265 | 25.407 | 44.193 | 20.579 | 81.810 | 47.171 | 74.358 | 38.602 | 70.377 | 101.713 | 10.374 | 0 | 73.337 | 35.735 | 40.261 | 64.851 |
| 15 | 132.588 | 64.269 | 80.133 | 94.243 | 103.543 | 78.575 | 31.396 | 119.291 | 8.687 | 107.609 | 72.007 | 166.042 | 68.947 | 73.337 | 0 | 104.408 | 109.176 | 16.765 |
| 16 | 33.866 | 38.572 | 76.609 | 29.224 | 44.743 | 37.142 | 103.862 | 26.447 | 103.454 | 29.660 | 65.482 | 69.927 | 42.372 | 35.735 | 104.408 | 0 | 41.083 | 92.484 |
| 17 | 36.211 | 49.025 | 88.454 | 23.575 | 33.741 | 32.612 | 117.697 | 21.321 | 110.518 | 22.225 | 94.118 | 75.040 | 41.506 | 40.261 | 109.176 | 41.083 | 0 | 103.651 |
| 18 | 123.403 | 54.149 | 68.021 | 83.888 | 97.100 | 69.551 | 27.890 | 111.768 | 23.307 | 99.424 | 57.621 | 158.339 | 61.578 | 64.851 | 16.765 | 92.484 | 103.651 | 0 |

**Table 8.6** Normalized Euclidean extended Ichino–Yaguchi dissimilarities: baseball data $r = 1$ (Example 8.13)

| u | 1 | {2,10} | 3 | 4 | 5 | 6 | 7 | 8 | 9 | 11 | 12 | 13 | 14 | 15 | 16 | 17 | 18 |
|---|---|--------|---|---|---|---|---|---|---|----|----|----|----|----|----|----|----|
| 1 | 0 | 85.827 | 80.666 | 45.445 | 58.953 | 141.677 | 132.593 | 132.593 | 155.742 | 89.012 | 71.468 | 86.998 | 144.919 | 136.208 | 40.098 | 125.013 | 131.231 |
| {2,10} | 85.827 | 0 | 66.774 | 43.788 | 50.345 | 60.773 | 104.190 | 67.952 | 86.379 | 113.130 | 90.762 | 53.863 | 60.896 | 107.847 | 55.111 | 56.759 | 88.250 |
| 3 | 80.666 | 66.774 | 0 | 51.719 | 45.333 | 101.455 | 63.546 | 111.670 | 74.914 | 70.767 | 110.659 | 49.700 | 107.102 | 64.439 | 67.738 | 98.309 | 52.815 |
| 4 | 45.445 | 43.788 | 51.719 | 0 | 33.891 | 104.400 | 95.985 | 105.613 | 108.555 | 78.269 | 77.844 | 50.363 | 106.066 | 103.541 | 34.588 | 95.621 | 89.450 |
| 5 | 58.953 | 50.345 | 45.333 | 33.891 | 0 | 82.053 | 110.230 | 80.845 | 107.392 | 102.305 | 64.648 | 41.374 | 89.165 | 107.468 | 52.724 | 71.400 | 95.524 |
| 6 | 141.677 | 60.773 | 101.455 | 104.400 | 82.053 | 0 | 142.772 | 38.793 | 91.217 | 165.494 | 110.517 | 74.162 | 20.577 | 129.400 | 112.040 | 31.216 | 108.393 |
| 7 | 132.593 | 104.190 | 63.546 | 95.985 | 110.230 | 142.772 | 0 | 168.651 | 70.222 | 65.142 | 170.421 | 82.896 | 139.949 | 21.793 | 103.312 | 154.543 | 34.115 |
| 8 | 132.593 | 67.952 | 111.670 | 105.613 | 80.845 | 38.793 | 168.651 | 0 | 127.984 | 177.291 | 82.326 | 88.919 | 46.683 | 157.492 | 110.111 | 12.932 | 139.495 |
| 9 | 155.742 | 86.379 | 74.914 | 108.555 | 107.392 | 91.217 | 70.222 | 127.984 | 0 | 125.263 | 165.200 | 77.512 | 86.734 | 52.170 | 119.115 | 113.645 | 37.110 |
| 11 | 89.012 | 113.130 | 70.767 | 78.269 | 102.305 | 165.494 | 65.142 | 177.291 | 125.263 | 0 | 152.634 | 98.674 | 166.783 | 82.652 | 76.638 | 168.259 | 87.394 |
| 12 | 71.468 | 90.762 | 110.659 | 77.844 | 64.648 | 110.517 | 170.421 | 82.326 | 165.200 | 152.634 | 0 | 99.276 | 116.787 | 166.037 | 79.571 | 80.354 | 158.034 |
| 13 | 86.998 | 53.863 | 49.700 | 50.363 | 41.374 | 74.162 | 82.896 | 88.919 | 77.512 | 98.674 | 99.276 | 0 | 73.860 | 73.998 | 50.318 | 74.435 | 59.023 |
| 14 | 144.919 | 60.896 | 107.102 | 106.066 | 89.165 | 20.577 | 139.949 | 46.683 | 86.734 | 166.783 | 116.787 | 73.860 | 0 | 126.279 | 111.502 | 39.250 | 105.388 |
| 15 | 136.208 | 107.847 | 64.439 | 103.541 | 107.468 | 129.400 | 21.793 | 157.492 | 52.170 | 82.652 | 166.037 | 73.998 | 126.279 | 0 | 105.372 | 141.328 | 19.211 |
| 16 | 40.098 | 55.111 | 67.738 | 34.588 | 52.724 | 112.040 | 103.312 | 110.111 | 119.115 | 76.638 | 79.571 | 50.318 | 111.502 | 105.372 | 0 | 101.458 | 94.193 |
| 17 | 125.013 | 56.759 | 98.309 | 95.621 | 71.400 | 31.216 | 154.543 | 12.932 | 113.645 | 168.259 | 80.354 | 74.435 | 39.250 | 141.328 | 101.458 | 0 | 123.283 |
| 18 | 131.231 | 88.250 | 52.815 | 89.450 | 95.524 | 108.393 | 34.115 | 139.495 | 37.110 | 87.394 | 158.034 | 59.023 | 105.388 | 19.211 | 94.193 | 123.283 | 0 |

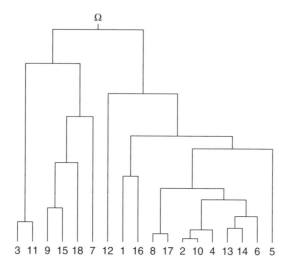

$\Omega$

3 11 9 15 18 7 12 1 16 8 17 2 10 4 13 14 6 5

**Figure 8.14** Adaptive agglomerative hierarchy based on normalized extended Ichina–Yaguchi dissimilarity: baseball data (Example 8.13).

also changed. This is because the normalizing terms (in Eqs. (4.2.31)–(4.2.34)) now have different values.

Then, the agglomerative algorithm can be applied once more. At the second $r = 2$ iteration, the observations $u = 8$ and $u = 17$ are merged into one cluster. Then, the prototype observation based on $u = 8$ and $u = 17$ is determined, and the new dissimilarities generated as before, ready for the next iteration. Proceeding in this way at each iteration, we obtain the hierarchy tree shown in Figure 8.14.

The two hierarchies of Figures 8.13 and 8.14 differ in some respects. Both contain three major sub-clusters, though their construction occurs at differing stages. One major sub-cluster consisting of observations $u = \{3, 7, 9, 11, 15, 18\}$ appears in both trees, but it is also clear from these hierarchies that their mergers occur at different iterations. These are teams that tend to have lower numbers for $Y_1 = $ # at-bats. The sub-cluster consisting of $\{1, 8, 16\}$ has more at-bats but still tends to have relatively low values for $Y_2 = $ # hits. The sub-cluster with observations $u = \{2, 4, 5, 6, 10, 13, 14, 17\}$ are teams which tend to have larger values for $Y_1 = $ # at-bats. The team $u = 12$ is set apart by having a large number of at-bats but scoring a relatively low number of hits. Important, however, is the fact that despite reasonably common major sub-clusters, the tree construction proceeds differently due to the adaptive nature of the algorithm through the calculation of the resulting prototypes at each stage. □

## 8.2   Pyramidal Clustering

Pure hierarchical clustering, by either divisive or agglomerative methods, produces non-overlapping clusters or partitions. Pyramid cluster methodology allows for overlapping clusters to occur. It is an agglomerative procedure. Pyramids for classical observations were first introduced by Bertrand (1986), Diday (1984, 1986b), and Diday and Bertrand (1986).

One approach to constructing a pyramid is based on generality degrees (defined in section 8.2.1) as explained in Brito (1991, 1994, 2000) and Brito and Diday (1990), and described in section 8.2.2. Pyramids can also be built along the lines of the principles of standard agglomerative methods based on dissimilarities (see section 8.2.3).

### 8.2.1   Generality Degree

A generality degree is a measure between observations that is particularly applicable to the construction of pyramids. More formally, let us define the notions of extent, completeness, and then generality degree in the following three definitions. An expanded discussion, along with numerous examples, can be found in Bock and Diday (2000) and Billard and Diday (2006a).

**Definition 8.9**   For observations $Y_1$ and $Y_2$ taking values in $\mathcal{Y} = \mathcal{Y}_1 \times \mathcal{Y}_2$, the **extent**, $Ext(\cdot)$, is the set of values in either or both $Y_1$ and $Y_2$. Furthermore, if the interest is in the descriptions $Y_1 \in \mathcal{Y}_1 \subseteq \mathcal{Y}$ and $Y_2 \in \mathcal{Y}_2 \subseteq \mathcal{Y}$, then the extent is the set of values in $\mathcal{Y}_1 \cup \mathcal{Y}_2 \subseteq \mathcal{Y}$.                                       □

**Example 8.14**   Suppose we have a multi-valued list variable taking values from $\mathcal{Y} = \{\text{coal}, \text{electricity}, \text{gas}, \text{wood}, \text{other}\}$, and suppose there are two observations, $Y_1 = \{\text{coal}, \text{electricity}\}$ and $Y_2 = \{\text{coal}, \text{gas}, \text{wood}\}$. Then the extent of $Y_1$ and $Y_2$ is

$$Ext(Y_1, Y_2) = \{\text{coal}, \text{electricity}\} \wedge \{\text{coal}, \text{gas}, \text{wood}\}$$
$$= \{\text{coal}, \text{electricity}, \text{gas}, \text{wood}\}.$$

□

**Example 8.15**   Suppose $Y$ is an interval-valued variable taking values from $\mathbb{R}$. Suppose there are two observations $Y_1 = [98, 108]$ and $Y_2 = [88, 125]$. Then the extent of $Y_1$ and $Y_2$ is

$$Ext(Y_1, Y_2) = [98,108] \wedge [88,125] = [88,125].$$

easo3ning

However, if interest is in the extent of the two descriptions $Y_1 \geq 100$ and $Y_2 \geq 95$, then the extent becomes

$$Ext(Y_1, Y_2) = (Y_1 \geq 100) \wedge (Y_2 \geq 95) = [100,108] \wedge [95,125] = [95,125].$$

$\square$

**Definition 8.10** Suppose we have a set of $m$ observation in $\Omega$, and suppose the cluster $C \subseteq \Omega$ has $m_c$ observations, described by the realizations $\mathbf{Y}_1, \dots, \mathbf{Y}_{m_c}$ where each $\mathbf{Y}_u = (Y_{u1}, \dots, Y_{up})$, $u = 1, \dots, m_c$, is a $p$-dimensional observation. Suppose the symbolic assertion object $s$ is defined by

$$s = f(C) = \bigwedge_{j=1}^{p} [Y_j \subseteq_{u \in C} Y_{uj}] \tag{8.2.1}$$

where the mapping $f(s) = Ext(s|C)$. Suppose $g = g(\Omega)$ is the mapping with $g(s) = Ext(s|\Omega)$. Let $h$ be the iterated mapping which satisfies

$$h = f \ o \ g \text{ gives } h(s) = f(g(s))$$

$$h' = g \ o \ f \text{ gives } h'(C) = g(f(C)).$$

Then, this symbolic $s$ is **complete** if and only if $h(s) = s$. $\square$

**Definition 8.11** Let $\mathbf{Y} = (Y_1, \dots, Y_p)$ be a $p$-dimensional observation with $Y_j$ taking values from $\mathcal{Y}_j$, $j = 1, \dots, p$. Suppose the object $s$ has description

$$s = \bigwedge_{j=1}^{p} [Y_j \in D_j], \ D_j \subseteq \mathcal{Y}_j, \ j = 1, \dots, p. \tag{8.2.2}$$

Then, the **generality degree** of $s$, $G(s)$, is

$$G(s) = \prod_{j=1}^{p} \frac{c(D_j)}{c(\mathcal{Y}_j)}; \tag{8.2.3}$$

if $Y_j$ is a multi-valued list observation from the list of possible categories $\mathcal{Y}_j$, then $c(A)$ is the cardinality of $A$, while if $Y_j$ is an interval-valued observation from $\mathbb{R}$, then $c(A)$ is the length of the interval $A$. $\square$

A generality degree is not a dissimilarity measure (since, e.g., $G(s,s) \neq 0$).

**Example 8.16** Consider the *E. coli* data set of Table 3.8 used in Examples 3.10 and 8.2–8.7. We want to calculate the generality degrees as well as identify which mergers of observations are complete.

The base of the pyramid consists of the six observations $u = 1, \dots, 6$. Obviously at the first step, two single observations are merged to form a pair. Then at the second step, three-wise merges become possible, and so on. All possible two-, three-, four- and five-wise merges are shown in Table 8.7. The second column Code (with values $u_1 u_2 \dots$, where $u_k = 1, \dots, 6$, $k = 1, 2, \dots$)

**Table 8.7** Extents of possible clusters for *E. coli* data set (Example 8.16)

| s | Code | $Y_1$ $[a_1, b_1]$ | $Y_2$ $[a_2, b_2]$ | $Y_3$ $[a_3, b_3]$ | $Y_4$ $[a_4, b_4]$ | Complete ✓ |
|---|------|-----------|-----------|----------|----------|---|
| 1 | 12 | [59, 108] | [91, 126] | [51, 92] | [96, 109] | ✓ |
| 2 | 13 | [88, 125] | [88, 154] | [69, 91] | [85, 112] | |
| 3 | 14 | [80,108] | [117, 133] | [68, 89] | [92, 109] | ✓ |
| 4 | 15 | [81, 108] | [108, 129] | [69, 86] | [93, 109] | ✓ |
| 5 | 16 | [98, 113] | [96, 126] | [67, 95] | [94, 109] | ✓ |
| 6 | 23 | [59, 125] | [88, 154] | [51, 92] | [85, 112] | |
| 7 | 24 | [59, 108] | [91, 133] | [51, 92] | [92, 99] | |
| 8 | 25 | [59, 108] | [91, 129] | [51, 92] | [93, 99] | |
| 9 | 26 | [59, 113] | [91, 107] | [51, 95] | [94, 99] | ✓ |
| 10 | 34 | [80, 125] | [88, 154] | [68, 91] | [85, 112] | |
| 11 | 35 | [81, 125] | [88, 154] | [76, 91] | [85, 112] | ✓ |
| 12 | 36 | [88, 125] | [88, 154] | [67, 95] | [85, 112] | |
| 13 | 45 | [80, 101] | [108, 133] | [68, 89] | [92, 97] | ✓ |
| 14 | 46 | [80, 113] | [96, 133] | [67, 95] | [92, 99] | |
| 15 | 56 | [81, 113] | [96, 129] | [67, 95] | [93, 99] | |
| 16 | 123 | [59, 125] | [88, 154] | [51, 92] | [85, 112] | |
| 17 | 124 | [59, 108] | [91, 133] | [51, 92] | [92, 109] | |
| 18 | 125 | [59, 108] | [91, 129] | [51, 92] | [93, 109] | |
| 19 | 126 | [59, 113] | [91, 126] | [51, 95] | [94, 109] | ✓ |
| 20 | 134 | [80, 125] | [88, 154] | [68, 91] | [85, 112] | |
| 21 | 135 | [81, 125] | [88, 154] | [69, 91] | [85, 112] | ✓ |
| 22 | 136 | [88, 125] | [88, 154] | [67, 95] | [85, 112] | ✓ |
| 23 | 145 | [80, 108] | [108, 133] | [68, 89] | [92, 109] | ✓ |
| 24 | 146 | [80, 113] | [96, 133] | [67, 95] | [92, 109] | |
| 25 | 156 | [81, 113] | [96, 129] | [67, 95] | [93, 109] | ✓ |
| 26 | 234 | [59, 125] | [88, 154] | [51, 92] | [85, 112] | |
| 27 | 235 | [59, 125] | [88, 154] | [51, 92] | [85, 112] | |
| 28 | 236 | [59, 125] | [88, 154] | [51, 95] | [85, 112] | |
| 29 | 245 | [59, 108] | [91, 133] | [51, 92] | [92, 99] | |
| 30 | 246 | [59, 113] | [91, 133] | [51, 95] | [92, 99] | |
| 31 | 256 | [59, 113] | [91, 129] | [51, 95] | [93, 99] | ✓ |
| 32 | 345 | [80, 125] | [88, 154] | [68, 91] | [85, 112] | |
| 33 | 346 | [80, 125] | [88, 154] | [67, 95] | [85, 112] | |
| 34 | 356 | [81, 125] | [88, 154] | [67, 95] | [85, 112] | |

*(Continued)*

**Table 8.7** (Continued)

| s | Code | $Y_1$ $[a_1, b_1]$ | $Y_2$ $[a_2, b_2]$ | $Y_3$ $[a_3, b_3]$ | $Y_4$ $[a_4, b_4]$ | Complete ✓ |
|---|------|-----------|-----------|----------|----------|---|
| 35 | 456 | [80, 113] | [96, 133] | [67, 95] | [92, 99] | |
| 36 | 1234 | [59, 125] | [88, 154] | [51, 92] | [85, 112] | |
| 37 | 1235 | [59, 125] | [88, 154] | [51, 92] | [85, 112] | |
| 38 | 1236 | [59, 125] | [88, 154] | [51, 95] | [85, 112] | |
| 39 | 1245 | [59, 108] | [91, 133] | [51, 92] | [92, 109] | ✓ |
| 40 | 1246 | [59, 113] | [91, 133] | [51, 95] | [92, 109] | |
| 41 | 1256 | [59, 113] | [91, 129] | [51, 95] | [93, 109] | ✓ |
| 42 | 1345 | [80, 125] | [88, 154] | [68, 91] | [85, 112] | ✓ |
| 43 | 1346 | [80, 125] | [88, 154] | [67, 95] | [85, 112] | |
| 44 | 1356 | [81, 125] | [88, 154] | [67, 95] | [85, 112] | ✓ |
| 45 | 1456 | [80, 113] | [96, 133] | [67, 95] | [92, 109] | ✓ |
| 46 | 2345 | [59, 125] | [88, 154] | [51, 92] | [85, 112] | |
| 47 | 2346 | [59, 125] | [88, 154] | [51, 95] | [85, 112] | |
| 48 | 2356 | [59, 125] | [88, 154] | [51, 95] | [85, 112] | |
| 49 | 2456 | [59, 113] | [91, 133] | [51, 95] | [92, 99] | ✓ |
| 50 | 3456 | [80, 125] | [88, 154] | [67, 95] | [85, 112] | |
| 51 | 23456 | [59, 125] | [88, 154] | [51, 95] | [85, 112] | |
| 52 | 13456 | [80, 125] | [88, 154] | [67, 95] | [85, 112] | ✓ |
| 53 | 12456 | [59, 113] | [91, 133] | [51, 95] | [92, 109] | ✓ |
| 54 | 12356 | [59, 125] | [88, 154] | [51, 95] | [85, 112] | |
| 55 | 12346 | [59, 125] | [88, 154] | [51, 95] | [85, 112] | |
| 56 | 12345 | [59, 125] | [88, 154] | [51, 92] | [85, 112] | ✓ |
| 57 | $\Omega$ | [59, 125] | [88, 154] | [51, 95] | [85, 112] | |

indicates which observations have been merged to produce the new symbolic observation/object $s$, along with the corresponding extents for each merger. For example, $s = 23$ has Code $= 145$, and results from merging the observations $u = 1$, $u = 4$, and $u = 5$. The extent for this merger is

$$
\begin{aligned}
Ext(s = 23) = Ext(\text{Code} = 145) &= \{[98,108], [120,126], [69,76], [100,109]\} \\
&\wedge \{[80,101], [117,133], [68,89], [92,97]\} \\
&\wedge \{[81,99], [108,129], [76,86], [93,95]\} \\
&= \{[80,108], [108,133], [68,89], [92,109]\}.
\end{aligned}
$$

$$(8.2.4)$$

Only extents that are complete are permissible merges in a pyramid. Those that are complete extents are indicted by ✓ in Table 8.7. To see how these are determined, we look at the detail in Table 8.8.

There are 57 different $s$ values resulting from all possible mergers, including the entire data set $\Omega$. The first two columns of Table 8.8 give the $s$ value, and the Code $u_1u_2 \ldots$ of merged observations corresponding to $s = 1, \ldots, 57$, respectively. The next four columns relate in turn to the four variables $Y_j$, $j = 1, \ldots, 4$. Within each $Y_j$ column, there are $m = 6$ sub-columns, one for each observation $u = 1, \ldots, 6$. If, for a given variable, observation $u$ is contained in the extent of $s$, then the $u$ value appears in the sub-column for $u$; if not, a '.' appears.

For example, take $s = 1$, Code $= 12$, which relates to merging observations $u = 1$ and $u = 2$. For the variable $Y_1$, the extent is $Y_1(s = 1) = Y_1(\text{Code} = 12) = [59,108]$ (see Table 8.7). It is clear that the observations $\{1, 2, 4, 5\}$ are contained in this extent, but observations $\{3, 6\}$ are not. Similarly, whether or not each observation $u$ is contained in the extent of $s = 1$ for all variables can be determined, as shown in Table 8.8. Thus we see in this case that although for variables $Y_j, j = 1, 2, 3$, the extent of $s = 1$ contains more observations than the merged object itself (i.e., $u = 1$ and $u = 2$), for $Y_4$ the only observations from $\Omega$ are $u = 1$ and $u = 2$. Hence, from Definition 8.10, $s = 1$, or equivalently Code $=12$, is complete. Note, however, had interest centered on variables $Y_j, j = 1, 2, 3$, only then the object $s = 1$ would not be a complete object.

As a second example, when $s = 53$, Code $= 12456$, we see that for $Y_3$ the extent of $s$ contains all observations in $\Omega$ and so is not complete relative to $Y_3$, while for variables $Y_j, j = 1, 2, 4$, the extent of $s$ contains observations $u = 1, 2, 4, 5, 6$ only. Since, clearly, if the extent of only one of the $Y_j$ values contains itself, then that observation is complete. Therefore, in this case, object $s = 53$ (i.e., the merged set $\{1, 2, 4, 5, 6\}$) is complete.

To illustrate the calculation of the generality degree, take $s = 1$, which represents the extent of the two observations $u = 1$ and $u = 2$. From Table 8.7, the extent is $Ext(s = 1) = ([59,108], [91,126], [51, 92], [96,109])$. Then, the cardinality of $Y_j, j = 1, \ldots, 4$, becomes $c(D_1) = c([59,108]) = 49$; likewise, $c(D_2) = 35$, $c(D_3) = 41$ and $c(D_4) = 13$, respectively. Also, we have that the cardinality of $\mathcal{Y}_j, j = 1, \ldots, 4$, is $c(\mathcal{Y}_1) = 66$, $c(\mathcal{Y}_2) = 66$, $c(\mathcal{Y}_3) = 44$ and $c(\mathcal{Y}_4) = 27$, respectively. Then, from Eq. (8.2.3),

$$G(s = 1) = \frac{49}{66} \times \frac{35}{66} \times \frac{41}{44} \times \frac{13}{27} = 0.642 \times 0.530 \times 0.932 \times 0.481$$
$$= 0.17664.$$

The right-most column of Table 8.8 provides the generality degree calculated from Eq. (8.2.3) for all $s = 1, \ldots, 57$. The generality degree $G_j(s)$, by variable $Y_j$ for each $s$ is given in Table 8.15.

**Table 8.8** Complete objects detail and generality degree $G(s)$ for *E. coli* data (Example 8.16)

| s | Code | Y1 u 1 | 2 | 3 | 4 | 5 | 6 | Y2 u 1 | 2 | 3 | 4 | 5 | 6 | Y3 u 1 | 2 | 3 | 4 | 5 | 6 | Y4 u 1 | 2 | 3 | 4 | 5 | 6 | Generality degree G(s) |
|---|------|----|---|---|---|---|---|----|---|---|---|---|---|----|---|---|---|---|---|----|---|---|---|---|---|------|
| 1 | 12 | 1 | 2 | . | 4 | 5 | . | 1 | 2 | . | . | . | 6 | 1 | 2 | 3 | 4 | 5 | . | 1 | 2 | . | . | . | . | 0.17664 |
| 2 | 13 | 1 | . | 3 | . | . | 6 | 1 | 2 | 3 | 4 | 5 | 6 | 1 | . | 3 | . | 5 | . | 1 | 2 | 3 | 4 | 5 | 6 | 0.28030 |
| 3 | 14 | 1 | . | . | 4 | 5 | . | 1 | . | . | 4 | . | . | 1 | . | . | 4 | 5 | . | 1 | 2 | 3 | 4 | 5 | 6 | 0.03091 |
| 4 | 15 | 1 | . | . | . | 5 | . | 1 | . | . | . | 5 | . | 1 | . | . | 4 | 5 | . | 1 | 2 | . | 4 | 5 | 6 | 0.02980 |
| 5 | 16 | 1 | . | . | . | . | 6 | 1 | . | . | . | . | 6 | 1 | . | . | 4 | 5 | 6 | 1 | 2 | . | . | 5 | 6 | 0.03652 |
| 6 | 23 | 1 | 2 | 3 | 4 | 5 | 6 | 1 | 2 | 3 | 4 | 5 | 6 | 1 | 2 | 3 | 4 | 5 | . | 1 | 2 | 3 | 4 | 5 | 6 | 0.93182 |
| 7 | 24 | 1 | 2 | . | 4 | 5 | . | 1 | 2 | . | 4 | 5 | 6 | 1 | 2 | 3 | 4 | 5 | . | 1 | 2 | . | 4 | 5 | 6 | 0.11414 |
| 8 | 25 | 1 | 2 | . | 4 | 5 | . | 1 | 2 | . | . | 5 | 6 | 1 | 2 | 3 | 4 | 5 | . | . | 2 | . | 4 | 5 | 6 | 0.08851 |
| 9 | 26 | 1 | 2 | . | 4 | 5 | 6 | 1 | 2 | . | . | . | 6 | 1 | 2 | 3 | 4 | 5 | 6 | . | 2 | . | . | 5 | 6 | 0.03673 |
| 10 | 34 | 1 | . | 3 | 4 | 5 | 6 | 1 | 2 | 3 | 4 | 5 | 6 | 1 | . | 3 | 4 | 5 | . | 1 | 2 | 3 | 4 | 5 | 6 | 0.35640 |
| 11 | 35 | 1 | . | 3 | . | 5 | 6 | 1 | 2 | 3 | 4 | 5 | 6 | 1 | . | 3 | . | 5 | . | 1 | 2 | 3 | 4 | 5 | 6 | 0.22727 |
| 12 | 36 | 1 | . | 3 | . | . | 6 | 1 | 2 | 3 | 4 | 5 | 6 | 1 | . | 3 | 4 | 5 | 6 | 1 | 2 | 3 | 4 | 5 | 6 | 0.35675 |
| 13 | 45 | . | . | . | 4 | 5 | . | 1 | . | . | 4 | 5 | . | 1 | . | . | 4 | 5 | . | . | . | . | 4 | 5 | . | 0.01065 |
| 14 | 46 | 1 | . | . | 4 | 5 | 6 | 1 | . | . | 4 | 5 | 6 | 1 | . | 3 | 4 | 5 | 6 | 1 | 2 | . | 4 | 5 | 6 | 0.04625 |
| 15 | 56 | 1 | . | . | . | 5 | 6 | 1 | . | . | . | 5 | 6 | 1 | . | 3 | 4 | 5 | 6 | . | . | . | . | 5 | 6 | 0.03428 |
| 16 | 123 | 1 | 2 | 3 | 4 | 5 | 6 | 1 | 2 | 3 | 4 | 5 | 6 | 1 | 2 | 3 | 4 | 5 | . | 1 | 2 | 3 | 4 | 5 | 6 | 0.93182 |
| 17 | 124 | 1 | 2 | . | 4 | 5 | . | 1 | 2 | . | 4 | 5 | 6 | 1 | 2 | 3 | 4 | 5 | . | 1 | 2 | . | 4 | 5 | 6 | 0.27719 |
| 18 | 125 | 1 | 2 | . | 4 | 5 | 6 | 1 | 2 | . | . | 5 | 6 | 1 | 2 | 3 | 4 | 5 | . | 1 | 2 | . | . | 5 | 6 | 0.23604 |
| 19 | 126 | 1 | 2 | . | 4 | 5 | 6 | 1 | 2 | . | . | . | 6 | 1 | 2 | 3 | 4 | 5 | 6 | 1 | 2 | . | . | . | 6 | 0.24105 |
| 20 | 134 | 1 | . | 3 | 4 | 5 | 6 | 1 | 2 | 3 | 4 | 5 | 6 | 1 | . | 3 | 4 | 5 | . | 1 | 2 | 3 | 4 | 5 | 6 | 0.35640 |

*(Continued)*

**Table 8.8** (Continued)

| s | Code | Y₁ u | | | | | | Y₂ u | | | | | | Y₃ u | | | | | | Y₄ u | | | | | | Generality degree G(s) |
|---|------|---|---|---|---|---|---|---|---|---|---|---|---|---|---|---|---|---|---|---|---|---|---|---|---|---|
| | | 1 | 2 | 3 | 4 | 5 | 6 | 1 | 2 | 3 | 4 | 5 | 6 | 1 | 2 | 3 | 4 | 5 | 6 | 1 | 2 | 3 | 4 | 5 | 6 | |
| 21 | 135 | 1 | . | 3 | . | 5 | 6 | 1 | 2 | 3 | 4 | 5 | 6 | 1 | . | 3 | . | 5 | . | 1 | 2 | 3 | 4 | 5 | 6 | 0.33333 |
| 22 | 136 | 1 | . | 3 | . | . | 6 | 1 | 2 | 3 | 4 | 5 | 6 | 1 | . | 3 | 4 | 5 | 6 | 1 | 2 | 3 | 4 | 5 | 6 | 0.35675 |
| 23 | 145 | 1 | . | . | 4 | 5 | . | 1 | . | . | 4 | 5 | . | 1 | . | . | 4 | 5 | . | 1 | 2 | . | 4 | 5 | 6 | 0.04829 |
| 24 | 146 | 1 | . | . | 4 | 5 | 6 | 1 | . | . | 4 | 5 | 6 | 1 | . | 3 | 4 | 5 | 6 | 1 | 2 | . | 4 | 5 | 6 | 0.11231 |
| 25 | 156 | 1 | . | . | . | 5 | 6 | 1 | . | . | . | 5 | 6 | 1 | . | 3 | 4 | 5 | 6 | 1 | 2 | . | . | 5 | 6 | 0.09142 |
| 26 | 234 | 1 | . | . | . | 5 | 6 | 1 | 2 | 3 | 4 | 5 | 6 | 1 | 2 | 3 | 4 | 5 | . | 1 | 2 | 3 | 4 | 5 | 6 | 0.93182 |
| 27 | 235 | 1 | 2 | 3 | 4 | 5 | 6 | 1 | 2 | 3 | 4 | 5 | 6 | 1 | 2 | 3 | 4 | 5 | . | 1 | 2 | 3 | 4 | 5 | 6 | 0.93182 |
| 28 | 236 | 1 | 2 | 3 | 4 | 5 | 6 | 1 | 2 | 3 | 4 | 5 | 6 | 1 | 2 | 3 | 4 | 5 | 6 | 1 | 2 | 3 | 4 | 5 | 6 | 1.00000 |
| 29 | 245 | 1 | 2 | 3 | 4 | 5 | 6 | 1 | 2 | 3 | 4 | 5 | 6 | 1 | 2 | 3 | 4 | 5 | . | 1 | 2 | 3 | 4 | 5 | 6 | 0.11414 |
| 30 | 246 | 1 | 2 | . | 4 | 5 | . | 1 | 2 | . | 4 | 5 | 6 | 1 | 2 | 3 | 4 | 5 | 6 | 1 | 2 | . | 4 | 5 | 6 | 0.13499 |
| 31 | 256 | 1 | 2 | . | 4 | 5 | 6 | 1 | 2 | . | 4 | 5 | 6 | 1 | 2 | 3 | 4 | 5 | 6 | 1 | 2 | . | . | 5 | 6 | 0.10468 |
| 32 | 345 | 1 | 2 | . | 4 | 5 | 6 | 1 | 2 | 3 | 4 | 5 | 6 | 1 | . | 3 | 4 | 5 | . | 1 | 2 | 3 | 4 | 5 | 6 | 0.35640 |
| 33 | 346 | 1 | . | 3 | 4 | 5 | 6 | 1 | 2 | 3 | 4 | 5 | 6 | 1 | . | 3 | 4 | 5 | 6 | 1 | 2 | 3 | 4 | 5 | 6 | 0.43388 |
| 34 | 356 | 1 | . | 3 | 4 | 5 | 6 | 1 | 2 | 3 | 4 | 5 | 6 | 1 | . | 3 | 4 | 5 | 6 | 1 | 2 | 3 | 4 | 5 | 6 | 0.42424 |
| 35 | 456 | 1 | . | 3 | . | 5 | 6 | 1 | . | . | 4 | 5 | 6 | 1 | . | 3 | 4 | 5 | 6 | 1 | 2 | . | 4 | 5 | 6 | 0.04625 |
| 36 | 1234 | 1 | . | . | 4 | 5 | 6 | 1 | 2 | 3 | 4 | 5 | 6 | 1 | 2 | 3 | 4 | 5 | . | 1 | 2 | 3 | 4 | 5 | 6 | 0.93182 |
| 37 | 1235 | 1 | 2 | 3 | 4 | 5 | 6 | 1 | 2 | 3 | 4 | 5 | 6 | 1 | 2 | 3 | 4 | 5 | . | 1 | 2 | 3 | 4 | 5 | 6 | 0.93182 |
| 38 | 1236 | 1 | 2 | 3 | 4 | 5 | 6 | 1 | 2 | 3 | 4 | 5 | 6 | 1 | 2 | 3 | 4 | 5 | 6 | 1 | 2 | 3 | 4 | 5 | 6 | 1.00000 |
| 39 | 1245 | 1 | 2 | 3 | 4 | 5 | 6 | 1 | 2 | 3 | 4 | 5 | 6 | 1 | 2 | 3 | 4 | 5 | . | 1 | 2 | 3 | 4 | 5 | 6 | 0.27719 |
| 40 | 1246 | 1 | 2 | . | 4 | 5 | 6 | 1 | 2 | . | 4 | 5 | 6 | 1 | 2 | 3 | 4 | 5 | 6 | 1 | 2 | . | 4 | 5 | 6 | 0.32782 |

*(Continued)*

**Table 8.8** (Continued)

| s | Code | Y₁ u 1 | 2 | 3 | 4 | 5 | 6 | Y₂ u 1 | 2 | 3 | 4 | 5 | 6 | Y₃ u 1 | 2 | 3 | 4 | 5 | 6 | Y₄ u 1 | 2 | 3 | 4 | 5 | 6 | Generality degree G(s) |
|---|---|---|---|---|---|---|---|---|---|---|---|---|---|---|---|---|---|---|---|---|---|---|---|---|---|---|
| 41 | 1256 | 1 | 2 | . | 4 | 5 | 6 | 1 | 2 | . | . | 5 | 6 | 1 | 2 | 3 | 4 | 5 | 6 | 1 | 2 | . | . | 5 | 6 | 0.27916 |
| 42 | 1345 | 1 | . | 3 | 4 | 5 | 6 | 1 | 2 | 3 | 4 | 5 | 6 | 1 | . | 3 | 4 | 5 | . | 1 | 2 | 3 | 4 | 5 | 6 | 0.35640 |
| 43 | 1346 | 1 | . | 3 | 4 | 5 | 6 | 1 | 2 | 3 | 4 | 5 | 6 | 1 | . | 3 | 4 | 5 | 6 | 1 | 2 | 3 | 4 | 5 | 6 | 0.43388 |
| 44 | 1356 | 1 | . | 3 | . | 5 | 6 | 1 | 2 | 3 | 4 | 5 | 6 | 1 | . | 3 | 4 | 5 | 6 | 1 | 2 | 3 | 4 | 5 | 6 | 0.42424 |
| 45 | 1456 | 1 | . | 3 | 4 | 5 | 6 | 1 | . | . | 4 | 5 | 6 | 1 | . | 3 | 4 | 5 | 6 | 1 | 2 | . | 4 | 5 | 6 | 0.11231 |
| 46 | 2345 | 1 | 2 | 3 | 4 | 5 | 6 | 1 | 2 | 3 | 4 | 5 | 6 | 1 | 2 | 3 | 4 | 5 | . | 1 | 2 | 3 | 4 | 5 | 6 | 0.93182 |
| 47 | 2346 | 1 | 2 | 3 | 4 | 5 | 6 | 1 | 2 | 3 | 4 | 5 | 6 | 1 | 2 | 3 | 4 | 5 | 6 | 1 | 2 | 3 | 4 | 5 | 6 | 1.00000 |
| 48 | 2356 | 1 | 2 | 3 | 4 | 5 | 6 | 1 | 2 | 3 | 4 | 5 | 6 | 1 | 2 | 3 | 4 | 5 | 6 | 1 | 2 | 3 | 4 | 5 | 6 | 1.00000 |
| 49 | 2456 | 1 | 2 | . | 4 | 5 | 6 | 1 | 2 | . | 4 | 5 | 6 | 1 | 2 | 3 | 4 | 5 | 6 | 1 | 2 | . | 4 | 5 | 6 | 0.13499 |
| 50 | 3456 | 1 | . | 3 | 4 | 5 | 6 | 1 | 2 | 3 | 4 | 5 | 6 | 1 | . | 3 | 4 | 5 | 6 | 1 | 2 | 3 | 4 | 5 | 6 | 0.43388 |
| 51 | 23456 | 1 | 2 | 3 | 4 | 5 | 6 | 1 | 2 | 3 | 4 | 5 | 6 | 1 | 2 | 3 | 4 | 5 | 6 | 1 | 2 | 3 | 4 | 5 | 6 | 1.00000 |
| 52 | 13456 | 1 | . | 3 | 4 | 5 | 6 | 1 | 2 | 3 | 4 | 5 | 6 | 1 | 2 | 3 | 4 | 5 | 6 | 1 | 2 | 3 | 4 | 5 | 6 | 0.43388 |
| 53 | 12456 | 1 | 2 | . | 4 | 5 | 6 | 1 | 2 | . | 4 | 5 | 6 | 1 | 2 | 3 | 4 | 5 | 6 | 1 | 2 | . | 4 | 5 | 6 | 0.32782 |
| 54 | 12356 | 1 | 2 | 3 | 4 | 5 | 6 | 1 | 2 | 3 | 4 | 5 | 6 | 1 | 2 | 3 | 4 | 5 | 6 | 1 | 2 | 3 | 4 | 5 | 6 | 1.00000 |
| 55 | 12346 | 1 | 2 | 3 | 4 | 5 | 6 | 1 | 2 | 3 | 4 | 5 | 6 | 1 | 2 | 3 | 4 | 5 | 6 | 1 | 2 | 3 | 4 | 5 | 6 | 1.00000 |
| 56 | 12345 | 1 | 2 | 3 | 4 | 5 | 6 | 1 | 2 | 3 | 4 | 5 | 6 | 1 | 2 | 3 | 4 | 5 | . | 1 | 2 | 3 | 4 | 5 | 6 | 0.93182 |
| 57 | Ω | 1 | 2 | 3 | 4 | 5 | 6 | 1 | 2 | 3 | 4 | 5 | 6 | 1 | 2 | 3 | 4 | 5 | 6 | 1 | 2 | 3 | 4 | 5 | 6 | 1.00000 |

As an aside, let us calculate the generality degree of a single observation, e.g., $u = 1$. It is easy to show that

$$G(u = 1) = \frac{10}{66} \times \frac{6}{66} \times \frac{7}{44} \times \frac{9}{27} = 0.642 \times 0.530 \times 0.932 \times 0.481$$
$$= 0.00073;$$

i.e, $G(u = 1, u = 1) \equiv G(u = 1) = 0.00073 \neq 0$, showing that $G(s)$ is not a dissimilarity. □

## 8.2.2 Pyramid Construction Based on Generality Degree

Construction of a pyramid for symbolic data, based on generality degrees, was introduced in Brito (1991, 1994) and Brito and Diday (1990).

We illustrate the pyramid construction methodology with two examples. The first example uses a data set in which there are no identical observations for any of the variables. In this case, the generality degree and the method itself cause no undue difficulties. The second example has some identical observed values for some (but not all) of the variables. This requires some extra care in comparing the generality degrees in constructing the pyramid.

**Example 8.17** Take the *E. coli* data set and the generality degrees calculated in Example 8.16. Tables 8.7 and 8.8 provide the input for constructing the pyramid for this data set. From these tables, we see that 22 of the 56 combinations of observations produce an extent that is complete. Only those $s$ that are complete are used in the construction of the pyramid. These combinations along with their generality degrees are summarized in Table 8.9. While at the outset all 22 possibilities are candidate mergers, some may subsequently not be possible because of the nature of preceding selections in the construction steps (see, e.g., Step 3 in the construction below).

Suppose the pyramid construction is based on the minimum generality degree of possible mergers at each stage. At the first step, only pairwise combinations pertain. Therefore, we seek the minimum value of $G(s)$ from among the first seven objects in Table 8.9, i.e., we find $s' \in S_1 = \{1, 3, 4, 5, 9, 11, 13\}$ such that

$$G(s') = \min_{s \in S_1} G(s)$$
$$= \min\{0.1766, 0.0309, 0.0298, 0.0365, 0.0367, 0.2273, 0.0106\}.$$

That is,

$$\min_{s \in S_1} G(s) = G(13) = 0.01065.$$

The $s = 13$ corresponds to Code = 45, therefore, at Step 1, the observations $u = 4$ and $u = 5$ are merged.

**Table 8.9** Permissible merges for *E. coli* data (Example 8.17)

| s | Code | $Y_1$ $[a_1, b_1]$ | $Y_2$ $[a_2, b_2]$ | $Y_3$ $[a_3, b_3]$ | $Y_4$ $[a_4, b_4]$ | Generality degree $G(s)$[a] |
|---|------|--------------|--------------|-------------|-------------|------------|
| 1 | 12 | [59, 108] | [91, 126] | [51, 92] | [96, 109] | 0.17664 |
| 3 | 14 | [80,108] | [117, 133] | [68, 89] | [92, 109] | 0.03091 |
| 4 | 15 | [81, 108] | [108, 129] | [69, 86] | [93, 109] | $\mathbf{0.02951^2}$ |
| 5 | 16 | [98, 113] | [96, 126] | [67, 95] | [94, 109] | $\mathbf{0.03652^3}$ |
| 9 | 26 | [59, 113] | [91, 107] | [51, 95] | [94, 99] | $\mathbf{0.03673^4}$ |
| 11 | 35 | [81, 125] | [88, 154] | [76, 91] | [85, 112] | 0.22727 |
| 13 | 45 | [80, 101] | [108, 133] | [68, 89] | [92, 97] | $\mathbf{0.01065^1}$ |
| 19 | 126 | [59, 113] | [91, 126] | [51, 95] | [94, 109] | $\mathbf{0.24105^8}$ |
| 21 | 135 | [81, 125] | [88, 154] | [69, 91] | [85, 112] | 0.33333 |
| 22 | 136 | [88, 125] | [88, 154] | [67, 95] | [85, 112] | 0.35675 |
| 23 | 145 | [80, 108] | [108, 133] | [68, 89] | [92, 109] | $\mathbf{0.04829^5}$ |
| 25 | 156 | [81, 113] | [96, 129] | [67, 95] | [93, 109] | $\mathbf{0.04829^6}$ |
| 31 | 256 | [59, 113] | [91, 129] | [51, 95] | [93, 99] | 0.10468 |
| 39 | 1245 | [59, 108] | [91, 133] | [51, 92] | [92, 109] | 0.27719 |
| 41 | 1256 | [59, 113] | [91, 129] | [51, 95] | [93, 109] | $\mathbf{0.27916^9}$ |
| 42 | 1345 | [80, 125] | [88, 154] | [68, 91] | [85, 112] | $\mathbf{0.35640^{11}}$ |
| 44 | 1356 | [81, 125] | [88, 154] | [67, 95] | [85, 112] | 0.42424 |
| 45 | 1456 | [80, 113] | [96, 133] | [67, 95] | [92, 109] | $\mathbf{0.11231^7}$ |
| 49 | 2456 | [59, 113] | [91, 133] | [51, 95] | [92, 99] | 0.13499 |
| 52 | 13456 | [80, 125] | [88, 154] | [67, 95] | [85, 112] | $\mathbf{0.43388^{12}}$ |
| 53 | 12456 | [59, 113] | [91, 133] | [51, 95] | [92, 109] | $\mathbf{0.32782^{10}}$ |
| 56 | 12345 | [59, 125] | [88, 154] | [51, 92] | [85, 112] | 0.93182 |
| 57 | $\Omega$ | [59, 125] | [88, 154] | [51, 95] | [85, 112] | $\mathbf{1.00000^{13}}$ |

a) $\mathbf{G^k}$ indicates minimum $G$ value at Step **k** in constructing the pyramid.

At the second step, possible merged objects correspond to elements in $S_1$ that were not merged at Step 1. In addition, three-wise mergers with the already merged pair $\{4, 5\}$ are now possible; for these data, only the triple $\{1, 4, 5\}$ gives a complete object (as seen in Table 8.9). Hence, we seek the $s' \in S_2 = \{1, 3, 4, 5, 9, 11, 23\}$ which minimizes the $G(s)$, namely,

$$G(s') = \min_{s \in S_2} G(s) = G(4) = 0.02980.$$

Therefore, at Step 2, the observations $u = 1$ and $u = 5$ are merged.

At Step 3, we seek $s' \in S_3 = \{1, 3, 5, 9, 21, 23, 25\}$, which minimizes $G(s)$. Notice that the pairwise merger of observations $u = 3$ and $u = 5$ is no longer possible given that observation $u = 5$ was merged with observation $u = 4$ at Step 1 and with observation $u = 1$ at Step 2 (see Figure 8.15). Then, we have that

$$G(s') = \min_{s \in S_3} G(s) = G(5) = 0.03652,$$

with observations $u = 1$ and $u = 6$ being merged at this stage.

We continue in this manner, to give at

Step 4  $S_4 = \{9, 19, 22, 23, 25\}$, $G(s') = \min_{s \in S_4} G(s) = G(9) = 0.03673$,

i.e., merge observations $u = 2$ and $u = 6$ to give $\{2, 6\}$

Step 5  $S_5 = \{19, 23, 25\}$,         $G(s') = \min_{s \in S_5} G(s) = G(23) = 0.04829$,

i.e., merge $\{1, 5\}$ and $\{4,5\}$ to give $\{1, 4, 5\}$

Step 6  $S_6 = \{19, 25, 42, 45\}$,     $G(s') = \min_{s \in S_6} G(s) = G(25) = 0.09142$,

i.e., merge $\{1, 5\}$ and $\{1, 6\}$ to give $\{1, 5, 6\}$

Step 7  $S_7 = \{19, 41, 42, 45\}$,     $G(s') = \min_{s \in S_7} G(s) = G(45) = 0.11231$,

i.e., merge $\{1, 4, 5\}$ and $\{1, 5, 6\}$ to give $\{1, 4, 5, 6\}$

Step 8  $S_8 = \{19, 41, 42, 52, 53\}$, $G(s') = \min_{s \in S_8} G(s) = G(19) = 0.24105$,

i.e., merge $\{1, 6\}$ and $\{2, 6\}$ to give $\{1, 2, 6\}$

Step 9  $S_9 = \{41, 42, 52, 53\}$,     $G(s') = \min_{s \in S_9} G(s) = G(41) = 0.27916$,

i.e., merge $\{1, 2, 6\}$ and $\{1, 5, 6\}$ to give $\{1, 2, 5, 6\}$

Step 10 $S_{10} = \{42, 52, 53\}$,       $G(s') = \min_{s \in S_{10}} G(s) = G(53) = 0.32782$,

i.e., merge $\{1, 2, 5, 6\}$ and $\{1,4,5,6\}$ to give $\{1,2,4,5,6\}$

Step 11 $S_{11} = \{42, 52, 57\}$,       $G(s') = \min_{s \in S_{11}} G(s) = G(42) = 0.35640$,

i.e., merge$\{1, 4, 5\}$and $u = 3$ to give $\{1, 3, 4, 5\}$

Step 12 $S_{12} = \{52, 57\}$,           $G(s') = \min_{s \in S_{12}} G(s) = G(52) = 0.43388$,

i.e., merge $\{1, 3, 4, 5\}$ and $\{1, 4, 5, 6\}$ to give $\{1, 3, 4, 5, 6\}$

Step 13 $S_{13} = \{57\}$,               $G(s') = \min_{s \in S_{13}} G(s) = G(57) = 1.00000$,

i.e., merge $\{1, 3, 4, 5, 6\}$ and $\{1, 2, 4, 5, 6\}$ to give $\Omega$.

The completed pyramid is shown in Figure 8.15.                                    □

**Example 8.18**  Consider now the routes data set displayed in Table 8.10. There are nine routes of interest, whereby goods are transported across these routes to and from regions of origin and destination regions. There are five variables, $Y_1 = $ population at the region of origin, $Y_2 = $ population at the destination region, $Y_3 = $ annual volume of imports to the destination region, $Y_4 = $ distance traveled between region of origin and destination region, and $Y_5 = $ volume of

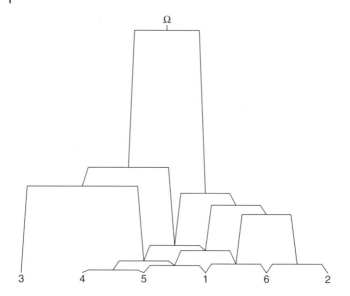

**Figure 8.15** Pyramid of *E. coli* data (Example 8.17).

**Table 8.10** Routes data (Example 8.18)

| u | $Y_1$<br>$[a_1, b_1]$ | $Y_2$<br>$[a_2, b_2]$ | $Y_3$<br>$[a_3, b_3]$ | $Y_4$<br>$[a_4, b_4]$ | $Y_5$<br>$[a_5, b_5]$ |
|---|---|---|---|---|---|
| 1 | [10980, 10980] | [1348, 2410] | [4200, 7201] | [101.28, 232.05] | [0.01, 935.86] |
| 2 | [3979, 3979] | [1348, 2410] | [4200, 7201] | [114.73, 423.19] | [0.01, 792.92] |
| 3 | [1305, 5525] | [3979, 3979] | [10465, 10465] | [594.00, 699.14] | [0.01, 201.71] |
| 4 | [254, 4422] | [3979, 3979] | [10465, 10465] | [479.98, 908.21] | [0.01, 466.37] |
| 5 | [1305, 5525] | [1623, 3133] | [4230, 8685] | [269.96, 791.86] | [0.01, 464.21] |
| 6 | [10980, 10980] | [714, 2851] | [1851, 8245] | [400.78, 655.11] | [0.01, 431.85] |
| 7 | [3979, 3979] | [10980, 10980] | [55574, 55574] | [206.89, 206.89] | [0.01, 874.33] |
| 8 | [1108, 2282] | [10980, 10980] | [55574, 55574] | [302.91, 419.28] | [0.01, 979.83] |
| 9 | [1623, 3133] | [10980, 10980] | [55574, 55574] | [293.63, 433.09] | [0.01, 973.65] |

goods transported over that route. For the present purpose, we consider the first $m = 6$ routes for the variables $Y_1, \dots, Y_4$.

The extents for all possible two-, three-, four- and five-wise mergers are shown in Table 8.11. Those extents that are obviously complete are indicated with a ✓. The detail for this determination is provided in Table 8.12. For example, from Table 8.12, for all four variables, the object $s = 1$ (i.e.,

**Table 8.11** Extents of possible clusters for routes data (Example 8.18)

| s | Code | $Y_1$ $[a_1, b_1]$ | $Y_2$ $[a_2, b_2]$ | $Y_3$ $[a_3, b_3]$ | $Y_4$ $[a_4, b_4]$ | Complete ✓ |
|---|---|---|---|---|---|---|
| 1 | 12 | [3979, 10980] | [1348, 2410] | [4200, 7201] | [101.28, 423.19] | ✓ |
| 2 | 13 | [1305, 10980] | [1348, 3979] | [4200, 10465] | [101.28, 699.14] | |
| 3 | 14 | [254, 10980] | [1348, 3979] | [4200, 10465] | [101.28, 908.21] | |
| 4 | 15 | [1305, 10980] | [1348, 3133] | [4200, 8685] | [101.28, 791.86] | x✓ |
| 5 | 16 | [10980, 10980] | [714, 2851] | [1851, 8245] | [101.28, 655.11] | ✓ |
| 6 | 23 | [1305, 5525] | [1348, 3979] | [4200, 10465] | [114.73, 699.14] | |
| 7 | 24 | [254, 4422] | [1348, 3979] | [4200, 10465] | [114.73, 908.21] | ✓ |
| 8 | 25 | [1305, 5525] | [1348, 3133] | [4200, 8685] | [114.73, 791.86] | x✓ |
| 9 | 26 | [3979, 10980] | [714, 2851] | [1851, 8245] | [114.73, 655.11] | ✓ |
| 10 | 34 | [254, 5525] | [3979, 3979] | [10465, 10465] | [479.98, 908.21] | ✓ |
| 11 | 35 | [1305, 5525] | [1623, 3979] | [4230, 10465] | [269.96, 791.86] | x✓ |
| 12 | 36 | [1305, 10980] | [714, 3979] | [1851, 10465] | [400.78, 699.14] | ✓ |
| 13 | 45 | [254, 5525] | [1623, 3979] | [4230, 10465] | [269.96, 908.21] | x✓ |
| 14 | 46 | [254, 10980] | [714, 3979] | [1851, 10465] | [400.78, 908.21] | |
| 15 | 56 | [1305, 10980] | [714, 3133] | [1851, 8685] | [269.96, 791.86] | |
| 16 | 123 | [1305, 10980] | [1348, 3979] | [4200, 10465] | [101.28, 699.14] | |
| 17 | 124 | [254, 10980] | [1348, 3979] | [4200, 10465] | [101.28, 908.21] | |
| 18 | 125 | [1305, 10980] | [1348, 3133] | [4200, 8685] | [101.28, 791.86] | ✓ |
| 19 | 126 | [3979, 10980] | [714, 2851] | [1851, 8245] | [101.28, 655.11] | ✓ |
| 20 | 134 | [254, 10980] | [1348, 3979] | [4200, 10465] | [101.28, 908.21] | |
| 21 | 135 | [1305, 10980] | [1348, 3979] | [4200, 10465] | [101.28, 791.86] | x✓ |
| 22 | 136 | [1305, 10980] | [714, 3979] | [1851, 10465] | [101.28, 699.14] | |
| 23 | 145 | [254, 10980] | [1348, 3979] | [4200, 10465] | [101.28, 908.21] | x✓ |
| 24 | 146 | [254, 10980] | [714, 3979] | [1851, 10465] | [101.28, 908.21] | |
| 25 | 156 | [1305, 10980] | [714, 3133] | [1851, 8685] | [101.28, 791.86] | x✓ |
| 26 | 234 | [254, 5525] | [1348, 3979] | [4200, 10465] | [114.73, 908.21] | |
| 27 | 235 | [1305, 5525] | [1348, 3979] | [4200, 10465] | [114.73, 791.86] | ✓ |
| 28 | 236 | [1305, 10980] | [714, 3979] | [1851, 10465] | [114.73, 699.14] | ✓ |
| 29 | 245 | [254, 5525] | [1348, 3979] | [4200, 10465] | [114.73, 908.21] | |
| 30 | 246 | [254, 10980] | [714, 3979] | [1851, 10465] | [114.73, 908.21] | |
| 31 | 256 | [1305, 10980] | [714, 3133] | [1851, 8685] | [114.73, 791.86] | |
| 32 | 345 | [254, 5525] | [1623, 3979] | [4230, 10465] | [269.96, 908.21] | ✓ |
| 33 | 346 | [254, 10980] | [714, 3979] | [1851, 10465] | [400.78, 908.21] | ✓ |
| 34 | 356 | [1305, 10980] | [714, 3979] | [1851, 10465] | [269.96, 791.86] | ✓ |

*(Continued)*

**Table 8.11** (Continued)

| s | Code | $Y_1$ $[a_1, b_1]$ | $Y_2$ $[a_2, b_2]$ | $Y_3$ $[a_3, b_3]$ | $Y_4$ $[a_4, b_4]$ | Complete ✓ |
|---|---|---|---|---|---|---|
| 35 | 456 | [254, 10980] | [714, 3979] | [1851, 10465] | [269.96, 908.21] | |
| 36 | 1234 | [254, 10980] | [1348, 3979] | [4200, 10465] | [101.28, 908.21] | |
| 37 | 1235 | [1305, 10980] | [1348, 3979] | [4200, 10465] | [101.28, 791.86] | x✓ |
| 38 | 1236 | [1305, 10980] | [714, 3979] | [1851, 10465] | [101.28, 699.14] | ✓ |
| 39 | 1245 | [254, 10980] | [1348, 3979] | [4200, 10465] | [101.28, 908.21] | x✓ |
| 40 | 1246 | [254, 10980] | [714, 3979] | [1851, 10465] | [101.28, 908.21] | |
| 41 | 1256 | [1305, 10980] | [714, 3133] | [1851, 8685] | [101.28, 791.86] | ✓ |
| 42 | 1345 | [254, 10980] | [1348, 3979] | [4200, 10465] | [101.28, 908.21] | |
| 43 | 1346 | [254, 10980] | [714, 3979] | [1851, 10465] | [101.28, 908.21] | |
| 44 | 1356 | [1305, 10980] | [714, 3979] | [1851, 10465] | [101.28, 791.86] | |
| 45 | 1456 | [254, 10980] | [714, 3979] | [1851, 10465] | [101.28, 908.21] | |
| 46 | 2345 | [254, 5525] | [1348, 3979] | [4200, 10465] | [114.73, 908.21] | ✓ |
| 47 | 2346 | [254, 10980] | [714, 3979] | [1851, 10465] | [114.73, 908.21] | |
| 48 | 2356 | [1305, 10980] | [714, 3979] | [1851, 10465] | [114.73, 791.86] | ✓ |
| 49 | 2456 | [254, 10980] | [714, 3979] | [1851, 10465] | [114.73, 908.21] | |
| 50 | 3456 | [254, 10980] | [714, 3979] | [1851, 10465] | [269.96, 908.21] | ✓ |
| 51 | 12345 | [254, 10980] | [1348, 3979] | [4200, 10465] | [101.28, 908.21] | ✓ |
| 52 | 12346 | [254, 10980] | [714, 3979] | [1851, 10465] | [101.28, 908.21] | |
| 53 | 12356 | [1305, 10980] | [714, 3979] | [1851, 10465] | [101.28, 791.86] | ✓ |
| 54 | 12456 | [254, 10980] | [714, 3979] | [1851, 10465] | [101.28, 908.21] | x✓ |
| 55 | 13456 | [254, 10980] | [714, 3979] | [1851, 10465] | [101.28, 908.21] | |
| 56 | 23456 | [254, 10980] | [714, 3979] | [1851, 10465] | [114.73, 908.21] | ✓ |
| 57 | Ω | [254, 10980] | [714, 3979] | [1851, 10465] | [101.28, 908.21] | |

Code = 12) is complete since for variable $Y_2$ the extent of $s$ is $s$ itself; this object is also complete relative to variables $Y_3$ and $Y_4$ but not relative to $Y_1$.

Other objects that at first appearance are not complete but in fact are complete are indicated by 'x ✓'. Consider the object $s = 4$ (Code = 15). Relative to the variable $Y_1$, we see from Table 8.12 that the extent contains observations {1, 2, 3, 5, 6} and so is not complete; likewise for variable $Y_4$. In contrast, for variable $Y_2$ we see that the extent appears to include the observations {1, 2, 5}, i.e., itself ({1, 5}) and also observation $u = 2$. However, $Y_{22} = Y_{12} = [1345, 2410]$. Therefore, the assertion or object $s$ that is the *most* specific is the extent {1, 5} rather than {1, 2, 5}. Therefore, relative to the variable $Y_2$, the object $s = 4$ is complete. Likewise, this is a complete object for variable $Y_4$ alone, and also for the set $Y_j, j = 1, \ldots, 4$.

**Table 8.12** Complete objects detail and generality degree $G(s)$ for routes data (Example 8.18)

| s | Code | Y1 u | | | | | | Y2 u | | | | | | Y3 u | | | | | | Y4 u | | | | | | Generality degree |
|---|---|---|---|---|---|---|---|---|---|---|---|---|---|---|---|---|---|---|---|---|---|---|---|---|---|---|
| | | 1 | 2 | 3 | 4 | 5 | 6 | 1 | 2 | 3 | 4 | 5 | 6 | 1 | 2 | 3 | 4 | 5 | 6 | 1 | 2 | 3 | 4 | 5 | 6 | $G(s)$ |
| 1 | 12 | 1 | 2 | . | . | . | 6 | 1 | 2 | . | . | . | . | 1 | 2 | . | . | . | . | 1 | 2 | . | . | . | . | 0.02951 |
| 2 | 13 | 1 | 2 | 3 | . | . | 6 | 1 | 2 | 3 | 4 | 5 | . | 1 | 2 | 3 | 4 | 5 | . | 1 | 2 | 3 | . | . | 6 | 0.39168 |
| 3 | 14 | 1 | 2 | 3 | 4 | . | 6 | 1 | 2 | 3 | 4 | 5 | . | 1 | 2 | 3 | 4 | 5 | . | 1 | 2 | 3 | 4 | 5 | 6 | 0.58608 |
| 4 | 15 | 1 | 2 | 3 | . | 5 | 6 | 1 | 2 | . | . | 5 | . | 1 | 2 | . | . | 5 | . | 1 | 2 | 3 | 4 | . | 6 | 0.21974 |
| 5 | 16 | 1 | . | . | . | 5 | 6 | 1 | 2 | . | . | . | 6 | 1 | 2 | . | . | . | 6 | 1 | 2 | 3 | . | 5 | 6 | 0.33345 |
| 6 | 23 | . | 2 | 3 | . | . | 6 | . | 2 | 3 | 4 | 5 | . | . | 2 | 3 | 4 | 5 | . | . | 2 | 3 | . | 5 | 6 | 0.16700 |
| 7 | 24 | . | 2 | . | 4 | . | 6 | . | 2 | 3 | 4 | 5 | . | . | 2 | 3 | 4 | 5 | . | . | 2 | 3 | 4 | 5 | 6 | 0.22395 |
| 8 | 25 | . | 2 | 3 | . | 5 | 6 | . | 2 | . | . | 5 | . | . | 2 | . | . | 5 | . | . | 2 | 3 | . | 5 | 6 | 0.09398 |
| 9 | 26 | 1 | 2 | . | . | . | 6 | 1 | 2 | . | . | . | 6 | 1 | 2 | . | . | . | 6 | . | 2 | 3 | . | . | 6 | 0.21236 |
| 10 | 34 | . | 2 | 3 | 4 | 5 | . | . | . | 3 | 4 | . | . | . | . | 3 | 4 | . | . | . | . | 3 | 4 | . | . | 0.26079 |
| 11 | 35 | . | 2 | 3 | . | 5 | 6 | . | . | 3 | 4 | 5 | . | . | . | 3 | 4 | 5 | . | . | . | 3 | . | 5 | 6 | 0.13291 |
| 12 | 36 | 1 | 2 | 3 | . | 5 | . | 1 | 2 | 3 | 4 | 5 | 6 | 1 | 2 | 3 | 4 | 5 | 6 | . | . | 3 | . | . | 6 | 0.33352 |
| 13 | 45 | . | 2 | 3 | 4 | 5 | . | . | . | 3 | 4 | 5 | . | . | . | 3 | 4 | 5 | . | . | . | 3 | 4 | 5 | 6 | 0.20302 |
| 14 | 46 | 1 | 2 | 3 | 4 | 5 | 6 | 1 | 2 | 3 | 4 | 5 | 6 | 1 | 2 | 3 | 4 | 5 | 6 | . | . | 3 | 4 | . | 6 | 0.62884 |
| 15 | 56 | 1 | 2 | 3 | . | 5 | 6 | 1 | 2 | . | . | 5 | 6 | 1 | 2 | . | . | 5 | 6 | . | . | 3 | . | 5 | 6 | 0.34292 |
| 16 | 123 | 1 | 2 | 3 | . | 5 | 6 | 1 | 2 | 3 | 4 | 5 | . | 1 | 2 | 3 | 4 | 5 | . | 1 | 2 | 3 | . | 5 | 6 | 0.39168 |
| 17 | 124 | 1 | 2 | 3 | 4 | 5 | 6 | 1 | 2 | 3 | 4 | 5 | . | 1 | 2 | 3 | 4 | 5 | . | 1 | 2 | 3 | 4 | 5 | 6 | 0.58608 |
| 18 | 125 | 1 | 2 | 3 | . | 5 | 6 | 1 | 2 | . | . | 5 | . | 1 | 2 | . | . | 5 | . | 1 | 2 | 3 | . | 5 | 6 | 0.21974 |
| 19 | 126 | 1 | 2 | . | . | . | 6 | 1 | 2 | . | . | . | 6 | 1 | 2 | . | . | . | 6 | 1 | 2 | . | . | . | 6 | 0.21765 |
| 20 | 134 | 1 | 2 | 3 | 4 | 5 | 6 | 1 | 2 | 3 | 4 | 5 | . | 1 | 2 | 3 | 4 | 5 | . | 1 | 2 | 3 | 4 | 5 | 6 | 0.58608 |

(Continued)

**Table 8.12** (Continued)

| s | Code | $Y_1$ | | | | | | $Y_2$ | | | | | | $Y_3$ | | | | | | $Y_4$ | | | | | | Generality degree $G(s)$ |
|---|---|---|---|---|---|---|---|---|---|---|---|---|---|---|---|---|---|---|---|---|---|---|---|---|---|---|
| | | 1 | 2 | 3 | 4 | 5 | 6 | 1 | 2 | 3 | 4 | 5 | 6 | 1 | 2 | 3 | 4 | 5 | 6 | 1 | 2 | 3 | 4 | 5 | 6 | |
| 21 | 135 | 1 | 2 | 3 | . | 5 | 6 | 1 | 2 | 3 | 4 | 5 | . | 1 | 2 | 3 | 4 | 5 | . | 1 | 2 | 3 | . | 5 | 6 | 0.45242 |
| 22 | 136 | 1 | 2 | 3 | . | 5 | 6 | 1 | 2 | 3 | 4 | 5 | 6 | 1 | 2 | 3 | 4 | 5 | 6 | 1 | 2 | 3 | . | . | 6 | 0.66831 |
| 23 | 145 | 1 | 2 | 3 | 4 | 5 | 6 | 1 | 2 | 3 | 4 | 5 | . | 1 | 2 | 3 | 4 | 5 | . | 1 | 2 | 3 | 4 | 5 | 6 | 0.58608 |
| 24 | 146 | 1 | 2 | 3 | 4 | 5 | 6 | 1 | 2 | 3 | 4 | 5 | 6 | 1 | 2 | 3 | 4 | 5 | 6 | . | 2 | 3 | 4 | 5 | 6 | 1.00000 |
| 25 | 156 | 1 | 2 | 3 | . | 5 | 6 | 1 | 2 | 3 | 4 | 5 | 6 | . | 2 | 3 | 4 | 5 | 6 | 1 | 2 | 3 | 4 | 5 | 6 | 0.45375 |
| 26 | 234 | . | 2 | 3 | 4 | 5 | . | 1 | 2 | 3 | 4 | 5 | . | 1 | 2 | 3 | 4 | 5 | . | . | 2 | 3 | 4 | 5 | . | 0.28321 |
| 27 | 235 | . | 2 | 3 | . | 5 | . | 1 | 2 | 3 | 4 | 5 | . | 1 | 2 | 3 | 4 | 5 | . | . | 2 | 3 | 4 | 5 | . | 0.19349 |
| 28 | 236 | 1 | 2 | 3 | . | 5 | 6 | 1 | 2 | 3 | 4 | 5 | 6 | 1 | 2 | 3 | 4 | 5 | 6 | . | 2 | 3 | . | 5 | . | 0.65327 |
| 29 | 245 | . | 2 | 3 | 4 | 5 | . | 1 | 2 | 3 | 4 | 5 | . | 1 | 2 | 3 | 4 | 5 | . | . | 2 | 3 | 4 | 5 | . | 0.28321 |
| 30 | 246 | 1 | 2 | 3 | 4 | 5 | 6 | 1 | 2 | 3 | 4 | 5 | 6 | 1 | 2 | 3 | 4 | 5 | 6 | . | 2 | 3 | 4 | 5 | 6 | 0.98333 |
| 31 | 256 | 1 | 2 | 3 | . | 5 | 6 | 1 | 2 | . | 4 | 5 | 6 | 1 | 2 | . | 4 | 5 | 6 | . | 2 | 3 | . | 5 | 6 | 0.44491 |
| 32 | 345 | . | 2 | 3 | 4 | 5 | . | . | 2 | 3 | 4 | 5 | . | . | . | 3 | 4 | 5 | . | . | . | 3 | 4 | 5 | . | 0.20302 |
| 33 | 346 | 1 | 2 | 3 | 4 | 5 | 6 | 1 | 2 | 3 | 4 | 5 | 6 | 1 | 2 | 3 | 4 | 5 | 6 | . | . | 3 | 4 | . | 6 | 0.62884 |
| 34 | 356 | 1 | 2 | 3 | . | 5 | 6 | 1 | 2 | 3 | 4 | 5 | 6 | 1 | 2 | 3 | 4 | 5 | 6 | . | . | 3 | 4 | . | 6 | 0.58340 |
| 35 | 456 | 1 | 2 | 3 | 4 | 5 | 6 | 1 | 2 | 3 | 4 | 5 | 6 | 1 | 2 | 3 | 4 | 5 | 6 | . | . | 3 | 4 | 5 | 6 | 0.79096 |
| 36 | 1234 | 1 | 2 | 3 | 4 | 5 | 6 | 1 | 2 | 3 | 4 | 5 | . | 1 | 2 | 3 | 4 | 5 | . | 1 | 2 | 3 | 4 | 5 | 6 | 0.58608 |
| 37 | 1235 | 1 | 2 | 3 | . | 5 | 6 | 1 | 2 | 3 | 4 | 5 | . | 1 | 2 | 3 | 4 | 5 | . | 1 | 2 | 3 | . | 5 | 6 | 0.45242 |
| 38 | 1236 | 1 | 2 | 3 | . | 5 | 6 | 1 | 2 | 3 | 4 | 5 | 6 | 1 | 2 | 3 | 4 | 5 | 6 | 1 | 2 | 3 | . | . | 6 | 0.66831 |

(Continued)

**Table 8.12** (Continued)

| s | Code | $Y_1$ | | | | | | $Y_2$ | | | | | | $Y_3$ | | | | | | $Y_4$ | | | | | | Generality degree |
|---|------|---|---|---|---|---|---|---|---|---|---|---|---|---|---|---|---|---|---|---|---|---|---|---|---|---|
| | | $u$ | | | | | | $u$ | | | | | | $u$ | | | | | | $u$ | | | | | | |
| | | 1 | 2 | 3 | 4 | 5 | 6 | 1 | 2 | 3 | 4 | 5 | 6 | 1 | 2 | 3 | 4 | 5 | 6 | 1 | 2 | 3 | 4 | 5 | 6 | $G(s)$ |
| 39 | 1245 | 1 | 2 | 3 | 4 | 5 | 6 | 1 | 2 | 3 | 4 | 5 | . | 1 | 2 | 3 | 4 | 5 | . | 1 | 2 | 3 | 4 | 5 | 6 | 0.58608 |
| 40 | 1246 | 1 | 2 | 3 | 4 | 5 | 6 | 1 | 2 | 3 | 4 | 5 | 6 | 1 | 2 | 3 | 4 | 5 | 6 | 1 | 2 | 3 | 4 | 5 | 6 | 1.00000 |
| 41 | 1256 | 1 | 2 | 3 | . | 5 | 6 | 1 | 2 | . | . | 5 | 6 | 1 | 2 | . | . | 5 | 6 | 1 | 2 | 3 | . | 5 | 6 | 0.45375 |
| 42 | 1345 | 1 | 2 | 3 | 4 | 5 | 6 | 1 | 2 | 3 | 4 | 5 | . | 1 | 2 | 3 | 4 | 5 | 6 | 1 | 2 | 3 | 4 | 5 | 6 | 0.58608 |
| 43 | 1346 | 1 | 2 | 3 | 4 | 5 | 6 | 1 | 2 | 3 | 4 | 5 | 6 | 1 | 2 | 3 | 4 | 5 | 6 | 1 | 2 | 3 | 4 | 5 | 6 | 1.00000 |
| 44 | 1356 | 1 | 2 | 3 | . | 5 | 6 | 1 | 2 | 3 | 4 | 5 | 6 | 1 | 2 | 3 | 4 | 5 | 6 | 1 | 2 | 3 | . | 5 | 6 | 0.77195 |
| 45 | 1456 | 1 | 2 | 3 | 4 | 5 | 6 | 1 | 2 | 3 | 4 | 5 | 6 | 1 | 2 | 3 | 4 | 5 | 6 | 1 | 2 | 3 | 4 | 5 | 6 | 1.00000 |
| 46 | 2345 | . | 2 | 3 | 4 | 5 | 6 | 1 | 2 | 3 | 4 | 5 | . | 1 | 2 | 3 | 4 | 5 | . | . | 2 | 3 | 4 | 5 | 6 | 0.28321 |
| 47 | 2346 | 1 | 2 | 3 | 4 | 5 | 6 | 1 | 2 | 3 | 4 | 5 | 6 | 1 | 2 | 3 | 4 | 5 | 6 | 1 | 2 | 3 | 4 | 5 | 6 | 0.98333 |
| 48 | 2356 | 1 | 2 | 3 | . | 5 | 6 | 1 | 2 | 3 | 4 | 5 | 6 | 1 | 2 | 3 | 4 | 5 | 6 | 1 | 2 | 3 | . | 5 | 6 | 0.75692 |
| 49 | 2456 | 1 | 2 | 3 | 4 | 5 | 6 | 1 | 2 | 3 | 4 | 5 | 6 | 1 | 2 | 3 | 4 | 5 | 6 | 1 | 2 | 3 | 4 | 5 | 6 | 0.98333 |
| 50 | 3456 | 1 | 2 | 3 | 4 | 5 | 6 | 1 | 2 | 3 | 4 | 5 | 6 | 1 | 2 | 3 | 4 | 5 | 6 | . | 2 | 3 | 4 | 5 | 6 | 0.79096 |
| 51 | 12345 | 1 | 2 | 3 | 4 | 5 | 6 | 1 | 2 | 3 | 4 | 5 | . | 1 | 2 | 3 | 4 | 5 | . | 1 | 2 | 3 | 4 | 5 | 6 | 0.58608 |
| 52 | 12346 | 1 | 2 | 3 | 4 | 5 | 6 | 1 | 2 | 3 | 4 | 5 | 6 | 1 | 2 | 3 | 4 | 5 | 6 | 1 | 2 | 3 | 4 | 5 | 6 | 1.00000 |
| 53 | 12356 | 1 | 2 | 3 | . | 5 | 6 | 1 | 2 | 3 | 4 | 5 | 6 | 1 | 2 | 3 | 4 | 5 | 6 | 1 | 2 | 3 | . | 5 | 6 | 0.77195 |
| 54 | 12456 | 1 | 2 | 3 | 4 | 5 | 6 | 1 | 2 | 3 | 4 | 5 | 6 | 1 | 2 | 3 | 4 | 5 | 6 | 1 | 2 | 3 | 4 | 5 | 6 | 1.00000 |
| 55 | 13456 | 1 | 2 | 3 | 4 | 5 | 6 | 1 | 2 | 3 | 4 | 5 | 6 | 1 | 2 | 3 | 4 | 5 | 6 | 1 | 2 | 3 | 4 | 5 | 6 | 1.00000 |
| 56 | 23456 | 1 | 2 | 3 | 4 | 5 | 6 | 1 | 2 | 3 | 4 | 5 | 6 | 1 | 2 | 3 | 4 | 5 | 6 | . | 2 | 3 | 4 | 5 | 6 | 0.98333 |
| 57 | Ω | 1 | 2 | 3 | 4 | 5 | 6 | 1 | 2 | 3 | 4 | 5 | 6 | 1 | 2 | 3 | 4 | 5 | 6 | 1 | 2 | 3 | 4 | 5 | 6 | 1.00000 |

Table 8.13 summarizes the extents and generality degree values for the 31 extents that are complete, plus that for the entire data set $\Omega$ ($s = 57$). This table contains the input for the construction of the pyramid. The construction is similar to that used in Example 8.17. The ensuing steps are:

Step 1    $S_1 = \{1, 4, 5, 7, 8, 9, 10, 11, 12, 13\}$, $G(s') = \min_{s \in S_1} G(s) = G(1) = 0.02951$,

      i.e., merge observations $u = 1$ and $u = 2$ to give $\{1, 2\}$

Step 2    $S_2 = \{4, 5, 7, 8, 9, 10, 11, 12, 13, 18, 19\}$, $G(s') = \min_{s \in S_2} G(s) = G(8) = 0.09398$,

      i.e., merge observations $u = 2$ and $u = 5$ to give $\{2, 5\}$

Step 3    $S_3 = \{5, 7, 10, 11, 12, 13, 18, 19, 27\}$,    $G(s') = \min_{s \in S_6} G(s) = G(11) = 0.13291$,

      i.e., merge observations $u = 3$ and $u = 5$ to give $\{3, 5\}$

Step 4    $S_4 = \{5, 10, 12, 18, 19, 27, 32, 34\}$, $G(s') = \min_{s \in S_4} G(s) = G(27) = 0.19349$,

      i.e., merge $\{2, 5\}$ and $\{3, 5\}$ to give $\{2, 3, 5\}$

Step 5    $S_5 = \{5, 10, 12, 18, 19, 32, 34, 37, 46\}$, $G(s') = \min_{s \in S_5} G(s) = G(32) = 0.20302$,

      i.e., merge $u = 4$ and $\{3, 5\}$ to give $\{3, 4, 5\}$

Step 6    $S_6 = \{5, 18, 19, 37, 46\}$,    $G(s') = \min_{s \in S_6} G(s) = G(19) = 0.21765$,

      i.e., merge $u = 6$ and $\{1, 2\}$ to give $\{1, 2, 6\}$

Step 7    $S_7 = \{18, 37, 41, 46\}$,    $G(s') = \min_{s \in S_7} G(s) = G(18) = 0.21974$,

      i.e., merge $\{1, 2\}$ and $\{2, 5\}$ to give $\{1, 2, 5\}$

Step 8    $S_8 = \{37, 41, 46\}$, $G(s') = \min_{s \in S_8} G(s) = G(46) = 0.28321$,

      i.e., merge $\{2, 3, 5\}$ and $\{3, 4, 5\}$ to give $\{2, 3, 4, 5\}$

Step 9    $S_9 = \{37, 41, 51\}$,    $G(s') = \min_{s \in S_9} G(s) = G(37) = 0.45242$,

      i.e., merge $\{1, 2, 5\}$ and $\{2, 3, 5\}$ to give $\{1, 2, 3, 5\}$

Step 10   $S_{10} = \{41, 51, 53\}$,    $G(s') = \min_{s \in S_{10}} G(s) = G(41) = 0.45375$,

      i.e., merge $\{1, 2, 6\}$ and $\{1, 2, 5\}$ to give $\{1, 2, 5, 6\}$

Step 11   $S_{11} = \{51, 53, 57\}$,    $G(s') = \min_{s \in S_{11}} G(s) = G(51) = 0.58608$,

      i.e., merge $\{1, 2, 3, 5\}$ and $\{2, 3, 4, 5\}$ to give $\{1, 2, 3, 4, 5\}$

Step 12   $S_{12} = \{53, 57\}$,    $G(s') = \min_{s \in S_{12}} G(s) = G(53) = 0.77195$,

      i.e., merge $\{1, 2, 3, 5\}$ and $\{1, 2, 5, 6\}$ to give $\{1, 2, 3, 5, 6\}$

Step 13   $S_{13} = \{57\}$,    $G(s') = \min_{s \in S_{13}} G(s) = G(57) = 1.00000$,

      i.e., merge$\{1, 2, 3, 5, 6\}$ and $\{1, 2, 3, 4, 5\}$ to give $\Omega$.

These are the adjusted generality degrees.

**Table 8.13** Permissible merges for routes data (Example 8.18)

| s | Code | $Y_1$ [$a_1, b_1$] | $Y_2$ [$a_2, b_2$] | $Y_3$ [$a_3, b_3$] | $Y_4$ [$a_4, b_4$] | Generality degree $G(s)$[a] |
|---|---|---|---|---|---|---|
| 1 | 12 | [3979, 10980] | [1348, 2410] | [4200, 7201] | [101.28, 423.19] | **0.02951**[1] |
| 4 | 15 | [1305, 10980] | [1348, 3133] | [4200, 8685] | [101.28, 791.86] | 0.21974 |
| 5 | 16 | [10980, 10980] | [714, 2851] | [1851, 8245] | [101.28, 655.11] | 0.33345 |
| 7 | 24 | [254, 4422] | [1348, 3979] | [4200, 10465] | [114.73, 908.21] | 0.22395 |
| 8 | 25 | [1305, 5525] | [1348, 3133] | [4200, 8685] | [114.73, 791.86] | **0.09398**[2] |
| 9 | 26 | [3979, 10980] | [714, 2851] | [1851, 8245] | [114.73, 655.11] | 0.21236 |
| 10 | 34 | [254, 5525] | [3979, 3979] | [10465, 10465] | [479.98, 908.21] | 0.26079 |
| 11 | 35 | [1305, 5525] | [1623, 3979] | [4230, 10465] | [269.96, 791.86] | **0.13291**[3] |
| 12 | 36 | [1305, 10980] | [714, 3979] | [1851, 10465] | [400.78, 699.14] | 0.33352 |
| 13 | 45 | [254, 5525] | [1623, 3979] | [4230, 10465] | [269.96, 908.21] | 0.20302 |
| 18 | 125 | [1305, 10980] | [1348, 3133] | [4200, 8685] | [101.28, 791.86] | **0.21974**[7] |
| 19 | 126 | [3979, 10980] | [714, 2851] | [1851, 8245] | [101.28, 655.11] | **0.21765**[6] |
| 21 | 135 | [1305, 10980] | [1348, 3979] | [4200, 10465] | [101.28, 791.86] | 0.45242 |
| 23 | 145 | [254, 10980] | [1348, 3979] | [4200, 10465] | [101.28, 908.21] | 0.58608 |
| 25 | 156 | [1305, 10980] | [714, 3133] | [1851, 8685] | [101.28, 791.86] | 0.45375 |
| 27 | 235 | [1305, 5525] | [1348, 3979] | [4200, 10465] | [114.73, 791.86] | **0.19349**[4] |
| 28 | 236 | [1305, 10980] | [714, 3979] | [1851, 10465] | [114.73, 699.14] | 0.65327 |
| 32 | 345 | [254, 5525] | [1623, 3979] | [4230, 10465] | [269.96, 908.21] | **0.20302**[5] |
| 33 | 346 | [254, 10980] | [714, 3979] | [1851, 10465] | [400.78, 908.21] | 0.62884 |
| 34 | 356 | [1305, 10980] | [714, 3979] | [1851, 10465] | [269.96, 791.86] | 0.58340 |
| 37 | 1235 | [1305, 10980] | [1348, 3979] | [4200, 10465] | [101.28, 791.86] | **0.45242**[9] |
| 38 | 1236 | [1305, 10980] | [714, 3979] | [1851, 10465] | [101.28, 699.14] | 0.66831 |
| 39 | 1245 | [254, 10980] | [1348, 3979] | [4200, 10465] | [101.28, 908.21] | 0.58608 |
| 41 | 1256 | [1305, 10980] | [714, 3133] | [1851, 8685] | [101.28, 791.86] | **0.45375**[10] |
| 46 | 2345 | [254, 5525] | [1348, 3979] | [4200, 10465] | [114.73, 908.21] | **0.28321**[8] |
| 48 | 2356 | [1305, 10980] | [714, 3979] | [1851, 10465] | [114.73, 791.86] | 0.75692 |
| 50 | 3456 | [254, 10980] | [714, 3979] | [1851, 10465] | [269.96, 908.21] | 0.79096 |
| 51 | 12345 | [254, 10980] | [1348, 3979] | [4200, 10465] | [101.28, 908.21] | **0.58608**[11] |
| 53 | 12356 | [1305, 10980] | [714, 3979] | [1851, 10465] | [101.28, 791.86] | **1.77195**[12] |
| 54 | 12456 | [254, 10980] | [714, 3979] | [1851, 10465] | [101.28, 908.21] | 1.00000 |
| 56 | 23456 | [254, 10980] | [714, 3979] | [1851, 10465] | [114.73, 908.21] | 0.98333 |
| 57 | $\Omega$ | [254, 10980] | [714, 3979] | [1851, 10465] | [101.28, 908.21] | **1.00000**[13] |

a) $\mathbf{G}^k$ indicates minimum $G$ value at Step **k** in constructing the pyramid.

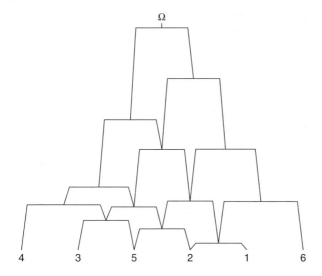

**Figure 8.16** Pyramid of routes data, $m = 6$ (Example 8.18).

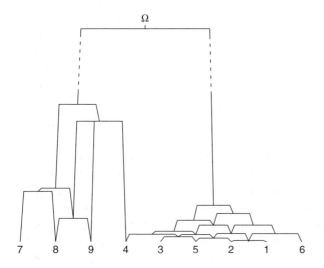

**Figure 8.17** Pyramid of all $m = 9$ routes data (Example 8.18).

**Table 8.14** Generality degree and merges all routes (Example 8.18)

| Step $s$ | Cluster | $G(s)$ | Step $s$ | Cluster | $G(s)$ |
|---|---|---|---|---|---|
| 1 | $\{1,2\}$ | 0.00150 | 11 | $\{1,2,3,4,5\}$ | 0.02989 |
| 2 | $\{2,5\}$ | 0.00479 | 12 | $\{8,9\}$ | 0.03263 |
| 3 | $\{3,5\}$ | 0.00678 | 13 | $\{1,2,3,5,6\}$ | 0.03937 |
| 4 | $\{2,3,5\}$ | 00987 | 14 | $\{1,2,3,4,5,6\}$ | 0.05099 |
| 5 | $\{3,4,5\}$ | 0.01035 | 15 | $\{7,8\}$ | 0.07045 |
| 6 | $\{1,2,6\}$ | 0.01110 | 16 | $\{7,8,9\}$ | 0.07503 |
| 7 | $\{1,2,5\}$ | 0.01121 | 17 | $\{4,9\}$ | 0.16947 |
| 8 | $\{3,4,5\}$ | 0.01444 | 18 | $\{4,8,9\}$ | 0.16947 |
| 9 | $\{1,2,3,5\}$ | 0.02307 | 19 | $\{4,7,8,9\}$ | 0.19339 |
| 10 | $\{1,2,5,6\}$ | 0.02314 | 20 | $\{1,\dots,9\}$ | 1.0 |

The pyramid thus obtained for these six observations is shown in Figure 8.16; the same pyramid emerges when using the unadjusted generality degree.

The corresponding pyramid when all $m = 9$ observations are considered is shown in Figure 8.17. A summary of the step-wise merges and the generality degrees is shown in Table 8.14. ▫

### 8.2.3 Pyramids from Dissimilarity Matrix

Pyramids can also be constructed using the principles covered in section 8.1 of agglomerative methods, except that now any one observation can appear in at most two clusters. This approach was first developed for classical observations by Bertrand (1986), Diday (1986b), and Diday and Bertrand (1986). This is illustrated for symbolic data briefly in the following example.

**Example 8.19** Consider the modal multi-valued household data of Table 4.3. Let us use the first $m = 4$ observations, and let us take the unweighted non-normalized Euclidean extended Ichino–Yaguchi distances for $\gamma = 0.25$, calculated in Example 4.6 and shown in Table 4.7. The distance matrix is

$$\mathbf{D} = \begin{bmatrix} 0 & 0.128 & 0.149 & 0.199 \\ 0.128 & 0 & 0.213 & 0.085 \\ 0.149 & 0.213 & 0 & 0.270 \\ 0.199 & 0.085 & 0.270 & 0 \end{bmatrix}. \tag{8.2.5}$$

Notice that, as written, this appears to be a non-Robinson matrix. However, if we re-order the observations $\{1, 2, 3, 4\}$ into $\{3, 1, 2, 4\}$, then we see that the distance matrix becomes

$$
\mathbf{D}_{(1)} =
\begin{bmatrix}
0 & 0.149 & 0.213 & 0.270 \\
0.149 & 0 & 0.128 & 0.199 \\
0.213 & 0.128 & 0 & \mathbf{0.085} \\
0.270 & 0.199 & \mathbf{0.085} & 0
\end{bmatrix}
\begin{matrix}
3 \\ 1 \\ 2 \\ 4
\end{matrix}
\qquad (8.2.6)
$$

which is a Robinson matrix; the column to the right of the matrix identifies the observation ($u$).

Suppose we build the hierarchy tree using the complete-link agglomerative algorithm adapted to pyramid construction. From Eq. (8.2.6), we see the minimum distance is the $d(2, 4) = 0.085$ between the observations $u = 2$ and $u = 4$. Merging these into one set $\{2, 4\}$, and using the maximum distances from Eq. (8.1.2), the distance matrix becomes

$$
\mathbf{D}_{(2)} =
\begin{bmatrix}
0 & 0.149 & 0.213 & 0.270 & 0.270 \\
0.149 & 0 & \mathbf{0.128} & 0.199 & 0.199 \\
0.213 & \mathbf{0.128} & 0 & - & - \\
0.270 & 0.199 & - & 0 & - \\
0.270 & 0.199 & - & - & 0
\end{bmatrix}
\begin{matrix}
3 \\ 1 \\ 2 \\ 4 \\ \{2, 4\}
\end{matrix}
$$
$$(8.2.7)$$

where the " $-$ " indicates a merger that is not possible at this stage, e.g., the first " $-$ " in the third row and fourth column corresponds to the merger of $u = 2$ and $u = 4$ which has already occurred and so is no longer possible again.

Next, we obtain the distance matrix $\mathbf{D}_{(3)}$ for the sets $\{3\}, \{1\}, \{1, 2\}, \{2, 4\}$, and $\{4\}$. This tells us that the next merge occurs between observations $u = 1$ and $u = 3$. At the fourth stage, the distance matrix $\mathbf{D}_{(4)}$ is calculated for the sets $\{1, 3\}, \{1, 2\}$, and $\{2, 4\}$, from which we obtain the merger of $\{1, 2\}$ and $\{2, 4\}$. We now have

$$
\mathbf{D}_4 =
\begin{bmatrix}
0 & \mathbf{0.213} & 0.270 \\
\mathbf{0.213} & 0 & 0.270 \\
0.270 & 0.270 & 0
\end{bmatrix}
\begin{matrix}
\{1, 3\} \\ \{1, 2\} \\ \{1, 2, 4\}
\end{matrix}
\qquad (8.2.8)
$$

Here, the minimum distance is the $d(\{1, 3\}, \{1, 2\}) = 0.213$ between the sets $\{1, 3\}$ and $\{1, 2\}$, to produce the set $\{1, 2, 3\}$. Hence, the final merge is between sets $\{1, 2, 3\}$ and $\{1, 2, 4\}$ with distance $d(\{1, 2, 3\}, \{1, 2, 4\}) = 0.270$.

The pyramid is displayed in Figure 8.18. The $y$-axis shows the distance measure at the stage of each merger. $\qquad\square$

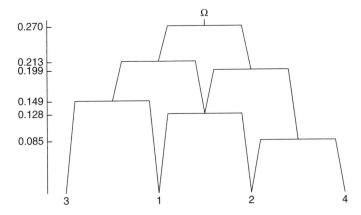

**Figure 8.18** Pyramid of $m = 4$ multi-valued households data based on dissimilarity matrix (Example 8.19).

**Example 8.20** As intimated in Chapter 5, hierarchy and pyramid trees can suggest partitions in a data set. This is illustrated with a subset of the routes data set of Table 8.10. A set of $m = 4$ observations produced the hierarchy tree of Figure 8.19 (a), while these same observations produced the pyramid of

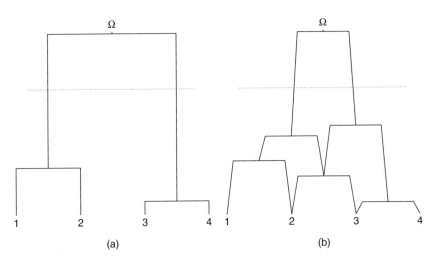

**Figure 8.19** Partitions cuts on hierarchy and pyramid trees (Example 8.20). (a) Hierarchy tree (b) Pyramid tree.

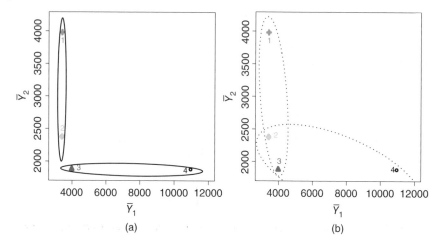

**Figure 8.20** Bi-plots $\bar{Y}_1 \times \bar{Y}_2$ for four routes (Example 8.20). (a) Partitions from hierarchy (b) Partitions from pyramid

Figure 8.19(b). (The observations $u = 1, 2, 3, 4$ in fact correspond, respectively, to the observations $4, 3, 1, 2$ of Table 8.10.) Figure 8.20 shows the biplot of the means for the variables $(Y_1, Y_2)$. In Figure 8.20(a), the two clusters identified from the hierarchy are shown; notice these are non-overlapping clusters. In contrast, the two clusters identified by the pyramid tree are overlapping clusters, as shown in Figure 8.20(b). □

### 8.2.4 Other Issues

Some pyramids can be weakly indexed or strictly indexed (see, e.g., Diday (1986b), Fichet (1987), Durand and Fichet (1988), Mirkin (1996), Bertrand (1995), Brito (2000), and Bertrand and Janowitz (2002)). Some theory is available showing the connection between pyramid construction and Galois lattices (see, e.g., Polaillon (1998, 2000), Polaillon and Diday (1996), Diday and Emilion (1998) and Brito and Polaillon (2005)).

All the hierarchies covered herein, be these divisive, agglomerative, or pyramidal trees, can be characterized as a structure drawn in a two-dimensional plane with the observations/clusters displayed along the $x$-axis. When the base is expanded from the single $x$-axis to a two-dimensional plane with the tree being constructed up the third dimension (i.e., the $z$-axis), then it becomes possible for observations to fall into three clusters instead of the limit of two clusters in standard pyramids. These "three-dimensional" pyramids are called spatial pyramids (see Diday (2004, 2008), Pak (2005) and Rahal (2010)).

## Exercises

**8.1** Refer to Example 8.1. Complete the details for the construction of the tree of Figure 8.1(b) by obtaining the dissimilarity matrix $\mathbf{D}_{(r)}$, for $r = 1, \dots, 11$ and applying the complete-link criterion of Definition 8.2.

**8.2** Build the hierarchy tree for the *E. coli* data of Table 3.8 using the Gowda–Diday dissimilarities as in Example 8.2, but now using the average-link agglomerative approach of Definition 8.3. Compare this tree with that in Figure 8.2. Explain the differences between them, if any.

**8.3** Take the unweighted normalized Euclidean distances based on the Gowda–Diday dissimilarities, calculated in Example 3.11 and shown in Table 3.10 column (*h*), for the *E. coli* data. Construct (i) a single-link tree, (ii) a compete-link tree, and (iii) an average-link tree, for this case. Compare your results.

**8.4** Refer to Example 8.4. Construct the trees of Figure 8.4 using the single-link and complete-link approach, respectively, for the *E. coli* data and the Ichino–Yaguchi dissimilarities for $Y_1$ only. What interpretation do you give to explain these trees?

**8.5** Refer to Example 8.4. Construct the single-link and complete-link hierarchies for the *E. coli* data using the variable $Y_4 = AT$ and the Ichino–Yaguchi dissimilarities. How do these trees compare with those of Exercise 8.4?

**8.6** Take at least one of the de Carvalho non-normalized Euclidean distances displayed in Table 3.14 for the *E. coli* data (calculated in Example 3.14). Build (a) a single-link tree, (b) a compete-link tree, and (c) an average-link tree, for this case. Compare your results.

**8.7** Repeat Exercise 8.6 for one of the de Carvalho normalized Euclidean distances shown in Table 3.14.

**8.8** Take the airline data for the $m = 8$ airlines used in Examples 8.8 and 8.9. Repeat the tree construction using flight time $Y_1$ only for the Ward criterion and for the single-link criterion. How do your trees compare with those in Figures 8.8 and 8.9(a), respectively, if at all, and why?

**8.9** Determine the underlying details in the construction of the hierarchy of Example 8.10 and Figure 8.10.

**8.10** Refer to the mixed variable census data of Example 8.11. Construct the hierarchy trees obtained from the single-link and complete-link agglomerative methodology. Compare these trees with those of Figure 8.11. What interpretation might you give to explain the differences, if any, between the trees? Can you give an interpretative meaning to the trees that were built by this methodology?

**8.11** The aggregation producing Table 3.8 involved an extreme outlier in the observation $u = 3$, lyt dsDNA. If that outlier is omitted, the data for this observation becomes $\mathbf{Y}_3 = (Y_{31}, Y_{32}, Y_{33}, Y_{34}) = ([88,104], [88,117], 82, 89], [85, 99])$. Values for the other observations remain as in Table 3.8. Construct the pyramid based on this new data set and compare with that obtained in Figure 8.15.

**8.12** Repeat Exercise 8.11 using the weights proportional to the respective $n_u$ values.

**8.13** Take the routes data for all $m = 9$ observations and Table 8.10 and Figure 8.17.
   (a) Complete the details in calculating the generality degrees and construction of the pyramid for these entities.
   (b) How do you explain or understand the conclusion that route $u = 4$ essentially connects the two distinct parts of Figure 8.17?

**8.14** For the routes data set of Table 8.10, construct the pyramid based on the variables (i) $Y_j, j = 1, 2, 3, 5$, and (ii) $Y_j, j = 1, \ldots, 5$, respectively.

# Appendix

Table 8.15 Generality degrees by variable, *E. coli* data (Example 8.16)

| s | Code | $G_1(s)$ | $G_2(s)$ | $G_3(s)$ | $G_4(s)$ | $G(s)$ |
|---|---|---|---|---|---|---|
| 1 | 12 | 0.74242 | 0.53030 | 0.93182 | 0.48148 | 0.17664 |
| 2 | 13 | 0.56061 | 1.00000 | 0.50000 | 1.00000 | 0.28030 |
| 3 | 14 | 0.42424 | 0.24242 | 0.47727 | 0.62963 | 0.03091 |
| 4 | 15 | 0.40909 | 0.31818 | 0.38636 | 0.59259 | 0.02980 |
| 5 | 16 | 0.22727 | 0.45455 | 0.63636 | 0.55556 | 0.03652 |
| 6 | 23 | 1.00000 | 1.00000 | 0.93182 | 1.00000 | 0.93182 |
| 7 | 24 | 0.74242 | 0.63636 | 0.93182 | 0.25926 | 0.11414 |
| 8 | 25 | 0.74242 | 0.57576 | 0.93182 | 0.22222 | 0.08851 |
| 9 | 26 | 0.81818 | 0.24242 | 1.00000 | 0.18519 | 0.03673 |
| 10 | 34 | 0.68182 | 1.00000 | 0.52273 | 1.00000 | 0.35640 |
| 11 | 35 | 0.66667 | 1.00000 | 0.34091 | 1.00000 | 0.22727 |
| 12 | 36 | 0.56061 | 1.00000 | 0.63636 | 1.00000 | 0.35675 |
| 13 | 45 | 0.31818 | 0.37879 | 0.47727 | 0.18519 | 0.01065 |
| 14 | 46 | 0.50000 | 0.56061 | 0.63636 | 0.25926 | 0.04625 |
| 15 | 56 | 0.48485 | 0.50000 | 0.63636 | 0.22222 | 0.03428 |
| 16 | 123 | 1.00000 | 1.00000 | 0.93182 | 1.00000 | 0.93182 |
| 17 | 124 | 0.74242 | 0.63636 | 0.93182 | 0.62963 | 0.27719 |
| 18 | 125 | 0.74242 | 0.57576 | 0.93182 | 0.59259 | 0.23604 |
| 19 | 126 | 0.81818 | 0.53030 | 1.00000 | 0.55556 | 0.24105 |
| 20 | 134 | 0.68182 | 1.00000 | 0.52273 | 1.00000 | 0.35640 |
| 21 | 135 | 0.66667 | 1.00000 | 0.50000 | 1.00000 | 0.33333 |
| 22 | 136 | 0.56061 | 1.00000 | 0.63636 | 1.00000 | 0.35675 |
| 23 | 145 | 0.42424 | 0.37879 | 0.47727 | 0.62963 | 0.04829 |
| 24 | 146 | 0.50000 | 0.56061 | 0.63636 | 0.62963 | 0.11231 |
| 25 | 156 | 0.48485 | 0.50000 | 0.63636 | 0.59259 | 0.09142 |
| 26 | 234 | 1.00000 | 1.00000 | 0.93182 | 1.00000 | 0.93182 |
| 27 | 235 | 1.00000 | 1.00000 | 0.93182 | 1.00000 | 0.93182 |
| 28 | 236 | 1.00000 | 1.00000 | 1.00000 | 1.00000 | 1.00000 |
| 29 | 245 | 0.74242 | 0.63636 | 0.93182 | 0.25926 | 0.11414 |
| 30 | 246 | 0.81818 | 0.63636 | 1.00000 | 0.25926 | 0.13499 |
| 31 | 256 | 0.81818 | 0.57576 | 1.00000 | 0.22222 | 0.10468 |
| 32 | 345 | 0.68182 | 1.00000 | 0.52273 | 1.00000 | 0.35640 |

*(Continued)*

**Table 8.15** (Continued)

| s | Code | $G_1(s)$ | $G_2(s)$ | $G_3(s)$ | $G_4(s)$ | $G(s)$ |
|---|---|---|---|---|---|---|
| 33 | 346 | 0.68182 | 1.00000 | 0.63636 | 1.00000 | 0.43388 |
| 34 | 356 | 0.6666 | 1.00000 | 0.63636 | 1.00000 | 0.42424 |
| 35 | 456 | 0.50000 | 0.56061 | 0.63636 | 0.25926 | 0.04625 |
| 36 | 1234 | 1.00000 | 1.00000 | 0.93182 | 1.00000 | 0.93182 |
| 37 | 1235 | 1.00000 | 1.00000 | 0.93182 | 1.00000 | 0.93182 |
| 38 | 1236 | 1.00000 | 1.00000 | 1.00000 | 1.00000 | 1.00000 |
| 39 | 1245 | 0.74242 | 0.63636 | 0.93182 | 0.62963 | 0.27719 |
| 40 | 1246 | 0.81818 | 0.63636 | 1.00000 | 0.6296 | 0.43388 |
| 44 | 1456 | 0.50000 | 0.56061 | 0.63636 | 0.6296 | 0.11231 |
| 46 | 2345 | 1.00000 | 1.00000 | 0.93182 | 1.00000 | 0.93182 |
| 47 | 2346 | 1.00000 | 1.00000 | 1.00000 | 1.00000 | 1.00000 |
| 48 | 2356 | 1.00000 | 1.00000 | 1.00000 | 1.00000 | 1.00000 |
| 49 | 2456 | 0.81818 | 0.63636 | 1.00000 | 0.25926 | 0.13499 |
| 50 | 3456 | 0.68182 | 1.00000 | 0.63636 | 1.0000 | 0.43388 |
| 51 | 23456 | 1.00000 | 1.00000 | 1.00000 | 1.00000 | 1.00000 |
| 52 | 13456 | 0.6818 | 1.0000 | 0.63636 | 1.00000 | 0.43388 |
| 53 | 12456 | 0.81818 | 0.63636 | 1.00000 | 0.62963 | 0.32782 |
| 54 | 12356 | 1.00000 | 1.00000 | 1.00000 | 1.00000 | 1.00000 |
| 55 | 12346 | 1.00000 | 1.00000 | 1.00000 | 1.00000 | 1.00000 |
| 56 | 12345 | 1.00000 | 1.00000 | 0.93182 | 1.00000 | 0.93182 |
| 57 | $\Omega$ | 1.00000 | 1.00000 | 1.00000 | 1.00000 | 1.00000 |

# References

Ali, M. M., Mikhail, N. N. and Haq, M. S. (1978). A class of bivariate distributions including the bivariate logistic. *Journal Multivariate Analysis* 8, 405–412.

Anderberg, M. R. (1973). *Cluster Analysis for Applications*. Academic Press, New York.

Bacão, F., Lobo, V. and Painho, M. (2005). Self-organizing maps as substitutes for K-means clustering. In: *Computational Science* (eds. V. S. Sunderam, G. D. van Albada, P. M. A. Sloot and J. J. Dongarra). Springer, Berlin, 476–483.

Banfield J. D. and Raftery, A. E. (1993). Model-based Gaussian and non-Gaussian clustering. *Biometrics* 49, 803–821.

Batagelj, V. (1988). Generalized Ward and related clustering problems. In: *Classification and Related Methods of Data Analysis* (ed. H. -H. Bock). North-Holland, Amsterdam, 67–74.

Batagelj, V. and Bren, M. (1995). Comparing resemblance measures. *Journal of Classification* 12, 73–90.

Batagelj, V., Doreian, P., Ferligoj, A. and Kejžar, N. (2014). *Understanding Large Temporal Networks and Spatial Networks: Exploration, Pattern Searching, Visualization and Network Evolution*. Wiley, Chichester.

Batagelj, V. and Ferligoj, A. (2000). Clustering relational data. In: *Data Analysis* (eds. W. Gaul, O. Opitz and M. Schader). Springer, New York, 3–16.

Batagelj, V., Kejžar, N. and Korenjak-Černe, S. (2015). Clustering of modal valued symbolic data. *Machine Learning arxiv:1507.06683*.

Bertrand, P. (1986). *Etude de la Représentation Pyramides*. Thèse de 3ème Cycle, Université de Paris, Dauphine.

Bertrand, P. (1995). Structural properties of pyramid clustering. In: *Partitioning Data Sets* (eds. I. J. Cox, P. Hansen and B. Julesz). DIMACS Series American Mathematical Society, Providence, 35–53.

Bertrand, P. and Goupil, F. (2000). Descriptive statistics for symbolic data. In: *Analysis of Symbolic Data: Exploratory Methods for Extracting Statistical Information from Complex Data* (eds. H. -H. Bock and E. Diday). Springer, Berlin, 103–124.

*Clustering Methodology for Symbolic Data*, First Edition. Lynne Billard and Edwin Diday.
© 2020 John Wiley & Sons Ltd. Published 2020 by John Wiley & Sons Ltd.

Bertrand, P. and Janowitz, M. F. (2002). Pyramids and weak hierarchies in the ordinal model for clustering. *Discrete Applied Mathematics* 122, 55–81.

Billard, L. (2008). Sample covariance functions for complex quantitative data. In: *Proceedings of the World Congress, International Association of Statistical Computing* (eds. M. Mizuta and J. Nakano). Japanese Society of Computational Statistics, Japan, 157–163.

Billard, L. (2011). Brief overview of symbolic data and analytic issues. *Statistical Analysis and Data Mining* 4, 149–156.

Billard, L. (2014). The past's present is now. What will the present's future bring? In: *Past, Present, and Future of Statistical Science* (eds. X. Lin, C. Genest, D. L. Banks, G. Molenberghs, D. W. Scott and J. -L. Wang). Chapman and Hall, New York, 323–334.

Billard, L. and Diday, E. (2003). From the statistics of data to the statistics of knowledge: Symbolic data analysis. *Journal American Statistical Association* 98, 470–487.

Billard, L. and Diday, E. (2006a). *Symbolic Data Analysis: Conceptual Statistics and Data Mining*. Wiley, Chichester.

Billard, L. and Diday, E. (2006b). Descriptive statistics for interval-valued observations in the presence of rules. *Computational Statistics* 21, 187–210.

Billard, L., Guo, H. -J. and Xu, W. (2016). Maximum likelihood estimators for bivariate interval-valued data. Manuscript.

Billard, L. and Kim, J. (2017). Hierarchical clustering for histogram data. *Wiley Interdisciplinary Reviews: Computational Statistics* 9, doi: 10.1002/wics.1405.

Billard, L. and Le-Rademacher, J. (2012). Principal component analysis for interval data. *Wiley Interdisciplinary Reviews: Computational Statistics* 4, 535–540; wics 1231.

Blaisdell, B. E., Campbell, A. M. and Karlin, S. (1996). Similarities and dissimilarities of phage genomes. *Proceedings of the National Academy of Sciences* 93, 5854–5859.

Bock, H. -H. (2003). Clustering algorithms and Kohonen maps for symbolic data. *New Trends in Computational Statistics with Biomedical Applications* 15, 217–229.

Bock, H. -H. (2007). Clustering methods: A history of $k$-means algorithms. In: *Selected Contributions in Data Analysis and Classification* (eds. P. Brito, P. Bertrand, G. Cucumel, and F. de Carvalho). Springer, Berlin, 161–172.

Bock, H. -H. and Diday, E. (eds.) (2000a). *Analysis of Symbolic Data: Exploratory Methods for Extracting Statistical Information from Complex Data*. Springer, Berlin.

Bock, H. -H. and Diday, E. (2000b). Symbolic objects. In: *Analysis of Symbolic Data: Exploratory Methods for Extracting Statistical Information from Complex Data* (eds. H. -H. Bock and E. Diday). Springer, Berlin, 55–77.

Breiman, L., Friedman, J., Olshen, R. and Stone, C. (1984). *Classification and Regression Trees*. Wadsworth, California.

Brito, M. P. (1991). *Analyse de Données Symbolique Pyramides d'Héritage*. Thèse de Doctorat, Université de Paris, Dauphine.

Brito, M. P. (1994). Use of pyramids in symbolic data analysis. In: *New Approaches in Classification and Data Analysis* (eds. E. Diday, Y. Lechevallier, M. Schader, P. Bertrand and B. Burtschy). Springer, Berlin, 378–386.

Brito, M. P. (2000). Hierarchical and pyramidal clustering with complete symbolic objects. In: *Analysis of Symbolic Data: Exploratory Methods for Extracting Statistical Information from Complex Data* (eds. H. -H. Bock and E. Diday). Springer, Berlin, 312–324.

Brito, M. P. and Diday, E. (1990). Pyramid representation of symbolic objects. In: *Knowledge, Data and Computer-Assisted Decisions* (eds. M. Schader and W. Gaul). Springer, Heidelberg, 3–16.

Brito, M. P. and Polaillon, G. (2005). Structuring probabilistic data by Galois lattices. *Mathematics and Social Sciences* 169, 77–104.

Caliński, T. and Harabasz, J. (1974). A dentrite method for cluster analysis. *Communications in Statistics* 3, 1–27.

Cariou, V. and Billard, L. (2015). Generalization method when manipulating relational databases. *Revue des Nouvelles Technologies de l'Information* 27, 59–86.

Celeux, G., Diday, E., Govaert, G., Lechevallier, Y. and Ralambondrainy, H. (1989). *Classification Automatique des Données*. Dunod, Paris.

Celeux, G. and Diebolt, J. (1985). The SEM algorithm: A probabilistic teacher algorithm derived from the EM algorithm for the mixture problem. *Computational Statistics Quarterly* 2, 73–82.

Celeux, G. and Govaert, G. (1992). A classification EM algorithm for clustering and two stochastic versions. *Computational Statistics and Data Analysis* 14, 315–332.

Celeux, G. and Govaert, G. (1995). Gaussian parsimonious clustering methods. *Pattern Recognition* 28, 781–793.

Census (2002). *California: 2000, Summary Population and Housing Characteristics*. US Department of Commerce Publishers.

Charles, C. (1977). *Regression Typologique et Reconnaissance des Formes*. Thèse de 3ème cycle, Université de Paris, Dauphine.

Chavent, M. (1997). *Analyse des Données Symbolique: Une Méthode Divisive de Classification*. Thèse de Doctorat, Université de Paris, Dauphine.

Chavent, M. (1998). A monothetic clustering algorithm. *Pattern Recognition Letters* 19, 989–996.

Chavent, M. (2000). Criterion-based divisive clustering for symbolic data. In: *Analysis of Symbolic Data: Exploratory Methods for Extracting Statistical Information from Complex Data* (eds. H. -H. Bock and E. Diday). Springer, Berlin, 299–311.

Chavent, M. and Lechevallier, Y. (2002). Dynamical clustering of interval data: Optimization of an adequacy criterion based on Hausdorff distance.

In: *Classification, Clustering, and Data Analysis* (eds. K. Jajuga, A. Sokolowski and H. -H. Bock). Springer, Berlin, 53–60.

Chavent, M., de Carvalho, F. A. T., Lechevallier, Y. and Verde, R. (2006). New clustering methods for interval data. *Computational Statistics* 21, 211–229.

Chen, Y. (2014). *Symbolic Data Regression and Clustering Methods.* Doctoral Dissertation, University of Georgia.

Chen, Y. and Billard, L. (2019). A study of divisive clustering with Hausdorff distances for interval data. *Pattern Recognition.*

Cheng, R. and Milligan, G. W. (1996). Measuring the influence of individual data points in cluster analysis. *Journal of Classification* 13, 315–335.

Choquet, G. (1954). Theory of capacities. *Annals Institute Fourier* 5, 131–295.

Clayton, D. G. (1978). A model for association in bivariate life tables and its application in epidemiological studies of familial tendency in chronic disease incidence. *Biometrika* 65, 141–151.

Consonni, V. and Todeschini, R. (2012). New similarity coefficients for binary data. *Communications in Mathematical and in Computer Chemistry* 68, 581–592.

Cormack, R. M. (1971). A review of classification. *Journal of the Royal Statistical Society A* 134, 321–367.

Davis, D. L. and Bouldin, D. W. (1979). A cluster separation measure. *IEEE Transactions on Pattern Analysis and Machine Intelligence* 1, 224–227.

De Carvalho, F. A. T. (1994). Proximity coefficients between Boolean symbolic objects. In: *New Approaches in Classification and Data Analysis* (eds. E. Diday, Y. Lechevallier, M. Schader, P. Bertrand and B. Burtschy). Springer, Berlin, 387–394.

De Carvalho, F. A. T. (1998). Extension based proximity coefficients between constrained Boolean symbolic objects. In: *Data Science, Classification, and Related methods* (eds. C. Hayashi, K. Yajima, H. -H. Bock, N. Ohsumi, Y. Tanaka, and Y. Baba). Springer, Berlin, 370–378.

De Carvalho, F. A. T., Brito, M. P. and Bock, H. -H. (2006). Dynamic clustering for interval data based on $L_2$ distance. *Computational Statistics* 21, 231–250.

De Carvalho, F. A. T. and Lechevallier, Y. (2009). Partitional clustering algorithms for symbolic interval data based on single adaptive distances. *Pattern Recognition* 42, 1223–1236.

De Carvalho, F. A. T., Lechevallier, Y. and Verde, R. (2008). Clustering methods in symbolic data analysis. In: *Symbolic Data Analysis and the SODAS Software* (eds. E. Diday and M. Noirhomme-Fraiture). Wiley, Chichester, 181–203.

De Carvalho, F. A. T., de Souza, R. M. C. R., Chavent, M. and Lechevallier, Y. (2006). Adaptive Hausdorff distances and dynamic clustering of symbolic interval data. *Pattern Recognition Letters* 27, 167–179.

De Carvalho, F. A. T., de Souza, R. M. C. R. and Silva, F. C. D. (2004). A clustering method for symbolic interval-type data using adaptive Chebyshev distances. In: *Advances in Artifical Intelligence, Lecture Notes in Computer Science*, vol. 3171. (eds. A. L. C. Bazzan and S. Labidi). Springer, Berlin, 266–275.

Dempster, A. P., Laird, N. M. and Rubin, D. B. (1977). Maximum likelihood for incomplete data via the EM algorithm. *Journal of the Royal Statistical Society B* 39, 1–38.

De Souza, R. M. C. R. and de Carvalho, F. A. T. (2004). Clustering of interval data based on city-block distances. *Pattern Recognition Letters* 25, 353–365.

De Souza, R. M. C. R., de Carvalho, F. A. T., Tenóio, C. P. and Lechevallier, Y. (2004). Dynamic cluster methods for interval data based on Mahalanobis distances. In: *Classification, Clustering, and Data Analysis* (eds. D. Banks, L. House, F. R. McMorris, P. Arabie and W. Gaul). Springer, Berlin, 251–360.

Diday E. (1971a). Une nouvelle méthode de classification automatique et reconnaissance des formes: La méthode des nuées dynamiques. *Revue de Statistique Appliquée* 2, 19–33.

Diday E. (1971b). La méthode des nuées dynamiques. *Revue de Statistique Appliquée* 19, 19–34.

Diday E. (1972a). Introduction à l'analyse factorielle typologique. *Rapport de Recherche 27*, INRIA, Rocquencourt.

Diday E. (1972b). Optimisation en classification automatique et reconnaisance des formes. *Revue Francaise d'Automatique, Informatique et Recherche Opérationelle* 6, 61–96.

Diday E. (1973). The dynamic clusters method in non-hierarchical clustering. *International Journal of Computer and Information Sciences* 2, 61–88.

Diday E. (1974). Classification automatique séquentielle pour grand tableaux. RAIRO, B-1, 29–61.

Diday E. (1978). Analyse canonique du point de vu de la classification automatique. *Rapport de Recherche 293*, INRIA, Rocquencourt.

Diday, E. (ed.) (1979). (with co-authors Bochi, S., Brossier, G., Celeux, G., Charles, C., Chifflet, R., Darcos, J., Diday, E., Diebolt, J., Fevre, P., Govaert, G., Hanani, O., Jacquet, D., Lechevallier, Y., Lemaire, J., Lemoine, Y., Molliere, J. L., Morisset, G., Ok-Sakun, Y., Rousseau, P., Sankoff, D., Schroeder, A., Sidi, J. and Taleng, F.). *Optimisation en Classification Automatique*, Tomes 1, 2. Institut National de Recherche en Informatique et en Automatique, Le Chesnay, France.

Diday, E. (1984). Une représentation visuelle des classes empietantes: les pyramides. *Rapport de Recherche 291*, INRIA, Rocquencourt.

Diday, E. (1986a). Canonical analysis from the automatic classification point of view. *Control and Cybernetics* 15, 115–137.

Diday, E. (1986b). Orders and overlapping clusters by pyramids. In: *Multidimensional Data Analysis* (eds. J. de Leeuw, W. Heiser, J. Meulman and F. Critchley). DSWO Press, Leiden, 201–234.

Diday, E. (1987). Introduction à l'approche symbolique en analyse des données. *Premier Jouneles Symbolique-Numerique*, CEREMADE, Université de Paris, Dauphine, 21–56.

Diday, E. (1989). Introduction à l'approche symbolique en analyse des données. *RAIRO Recherche Opérationnelle/Operations Research* 23, 193–236.

Diday, E. (1995). Probabilist, possibilist, and belief objects for knowledge analysis. *Annals of Operational Research* 55, 227–276.

Diday, E. (2004). Spatial pyramidal clustering based on a tessellation. In: *Classification, Clustering, and Data Mining Applications* (eds. D. Banks, F. R. McMorris, P. Arabie and W. Gaul). Springer, Berlin, 105–120.

Diday, E. (2008). Spatial classification. *Discrete Applied Mathematics* 156, 1271–1294.

Diday, E. (2016). Thinking by classes in data science: The symbolic data analysis paradigm. *WIREs Computational Statistics* 8, 172–205.

Diday E. and Bertrand, P. (1986). An extension of hierarchical clustering: the pyramid presentation. In: *Pattern Recognition in Practice II* (eds. E. S. Gelsema and L. N. Kanal). North-Holland, Amsterdam, 411–423.

Diday, E. and Emilion, R. (1998). Treillis de Galois et capacités de Choquet. *Compte Rendu Acadamey of Science Paris, Analyse Mathématique* 324, 261–266.

Diday, E. and Emilion, R. (2003). Maximal stochastic Galois lattices. *Journal of Discrete Applied Mathematics* 127, 271–284.

Diday, E. and Govaert, G. (1977). Classification avec adaptive intervals. *Revue Francaise d'Automatique, Informatique, et Recherche Opérationnelle, Information Computer Science* 11, 329–349.

Diday, E. and Noirhomme-Fraiture, M. (eds.) (2008). *Symbolic Data Analysis and the SODAS Software.* Wiley, Chichester.

Diday E. and Schroeder A. (1976). A new approach in mixed distribution detection. *Revue Francaise d'Automatique, Informatique, et Recherche Opérationnelle* 10, 75–106.

Diday E., Schroeder A. and Ok, Y. (1974). The dynamic clusters method in pattern recognition. In: *Proceedings of International Federation for Information Processing Congress* (ed. J. L. Rosenfeld). Elsevier, New York, 691–697.

Diday E. and Simon, J. C. (1976). Clustering analysis. In: *Digital Pattern Recognition* (ed. K. S. Fu). Springer, Berlin, 47–94.

Doreian, P., Batagelj, V. and Ferligoj, A. (2005). *Generalized Blockmodeling.* Cambridge University Press.

Dubes, R. and Jain, A. K. (1979). Validity studies in clustering methodologies. *Pattern Recognition* 11, 235–254.

Dunn, J. C. (1974). Well separated clusters and optimal fuzzy partitions. *Journal of Cybernetica* 4, 95–104.

Durand, C. and Fichet, B. (1988). One-to-one correspondence in pyramidal representation: a unified approach. In: *Classification and Related Methods of Data Analysis* (ed. H. -H. Bock). North-Holland, Amsterdam, 85–90.

Falduti, N. and Taibaly, H. (2004). Etude des retards sur les vols des compagnies aériennes. Report CEREMADE, Université de Paris, Dauphine, 63 pages.

Ferligoj, A. and Batagelj, V. (1982). Clustering with relational constraint. *Psychometrika* 47, 413–426.

Ferligoj, A. and Batagelj, V. (1983). Some types of clustering with relational constraint. *Psychometrika* 48, 541–552.

Ferligoj, A. and Batagelj, V. (1992). Direct multicriteria clustering algorithms. *Journal of Classification* 9, 43–61.

Fichet, B. (1987). Data analysis: geometric and algebraic structures. In: *Proceedings First World Congress of Bernoulli Society* (eds. Y. Prochorov and V. V. Sazanov). VNU Science Press, Utrecht, 123–132.

Fisher, R. A. (1936). The use of multiple measurements in taxonomic problems. *Annals of Eugenics* 7, 179–188.

Forgy, E. W. (1965). Cluster analysis of multivariate data: Efficiency versus interpretability of classifications. *Biometrics* 21, 768.

Foulkes, E. B., Gnanadesikan, R. and Kettenring, J. R. (1988). Variable selection in clustering. *Journal of Classification* 5, 205–228.

Fraley, C. and Raftery, A. E. (1998). How many clusters? Which clustering method? Answers via model-based cluster analysis. *The Computer Journal* 41, 578–588.

Frank, M. J. (1979). On the simultaneous associativity of $F(x, y)$ and $x + y - F(x, y)$. *Aequationes Mathematicae* 19, 194–226.

Fräti, P., Rezaei, M. and Zhao, Q. (2014). Centroid index: Cluster level similarity measure. *Pattern Recognition* 47, 3034–3045.

Friedman, J. H. (1987). Exploratory projection pursuit. *Journal of the American Statistical Association* 82, 249–266.

Friedman, J. H. and Rubin, J. (1967). On some variant criteria for grouping data. *Journal of the American Statistical Association* 62, 1159–1178.

Gaul, W. and Schader, M. (1994). Pyramidal classification based on incomplete dissimilarity data. *Journal of Classification* 11, 171–193.

Genest, C. and Ghoudi, K. (1994). Une famile de lois bidimensionnelles insolite. *Compte Rendu Academy Science Paris, Série I* 318, 351–354.

Goodall, D. W. (1966). Classification, probability and utility. *Nature* 211, 53–54.

Gordon, A. D. (1987). A review of hierarchical classification. *Journal of the Royal Statistical Society A* 150, 119–137.

Gordon, A. D. (1999). *Classification* (2nd edn). Chapman and Hall, Boca Raton.

Gowda, K. C. and Diday, E. (1991). Symbolic clustering using a new dissimilarity measure. *Pattern Recognition* 24, 567–578.

Gowda, K. C. and Diday, E. (1992). Symbolic clustering using a new similarity measure. *IEEE Transactions on Systems, Man, and Cybernetics* 22, 368–378.

Gower, J. C. (1971). Statistical methods of comparing different multivariate analyses of the same data. In: *Anglo-Romanian Conference on Mathematics in Archeology and Historical Sciences* (eds. F. R. Hodson, D. G. Kendall and P. Tautu). Edinburgh University Press, Edinburgh, 138–149.

Hajjar, C. and Hamdan, H. (2013). Interval data clustering using self-organizing maps based on adaptive Mahalanobis distances. *Neural Networks* 46, 124–132.

Hardy, A. (2007). Validation in unsupervised symbolic classification. In: *Selected Contributions in Data Analysis and Classification* (eds. P. Brito, P. Bertrand, G. Cucumel and F. de Carvalho). Springer, Berlin, 379–386.

Hardy, A. and Baume, J. (2007). Clustering and validation of interval data. In: *Selected Contributions in Data Analysis and Classification* (eds. P. Brito, P. Bertrand, G. Cucumel and F. de Carvalho). Springer, Berlin, 69–81.

Hausdorff, F. (1937). *Set Theory* [Translated into English by J. R. Aumann, 1957]. Chelsey, New York.

Hubert, L. and Arabie, P. (1985). Comparing partitions. *Journal of Classification* 2, 193–218.

Ichino, M. (1988). General metrics for mixed features – The Cartesian space theory for pattern recognition. In: *Proceedings of the 1988 Conference on Systems, Man, and Cybernetics.* International Academic Publisher, Pergaman, 494–497.

Ichino, M. and Yaguchi, H. (1994). Generalized Minkowski metrics for mixed feature type data analysis. *IEEE Transactions on Systems, Man and Cybernetics* 24, 698–708.

Irpino, A. and Romano, E. (2007). Optimal histogram representation of large data sets: Fisher vs piecewise linear approximation. *Revue des Nouvelles Technologies de l'Information* E-9, 99–110.

Irpino, A. and Verde, R. (2006). A new Wasserstein based distance for the hierarchical clustering of histogram symbolic data. In: *Proceedings COMPSTAT* (eds. A. Rizzi and M. Vichi). Physica-Verlag, Berlin, 869–876.

Irpino, A. and Verde, R. (2008). Dynamic clustering of interval data using a Wasserstein-based distance. *Pattern Recognition Letters* 29, 1648–1658.

Irpino, A., Verde, R. and de Carvalho, F. A. T. (2014). Dynamic clustering of histogram data based on adaptive squared Wasserstein distances. *Expert Systems with Applications* 41, 3351–3366.

Irpino, A., Verde, R. and Lechevallier, Y. (2006). Dynamic clustering of histograms using Wasserstein metric. In: *COMPSTAT 2006* (eds. A. Rizzi and M. Vichi). Physica-Verlag, Berlin, 869–876.

Jaccard, P. (1907). La distribution de la flore dans la zone alpine. *Revue Générale des Sciences* 18, 961–967. [Translated into English in 1912 in *New Phytologist* 11, 37–50.]

Jain, A. K. (2010). Data clustering: 50 years beyond $K$-means. *Pattern Recognition Letters* 31, 651–666.

Jain, A. K. and Dubes, R. C. (1988). *Algorithms for Clustering Data.* Prentice Hall, New Jersey.

Jain, A. K., Murty, M. N. and Flynn, P. J. (1999). Data clustering: A review. *ACM Computing Surveys* 31, 263–323.

Jardine, N. and Sibson, R. (1971). *Mathematical Taxonomy.* Wiley, New York.

Johnson, S. C. (1967). Hierarchical clustering schemes. *Psychometrika* 32, 241–254.

Kaufman, L. and Rousseeuw, P. J. (1986). Clustering large data sets (with Discussion). In: *Pattern Recognition in Practice II* (eds. E. S. Gelsema and L. N. Kanal). North-Holland, Amsterdam, 425–437.

Kaufman, L. and Rousseeuw, P. J. (1987). Clustering by means of medoids. In: *Statistical Data Analysis Based on the $\mathcal{L}_1$-Norm and Related Methods* (ed. Y. Dodge). North-Holland, Berlin, 405–416.

Kaufman, L. and Rousseeuw, P. L. (1990). *Finding Groups in Data: An Introduction to Cluster Analysis.* John Wiley, New York.

Kim, J. (2009). *Dissimilarity Measures for Histogram-valued Data and Divisive Clustering of Symbolic Objects.* Doctoral Dissertation, University of Georgia.

Kim, J. and Billard, L. (2011). A polythetic clustering process for symbolic observations and cluster validity indexes. *Computational Statistics and Data Analysis* 55, 2250–2262.

Kim, J. and Billard, L. (2012). Dissimilarity measures and divisive clustering for symbolic multimodal-valued data. *Computational Statistics and Data Analysis* 56, 2795–2808.

Kim, J. and Billard, L. (2013). Dissimilarity measures for histogram-valued observations. *Communications in Statistics: Theory and Methods* 42, 283–303.

Kim, J. and Billard, L. (2018). Double monothetic clustering for histogram-valued data. *Communications for Statistical Applications and Methods* 25, 263–274.

Kohonen, T. (2001). *Self-Organizing Maps.* Springer, Berlin, Heidelberg.

Korenjak-Černe, S., Batagelj, V. and Pavešić, B. J. (2011). Clustering large data sets described with discrete distributions and its application on TIMSS data set. *Statistical Analysis and Data Mining* 4, 199–215.

Košmelj, K. and Billard, L. (2011). Clustering of population pyramids using Mallows' $L^2$ distance. *Metodološki Zvezki* 8, 1–15.

Košmelj, K. and Billard, L. (2012). Mallows'$L^2$ distance in some multivariate methods and its application to histogram-type data. *Metodološki Zvezki* 9, 107–118.

Kuo, R. J., Ho, L. M. and Hu, C. M. (2002). Cluster analysis in industrial market segmentation through artificial neural network. *Computers and Industrial Engineering* 42, 391–399.

Lance, G. N. and Williams, W. T. (1967a). A general theory of classificatory sorting strategies. II. Clustering systems. *The Computer Journal* 10, 271–277.

Lance, G. N. and Williams, W. T. (1967b). A general theory of classificatory sorting strategies. I. Hierarchical systems. *The Computer Journal* 9, 373–380.

Lechevallier, Y., de Carvalho, F. A. T., Despeyroux, T. and de Melo, F. (2010). Clustering of multiple dissimilarity data tables for documents categorization. In: *Proceedings COMPSTAT* (eds. Y. Lechevallier and G. Saporta) 19, 1263–1270.

Le-Rademacher, J. and Billard, L. (2011). Likelihood functions and some maximum likelihood estimators for symbolic data. *Journal of Statistical Planning and Inference* 141, 1593–1602.

Leroy, B., Chouakria, A., Herlin, I. and Diday, E. (1996). Approche géométrique et classification pour la reconnaissance de visage. *Reconnaissance des Forms et Intelligence Artificelle*, INRIA and IRISA and CNRS, France, 548–557.

Levina, E. and Bickel, P. (2001). The earth mover's distance is the Mallows' distance. Some insights from statistics. *Proceedings IEEE International Conference Computer Vision*. Publishers IEEE Computer Society, 251–256.

Limam, M. (2005). *Méthodes de Description de Classes Combinant Classification et Discrimination en Analyse de Données Symboliques*. Thèse de Doctorat, Université de Paris, Dauphine.

Limam, M. M., Diday, E. and Winsberg, S. (2004). Probabilist allocation of aggregated statistical units in classification trees for symbolic class description. In: *Classification, Clustering and Data Mining Applications* (eds. D. Banks, L. House, F. R. McMorris, P. Arabie and W. Gaul). Springer, Heidelberg, 371–379.

Lisboa, P. J. G., Etchells, T. A., Jarman, I. H. and Chambers, S. J. (2013). Finding reproducible cluster partitions for the $k$-means algorithm. *BMC Bioinformatics* 14, 1–19.

Liu, F. (2016). *Cluster Analysis for Symbolic Interval Data Using Linear Regression Method*. Doctoral Dissertation, University of Georgia.

MacNaughton-Smith, P., Williams, W. T., Dale, M. B. and Mockett, L. G. (1964). Dissimilarity analysis: a new technique of hierarchical division. *Nature* 202, 1034–1035.

MacQueen, J. (1967). Some methods for classification and analysis of multivariate observations. In: *Proceedings of the 5th Berkeley Symposium on Mathematical Statistics and Probability* (eds. L. M. LeCam and J. Neyman). University of California Press, Berkeley, 1, 281–299.

Mahalanobis, P. C. (1936). On the generalized distance in statistics. *Proceedings of the National Institute of Science India* 2, 49–55.

Mallows, C. L. (1972). A note on asymptotic joint normality. *Annals of Mathematical Statistics* 43, 508–515.

Malerba, D., Esposito, F., Gioviale, V. and Tamma, V. (2001). Comparing dissimilarity measures for symbolic data analysis. In: *Proceedings of Exchange of Technology and Know-how and New Techniques and Technologies for Statistics*, Crete (eds. P. Nanopoulos and D. Wilkinson). Publisher European Communities Rome, 473–481.

McLachlan G. and Basford, K. E. (1988). *Mixture Models: Inference and Applications to Clustering*. Marcel Dekker, New York.

McLachlan G. and Peel D. (2000). *Finite Mixture Models*. Wiley, New York.

McQuitty, L. L. (1967). Expansion of similarity analysis by reciprocal pairs for discrete and continuous data. *Education Psychology Measurement* 27, 253–255.

Milligan, G. W. and Cooper, M. (1985). An examination of procedures for determining the number of clusters in a data set. *Psychometrika* 50, 159–179.

Mirkin, B. (1996). *Mathematical Classification and Clustering*. Kluwer, Dordrecht.

Murtagh, F. and Legendre, P. (2014). Ward's hierarchical clustering method: Clustering criterion and agglomerative algorithm. *Journal of Classification* 31, 274–295.

Nelsen, R. B. (2007). *An Introduction to Copulas*. Springer, New York.

Noirhomme-Fraiture, M. and Brito, M. P. (2011). Far beyond the classical data models: Symbolic data analysis. *Statistical Analysis and Data Mining* 4, 157–170.

Pak, K. K. (2005). *Classifications Hiérarchique et Pyramidale Spatiale*. Thése de Doctorat, Université de Paris, Dauphine.

Parks, J. M. (1966). Cluster analysis applied to multivariate geologic problems. *Journal of Geology* 74, 703–715.

Pearson, K. (1895). Note on regression and inheritance in the case of two parents. *Proceedings of the Royal Society of London* 58, 240–242.

Périnel, E. (1996). *Segmentation et Analyse des Données Symbolique: Applications á des Données Probabilities Imprécises*. Thèse de Doctorat, Université de Paris, Dauphine.

Polaillon, G. (1998). *Organisation et Interprétation par les Treis de Galois de Données de Type Multivalué, Intervalle ou Histogramme*. Thése de Doctorat, Université de Paris, Dauphine.

Polaillon, G. (2000). Pyramidal classification for interval data using Galois lattice reduction. In: *Analysis of Symbolic Data: Exploratory Methods for Extracting Statistical Information from Complex Data* (eds. H. -H. Bock and E. Diday). Springer, Berlin, 324–341.

Polaillon, G. and Diday, E. (1996). Galois lattices construction and application in symbolic data analysis. Report CEREMADE 9631, Université de Paris, Dauphine.

Punj, G. and Stewart, D. W. (1983). Cluster analysis in marketing research: Review and suggestions for application. *Journal of Marketing Research* 20, 134–148.

Quantin, C., Billard, L., Touati, M., Andreu, N., Cottin, Y., Zeller, M., Afonso, F., Battaglia, G., Seck, D., LeTeuff, G. and Diday, E. (2011). Classification and regression trees on aggregate data modeling: An application in acute myocardial infarction. *Journal of Probability and Statistics*, ID 523937, doi:10.1155/2011/523937.

Rahal, M. C. (2010). *Classification Pyramidale Spatiale: Nouveaux Algorithmes et Aide à l'Interprétation*. Thése de Doctorat, Université de Paris, Dauphine.

Ralambondrainy, H. (1995). A conceptual version of the $K$-means algorithm. *Pattern Recognition Letters* 16, 1147–1157.

Reynolds, A. P., Richards, G., de la Iglesia, B. and Rayward-Smith, V. J. (2006). Clustering rules: A comparison of partitioning and hierarchical clustering algorithms. *Journal of Mathematical Modelling and Algorithms* 5, 475–504.

Robinson, W. S. (1951). A method for chronologically ordering archaeological deposits. *American Antiquity* 16, 293–301.

Rüschendorf, L. (2001). Wasserstein metric. In: *Encyclopedia of Mathematics* (ed. M. Hazewinkel). Springer, Dordrecht, 631.

Schroeder, A. (1976). Analyse d'un mélange de distributions de probabilité de même type. *Revue de Statistiques Appliquées* 24, 39–62.

Schwarz, G. (1978). Estimating the dimension of a model. *Annals of Statistics* 6, 461–464.

Schweizer, B. (1984). Distributions are the numbers of the future. In: *Proceedings The Mathematics of Fuzzy Systems Meeting* (eds. A. di Nola and A. Ventes). University of Naples, Naples Italy, 137–149.

Scott, A. J. and Symons, M. J. (1971). Clustering methods based on likelihood ratio criteria. *Biometrics* 27, 387–397.

Seck, D. A. N. (2012). *Arbres de Décision Symboliques, Outils de Validation et d'Aide à l'Interprétation.* Thèse de Doctorat, Université de Paris, Dauphine.

Seck, D., Billard, L., Diday, E. and Afonso, F. (2010). A decision tree for interval-valued data with modal dependent variable. In: *Proceedings COMPSTAT 2010* (eds. Y. Lechevallier and G. Saporta). Springer, 19, 1621–1628.

Silverman, B. W. (1986). *Density Estimation for Statistics and Data Analysis.* Chapman and Hall, London.

Sklar, A. (1959.) Fonction de répartition à $n$ dimensions et leurs marges. *Institute Statistics Université de Paris* 8, 229–231.

Sneath, P. H. and Sokal, R. R. (1973). *Numerical Taxonomy.* Freeman, San Francisco.

Sokal, R. R. (1963). The principles and practice of numerical taxonomy. *Taxon* 12, 190–199.

Sokal, R. R. and Michener, C. D. (1958). A statistical method for evaluating systematic relationships. *University of Kansas Science Bulletin* 38, 1409–1438.

Sokal, R. R. and Sneath, P. H. (1963). *Principles of Numerical Taxonomy.* Freeman, San Francisco.

Steinley, D. (2006). $K$-means clustering: A half-century synthesis. *British Journal of Mathematical and Statistical Psychology* 59, 1–34.

Stéphan, V. (1998). *Construction d' Objects Symboliques par Synthèse des Résultats de Requêtes SQL.* Thèse de Doctorat, Université de Paris, Dauphine.

Stéphan, V., Hébrail, G. and Lechevallier, Y. (2000). Generation of symbolic objects from relational databases. In: *Analysis of Symbolic Data: Exploratory Methods for Extracting Statistical Information from Complex Data* (eds. H. -H. Bock and E. Diday). Springer, Berlin, 78–105.

Symons, M. J. (1981). Clustering criteria and multivariate normal mixtures. *Biometrics* 37, 35–43.

Tibshirani, R. and Walther, G. (2005). Cluster validation by prediction strength. *Journal of Computational and Graphical Statistics* 14, 511–528.

Vanessa, A. and Vanessa, L. (2004). La meilleure équipe de baseball. Report CEREMADE, Université de Paris, Dauphine.

Verde, R. and Irpino, A. (2007). Dynamic clustering of histogram data: Using the right metric. In: *Selected Contributions in Data Analysis and Classification* (eds. P. Brito, P. Bertrand, G. Cucumel and F. de Carvalho). Springer, Berlin, 123–134.

Vrac, M. (2002). *Analyse et Modelisation de Données Probabilistes Par Décomposition de Mélange de Copules et Application à une Base de Données Climatologiques.* Thèse de Doctorat, Université de Paris, Dauphine.

Vrac, M., Billard, L., Diday E. and Chédin, A. (2012). Copula analysis of mixture models. *Computational Statistics* 27, 427–457.

Ward, J. H. (1963). Hierarchical grouping to optimize an objective function. *Journal of the American Statistical Association* 58, 236–244.

Wei, G. C. G. and Tanner, M. A. (1990). A Monte Carlo implementation of the EM algorithm and the poor man's data augmentation algorithms. *Journal of the American Statistical Association* 85, 699–704.

Winsberg, S., Diday, E. and Limam, M. M. (2006): A tree structured classifier for symbolic class description. In: *Proceedings COMPSTAT* (eds. A. Rizzi and M. Vichi). Springer, 927–936.

Wishart, D. (1969). An algorithm for hierarchical classification. *Biometrics* 25, 165–170.

Xu, W. (2010). *Symbolic Data Analysis: Interval-valued Data Regression.* Doctoral Dissertation, University of Georgia.

# Index

*Clustering Methodology for Symbolic Data*, First Edition. Lynne Billard and Edwin Diday.
© 2020 John Wiley & Sons Ltd. Published 2020 by John Wiley & Sons Ltd.